Student Solutions Manual for

Technical Calculus with Analytic Geometry

Third Edition

Allyn J. Washington

Dutchess Community College

The Benjamin/Cummings Publishing Company
Menlo Park, California ● Reading, Massachusetts
Don Mills, Ontario ● Wokingham, U.K. ● Amsterdam
Sydney ● Singapore ● Tokyo ● Madrid
Bogota ● Santiago ● San Juan

Sponsoring Editor: Craig Bartholomew
Production Editor: Connie Sorensen
Cover Artist: Becky Felt

ISBN 0-8053-9514-8

7 8 9 10 -AL- 95 94 93 92 91

The Benjamin/Cummings Publishing Company, Inc.
2727 Sand Hill Road
Menlo Park, California 94025

Contents

This solutions manual contains the solutions to
all of the odd-numbered exercises in
**Technical Calculus with Analytic Geometry,
Third Edition.**

CHAPTER 1 **PLANE ANALYTIC GEOMETRY**

Exercises 1-1, p. 4

1. (a) $\frac{5}{4}$ is rational (it is written as the ratio of integers); $\frac{5}{4}$ is real (not the square root of a negative number). (b) $-\pi$ is irrational (cannot be written as the ratio of integers); $-\pi$ is real__(c) $\sqrt{-4}$ is imaginary (it is the square root of a negative number). (d) $6\sqrt{-1}$ is imaginary (it is a multiple of the square root of a negative number).

3. (a) $|3| = 3$ (b) $\left|\frac{7}{2}\right| = \frac{7}{2}$ (c) $|-4| = 4$

5. (a) $6 < 8$ (6 is to the left of 8)   (b) $7 > -5$ (7 is to the right of -5)
   (c) $\pi > -1$ (π is to the right of -1)

7.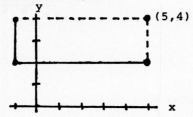

9. The points lie on a straight line.

11. Abscissas are 1 means x-coordinates are 1. Points are on a line parallel to the y-axis, one unit to the right.

13. On the y-axis, the abscissa (x-coordinate) of all points is 0.

15. If $x > 0$, a point is to the right of the y-axis.

17. If $|y| > 2$, a point is more than two units from the x-axis. It is above a line parallel to the x-axis and two units above, or it is below a line parallel to the x-axis and two units below.

19. If $x < 0$, a point is the the left of the y-axis. If $y > 1$, a point is above a line parallel to the x-axis and one unit above. For both conditions, a point must be in the second quadrant, above a line parallel to the x-axis and one unit above.

21. In the first quadrant both coordinates are positive and y/x is positive. In the third quadrant both coordinates are negative and y/x is positive. In the second and fourth quadrants, one coordinate is negative and the other is positive and y/x is negative.

23. In a rectangle, opposite sides are equal and parallel. From the figure we see that the fourth vertex is (5,4).

25. (a) b-a is a positive integer since b>a.
   (b) a-b is a negative integer since b>a.
   (c) $\frac{b-a}{b+a}$ is a positive rational number (less than 1) since b-a and b+a are positive integers (and b+a>b-a).

27.

1

Exercises 1-2, p. 9

1. $y = 3x$

x	y
-2	-6
-1	-3
0	0
1	3
2	6

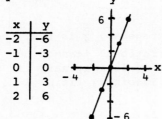

3. $y = 2x - 4$

x	y
-1	-6
0	-4
1	-2
2	0
3	2
4	4
5	6

5. $y = 7 - 2x$

x	y
-1	9
0	7
1	5
3	1
5	-3
7	-7

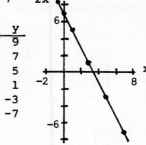

7. $y = \frac{1}{2}x - 2$

x	y
-4	-4
-2	-3
0	-2
2	-1
4	0
6	1
8	2

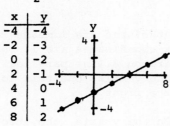

9. $y = x^2$

x	y
-3	9
-2	4
-1	1
0	0
1	1
2	4
3	9

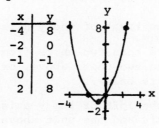

11. $y = 3 - x^2$

x	y
-3	-6
-2	-1
-1	2
0	3
1	2
2	-1
3	-6

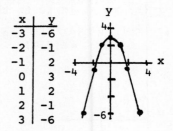

13. $y = \frac{1}{2}x^2 + 2$

x	y
-4	10
-2	4
0	2
2	4
4	10

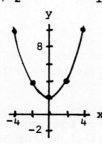

15. $y = x^2 + 2x$

x	y
-4	8
-2	0
-1	-1
0	0
2	8

17. $y = x^2 - 3x + 1$

x	y
-2	11
-1	5
0	1
1	-1
1.5	-1.25
2	-1
3	1
4	5
5	11

19. $y = x^3$

x	y
-2	-8
-1	-1
0	0
1	1
2	8

21. $y = x^3 - x^2$

x	y
-2	-12
-1	-2
0	0
0.5	-0.125
1	0
2	4
3	18

23. $y = x^4 - 4x^2$

x	y
-3	45
-2	0
-1	-3
0	0
1	-3
2	0
3	45

25. $y = \frac{1}{x}$

x	y	x	y	x	y
-4	-0.25	-0.2	-5	1	1
-2	-0.5	0	undef	2	0.5
-1	-1	0.2	5	4	0.25
-0.5	-2	0.5	2		

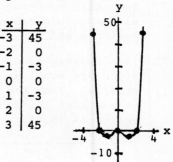

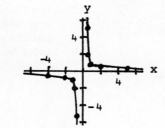

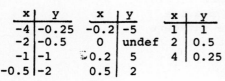

27. $y = \dfrac{1}{x^2}$

x	y
-2	0.25
-1	1
-0.5	4
0	undef
0.5	4
1	1
2	0.25

29. $y = \sqrt{x}$

x	y
neg	imag
0	0
1	1
2	1.4
4	2
6	2.4
9	3

31. $y = \sqrt{16 - x^2}$

x	y
-5	imag
-4	0
-2	3.5
0	4
2	3.5
4	0
5	imag

33. (a) $-y = x^2 - 3$ is not the same as $y = x^2 - 3$; not symmetric to x-axis
       $y = (-x)^2 - 3 = x^2 - 3$ is the same as $y = x^2 - 3$; symmetric to y-axis
    (b) $-y = 2x - x^2$ is not the same as $y = 2x - x^2$; not symmetric to x-axis
       $y = 2(-x) - (-x)^2 = -2x - x^2$ is not the same as $y = 2x - x^2$; not symm. to y-axis
    (c) $-y = \sqrt{x}$ is not the same as $y = \sqrt{x}$; not symmetric to x-axis
       $y = \sqrt{-x}$ is not the same as $y = \sqrt{x}$; not symmetric to y-axis

35. (a) $y = x^3$; no vertical asymptotes ; no variables in denominator
    (b) $y = 1/x$; $x = 0$ is vertical asymptote ; $x = 0$ makes denominator zero
    (c) $y = 1/(x^2 + 1)$; no vertical asymptotes ; $x^2 + 1$ always positive

37. $v = 100 - 32t$

t(s)	v(ft/s)
0	100
1	68
2	36
3	4
4	-28

39. $c = 0.011r + 4.0$
    (r in r/min, c in L/h)

r	c
500	9.5
1000	15.0
2000	26.0
3000	37.0

41. $r = 0.42v^2$

$v(\frac{mi}{h})$	r(ft)
0	0
20	170
40	670
60	1500

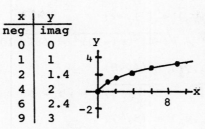

43. $A = 300w - w^2$

w(m)	A(m^2)
0	0
50	12500
100	20000
150	22500
200	20000
250	12500
300	0

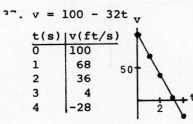

45. $I = \dfrac{400}{r^2}$
    (r in meters
     I in lumens/m^2)

r	I
0	undef
1	400
2	100
4	25
10	4

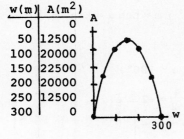

47. $P = 4x^3 - 18x^2 - 120x - 200$
    (x in units produced
     P in dollars)

x	P
0	-200
2	-480
4	-712
5	-750
6	-704
8	-264
10	800

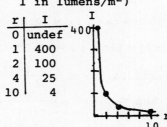
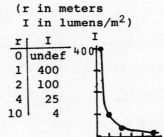

49. $y = x$ $y = |x|$

x	y
-2	-2
-1	-1
0	0
1	1
2	2

x	y
-2	2
-1	1
0	0
1	1
2	2

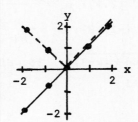

Exercises 1-3, p. 16

1. $(x_1, y_1) = (3,8)$, $(x_2, y_2) = (-1,-2)$

   $d = \sqrt{(-1-3)^2 + (-2-8)^2} = \sqrt{(-4)^2 + (-10)^2} = \sqrt{16 + 100} = \sqrt{116} = 2\sqrt{29}$

3. $(x_1, y_1) = (4,-5)$, $(x_2, y_2) = (4,-8)$

   $d = \sqrt{(4-4)^2 + [-8-(-5)]^2} = \sqrt{0^2 + (-3)^2} = \sqrt{9} = 3$

5. $(x_1, y_1) = (-1,0)$, $(x_2, y_2) = (5,-7)$

   $d = \sqrt{[5-(-1)]^2 + (-7-0)^2} = \sqrt{6^2 + (-7)^2} = \sqrt{36 + 49} = \sqrt{85}$

7. $(x_1, y_1) = (-4,-3)$, $(x_2, y_2) = (3,-3)$

   $d = \sqrt{[3-(-4)]^2 + [-3-(-3)]^2} = \sqrt{7^2 + 0^2} = \sqrt{49} = 7$

9. $(x_1, y_1) = (1.22,-3.45)$, $(x_2, y_2) = (-1.07,-5.16)$

   $d = \sqrt{(-1.07-1.22)^2 + [-5.16 - (-3.45)]^2} = \sqrt{(-2.29)^2 + (-1.71)^2} = \sqrt{8.168} = 2.86$

11. $(x_1, y_1) = (3,8)$, $(x_2, y_2) = (-1,-2)$

    $m = \dfrac{-2-8}{-1-3} = \dfrac{-10}{-4} = \dfrac{5}{2}$

13. $(x_1, y_1) = (4,-5)$, $(x_2, y_2) = (4,-8)$

    $m = \dfrac{-8-(-5)}{4-4} = \dfrac{-3}{0}$ Since denominator is zero, slope is undefined.

15. $(x_1, y_1) = (-1,0)$, $(x_2, y_2) = (5,-7)$

    $m = \dfrac{-7-0}{5-(-1)} = \dfrac{-7}{6} = -\dfrac{7}{6}$

17. $(x_1, y_1) = (-4,-3)$, $(x_2, y_2) = (3,-3)$

    $m = \dfrac{-3-(-3)}{3-(-4)} = \dfrac{0}{7} = 0$

19. $(x_1, y_1) = (1.22,-3.45)$, $(x_2, y_2) = (-1.07,-5.16)$

    $m = \dfrac{-5.16-(-3.45)}{-1.07-1.22} = \dfrac{-1.71}{-2.29} = 0.747$

21. $\alpha = 30°$; $m = \tan \alpha = \tan 30° = \dfrac{1}{3}\sqrt{3}$

23. $\alpha = 150°$; $m = \tan \alpha = \tan 150° = -\dfrac{1}{3}\sqrt{3}$

25. $m = 0.3640$; $m = \tan \alpha$; $0.3640 = \tan \alpha$; $\alpha = 20.0°$

27. $m = -6.691$; $m = \tan \alpha$; $-6.691 = \tan \alpha$; $\alpha = 98.5°$ $(0 \le \alpha < 180°)$

29. For line through $(6,-1)$ and $(4,3)$: $m_1 = \dfrac{3-(-1)}{4-6} = \dfrac{4}{-2} = -2$

    For line through $(-5,2)$ and $(-7,6)$: $m_2 = \dfrac{6-2}{-7-(-5)} = \dfrac{4}{-2} = -2$

    Since $m_1 = m_2$, the lines are parallel.

31. For line through $(-1,-4)$ and $(2,3)$: $m_1 = \dfrac{3-(-4)}{2-(-1)} = \dfrac{7}{3}$

    For line through $(-5,2)$ and $(-19,8)$: $m_2 = \dfrac{8-2}{-19-(-5)} = \dfrac{6}{-14} = -\dfrac{3}{7}$

    Since $m_2 = -\dfrac{1}{m_1}$, the lines are perpendicular.

33. $13 = \sqrt{[11-(-1)]^2 + (k-3)^2} = \sqrt{144 + (k-3)^2}$

    $169 = 144 + (k-3)^2$; $(k-3)^2 = 25$

    $k - 3 = \pm 5$; $k = 3 \pm 5 = 8, -2$

35. Since points are on same line, the slope through $(6,-1)$ and $(3,k)$ is the same as the slope through $(3,k)$ and $(-3,-7)$.

    $\dfrac{k-(-1)}{3-6} = \dfrac{-7-k}{-3-3}$; $\dfrac{k+1}{-3} = \dfrac{-7-k}{-6}$

    $2k + 2 = -7 - k$; $3k = -9$; $k = -3$

37. For (2,3) and (4,9): $d_1 = \sqrt{(4-2)^2 + (9-3)^2} = \sqrt{40} = 2\sqrt{10}$ Since $d_1 = d_2$, the
    For (4,9) and (-2,7): $d_2 = \sqrt{(-2-4)^2 + (7-9)^2} = \sqrt{40} = 2\sqrt{10}$ triangle is isosceles.

39. For (3,2) and (7,3): $m_1 = \frac{3-2}{7-3} = \frac{1}{4}$ For (-1,-3) and (3,2): $m_3 = \frac{2-(-3)}{3-(-1)} = \frac{5}{4}$

    For (-1,-3) and (3,-2): $m_2 = \frac{-2-(-3)}{3-(-1)} = \frac{1}{4}$ For (3,-2) and (7,3): $m_4 = \frac{3-(-2)}{7-3} = \frac{5}{4}$

    Since slopes of opposite sides are equal, both pairs of opposite sides are
    parallel and the figure is a parallelogram. (Other methods of proof involving
    the lengths of the sides can also be used.)

41. For (-1,3) and (3,5): $d_1 = \sqrt{[3-(-1)]^2 + (5-3)^2} = \sqrt{20}$; $m_1 = \frac{5-3}{3-(-1)} = \frac{1}{2}$

    For (3,5) and (5,1): $d_2 = \sqrt{(5-3)^2 + (1-5)^2} = \sqrt{20}$; $m_2 = \frac{1-5}{5-3} = -2$

    Since $m_1 = -\frac{1}{m_2}$, side 1 $\perp$ side 2, and they are the sides of a right triangle.
    $A = \frac{1}{2}(\sqrt{20})(\sqrt{20}) = \frac{1}{2}(20) = 10$

43. $(x_1,y_1) = (-4,9)$, $(x_2,y_2) = (6,1)$; midpoint is $(\frac{-4+6}{2}, \frac{9+1}{2}) = (1,5)$

Exercises 1-4, p. 21

1. $(x_1,y_1) = (-3,8)$, $m = 4$
   $y - y_1 = m(x - x_1)$
   $y - 8 = 4(x + 3)$
   $\quad\;\; = 4x + 12$
   $4x - y + 20 = 0$

3. $(x_1,y_1) = (-2,-5)$
   $(x_2,y_2) = (4,2)$
   $m = \frac{2+5}{4+2} = \frac{7}{6}$
   $y - y_1 = m(x - x_1)$
   $y + 5 = \frac{7}{6}(x + 2)$
   $6y + 30 = 7x + 14$
   $7x - 6y - 16 = 0$

5. $(x_1,y_1) = (1,3)$, $\alpha = 45°$
   $m = \tan 45° = 1$
   $y - y_1 = m(x - x_1)$
   $y - 3 = 1(x - 1)$
   $\quad\;\; = x - 1$
   $x - y + 2 = 0$

7. $(x_1,y_1) = (6,-3)$
   Parallel to x-axis
   means $m = 0$
   $y = y_1$, $y = -3$

9. Parallel to y-axis,
   3 units to left
   means all points
   have x-coordinates
   of -3
   $x = -3$

11. x-int. of 4 means point
    is (4,0)
    y-int. of -6 means b=-6
    or point is (0,-6)
    $m = \frac{-6-0}{0-4} = \frac{3}{2}$
    $y = mx + b$
    $y = \frac{3}{2}x - 6$, $2y = 3x - 12$
    $3x - 2y - 12 = 0$

13. Perpendicular to line with slope of 3
    means $m = -1/3$; $(x_1,y_1) = (1,-2)$
    $y - y_1 = m(x - x_1)$, $y + 2 = -\frac{1}{3}(x - 1)$
    $3y + 6 = -x + 1$, $x + 3y + 5 = 0$

15. Parallel to line with slope of $-1/2$
    means $m = -\frac{1}{2}$
    x-int. of 4 means point is (4,0)
    $y - y_1 = m(x - x_1)$
    $y - 0 = -\frac{1}{2}(x - 4)$
    $2y = -x + 4$, $x + 2y - 4 = 0$

17. Parallel to line through (7,-1)
    and (4,3) means
    $m = \frac{3+1}{4-7} = -\frac{4}{3}$
    y-int. of -2 means $b = -2$
    $y = mx + b$, $y = -\frac{4}{3}x - 2$
    $3y = -4x - 6$, $4x + 3y + 6 = 0$

19. Perpendicular to $5x-2y-3=0$

$2y = 5x - 3$, $y = \frac{5}{2}x - \frac{3}{2}$, $m = \frac{5}{2}$

$m_{line} = -\frac{2}{5}$; $(x_1, y_1) = (3, -4)$

$y - y_1 = m(x - x_1)$

$y + 4 = -\frac{2}{5}(x - 3)$

$5y + 20 = -2x + 6$

$2x + 5y + 14 = 0$

21. $4x - y = 8$

$y = 4x - 8$

$m = 4$, $b = -8$

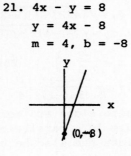

$(0, -8)$

23. $3x + 5y - 10 = 0$

$5y = -3x + 10$

$y = -\frac{3}{5}x + 2$

$m = -\frac{3}{5}$, $b = 2$

$(0, 2)$

25. $3x - 2y - 1 = 0$

$2y = 3x - 1$

$y = \frac{3}{2}x - \frac{1}{2}$

$m = \frac{3}{2}$, $b = -\frac{1}{2}$

27. $5x - 2y + 5 = 0$

$2y = 5x + 5$

$y = \frac{5}{2}x + \frac{5}{2}$

$m = \frac{5}{2}$, $b = \frac{5}{2}$

29. $4x - ky = 6 \qquad 6x + 3y + 2 = 0$

$ky = 4x + 6 \qquad 3y = -6x - 2$

$y = \frac{4}{k}x + \frac{6}{k} \qquad y = -2x - \frac{2}{3}$

$m_1 = \frac{4}{k}$, $m_2 = -2$

$m_1 = m_2$; $\frac{4}{k} = -2$; $k = -2$

31. $3x - y = 9 \qquad kx + 3y = 5$

$y = 3x - 9 \qquad 3y = -kx + 5$

$\qquad\qquad\qquad y = -\frac{k}{3}x + \frac{5}{3}$

$m_1 = 3$, $m_2 = -\frac{k}{3}$

$m_1 = -\frac{1}{m_2}$; $3 = \frac{3}{k}$; $k = 1$

33. $3x - 2y + 5 = 0 \qquad 4y = 6x - 1$

$2y = 3x + 5 \qquad\qquad y = \frac{3}{2}x - \frac{1}{4}$

$y = \frac{3}{2}x + \frac{5}{2} \qquad\qquad m_2 = \frac{3}{2}$

$m_1 = \frac{3}{2}$

$m_1 = m_2$ (lines parallel)

35. $6x - 3y - 2 = 0 \qquad x + 2y - 4 = 0$

$3y = 6x - 2 \qquad\qquad 2y = -x + 4$

$y = 2x - \frac{2}{3} \qquad\qquad y = -\frac{1}{2}x + 2$

$m_1 = 2 \qquad\qquad\qquad m_2 = -\frac{1}{2}$

$m_1 = -\frac{1}{m_2}$ (lines perpendicular)

37. x-int. of 4 means point is $(4,0)$

$2y - 3x - 4 = 0$

$y = \frac{3}{2}x + 2$, $b = 2$

lines passes through $(4,0)$ and $(0,2)$

$m = \frac{2-0}{0-4} = -\frac{1}{2}$; $y = mx + b$

$y = -\frac{1}{2}x + 2$, $2y = -x + 4$, $x+2y-4=0$

39. $5x - y = 6$

$\underline{x + y = 12}$

$6x = 18$

$x = 3$

$y = 9$

Point of intersection is $(3,9)$

$m = -3$

$y - y_1 = m(x - x_1)$

$y - 9 = -3(x - 3)$

$y - 9 = -3x + 9$

$3x + y - 18 = 0$

41. $\ell = 8 + 2c$

$= 2c + 8$

43. 50x characters printed by first microcomputer

60y characters printed by second microcomputer

$50x + 60y = 12200$

$5x + 6y - 1220 = 0$

45. $v_{av} = \frac{s - s_1}{t - t_1} = 50$ m/s ; $(t_1, s_1) = (0, 10)$; $\frac{s - 10}{t - 0} = 50$; $s = 50t + 10$

47. Let L = length, F = force,
   L_0 = natural length
   $L - L_0 = kF$, L_0 = 15 in.
   $L - L_0$ = 2 in. for F = 3 lb
   $2 = k(3)$, $k = \frac{2}{3}$ in./lb
   $L = \frac{2}{3}F + 15$

49. Let H = amount of heat, T = temperature
   $H - H_1 = k(T - T_1)$
   $1.26 = k(50 - 20)$, k = 0.042 kJ/°C
   $4.19 - H_0 = 0.042(20 - 0)$
   $4.19 - H_0 = 0.84$
   $H_0 = 4.19 - 0.84 = 3.35$ kJ

51. n_1 = 45 cars/min for t_1 = 30 min
   n_2 = 115 cars/min for t_2 = 90 min
   $n - n_1 = k(t - t_1)$
   $n - 45 = k(t - 30)$, $n - kt = 45 - 30k$
   $n - 115 = k(t - 90)$, $\underline{n - kt = 115 - 90k}$
   $$0 = 70 - 60k$$
   $n - 45 = \frac{7}{6}(t - 30)$ $k = 7/6$
   $n = \frac{7}{6}t + 10$

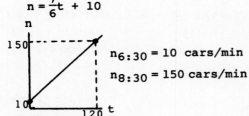

   $n_{6:30}$ = 10 cars/min
   $n_{8:30}$ = 150 cars/min

53. $y = 1 + \sqrt{x}$

x	$\sqrt{x}$	y
0	0.0	1.0
1	1.0	2.0
2	1.4	2.4
4	2.0	3.0
6	2.4	3.4
8	2.8	3.8
9	3.0	4.0

55. $h = 300 + 2t^{3/2}$

t	$t^{3/2}$	h
0	0.0	300
20	89.4	479
40	253	806
60	465	1230
80	716	1732
100	1000	2300

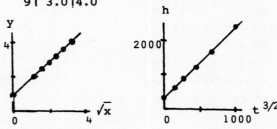

57. $y = 3x^4$

x	y
1	3
2	48
3	243
4	768

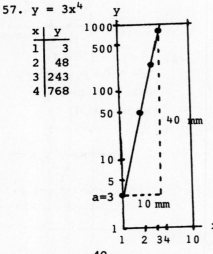

slope = $\frac{40}{10} = 4 = n$

$y = ax^n = 3x^4$

59.

v (m/s)	16.0	32.0	45.3	55.4	64.0
T (N)	0.100	0.400	0.800	1.20	1.60

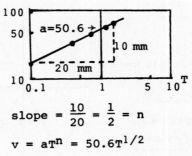

slope = $\frac{10}{20} = \frac{1}{2} = n$

$v = aT^n = 50.6T^{1/2}$

Exercises 1-5, p. 28

1. $(x - 2)^2 + (y - 1)^2 = 25$
   $h = 2, \; k = 1, \; r = \sqrt{25} = 5$
   Center is $(2,1)$

3. $(x + 1)^2 + y^2 = 4$
   $[x-(-1)]^2 + (y - 0)^2 = 2^2$
   $h = -1, \; k = 0, \; r = 2$
   Center is $(-1,0)$

5. Center $(0,0), \; r = 3$
   $x^2 + y^2 = r^2$
   $x^2 + y^2 = 9$

7. Center $(2,2), \; r = 4$
   $(x-h)^2 + (y-k)^2 = r^2$
   $(x-2)^2 + (y-2)^2 = 16$
   $x^2 - 4x + 4 + y^2 - 4y + 4 = 16$
   $x^2 + y^2 - 4x - 4y - 8 = 0$

9. Center $(-2,5), \; r = \sqrt{5}$
   $(x-h)^2 + (y-k)^2 = r^2$
   $[x-(-2)]^2 + (y-5)^2 = (\sqrt{5})^2$
   $(x + 2)^2 + (y - 5)^2 = 5$
   $x^2 + 4x + 4 + y^2 - 10y + 25 = 5$
   $x^2 + y^2 + 4x - 10y + 24 = 0$

11. Center $(-3,0), \; r = \frac{1}{2}$
    $(x-h)^2 + (y-k)^2 = r^2$
    $[x-(-3)]^2 + (y-0)^2 = (\frac{1}{2})^2$
    $(x+3)^2 + y^2 = \frac{1}{4}$
    $x^2 + 6x + 9 + y^2 = \frac{1}{4}$
    $4x^2 + 4y^2 + 24x + 35 = 0$

13. Center $(2,1)$
    Passes through $(4,-1)$
    $(x-h)^2 + (y-k)^2 = r^2$
    $(x-2)^2 + (y-1)^2 = r^2$
    $(4-2)^2 + (-1-1)^2 = r^2$
    $r^2 = 8$
    $(x-2)^2 + (y-1)^2 = 8$
    $x^2 - 4x + 4 + y^2 - 2y + 1 = 8$
    $x^2 + y^2 - 4x - 2y - 3 = 0$

15. Center $(-3,5)$
    Tangent to x-axis means
    $r = |k| = 5$
    $(x-h)^2 + (y-k)^2 = r^2$
    $[x-(-3)]^2 + (y-5)^2 = 5^2$
    $(x+3)^2 + (y-5)^2 = 25$
    $x^2 + 6x + 9 + y^2 - 10y + 25 = 25$
    $x^2 + y^2 + 6x - 10y + 9 = 0$

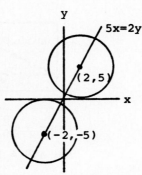

17. Tangent to both axes
    and lines y=4 and
    x=4 means center is
    (2,2) amd radius is 2.
    $(x-h)^2 + (y-k)^2 = r^2$
    $(x-2)^2 + (y-2)^2 = 2^2$
    $x^2 - 4x + 4 + y^2 - 4y + 4 = 4$
    $x^2 + y^2 - 4x - 4y + 4 = 0$

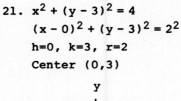

19. Center on line 5x=2y
    means 5h=2k ; r=5
    Tangent to x-axis
    means $r = |k| = 5$
    For k=5, 5h=2(5), h=2
    For k=-5, 5h=2(-5), h=-2
    $(x-h)^2 + (y-k)^2 = r^2$
    $(x-2)^2 + (y-5)^2 = 5^2$
    $x^2 - 4x + 4 + y^2 - 10y + 25 = 25$
    $x^2 + y^2 - 4x - 10y + 4 = 0$

    $(x+2)^2 + (y+5)^2 = 5^2$
    $x^2 + 4x + 4 + y^2 + 10y + 25 = 25$
    $x^2 + y^2 + 4x + 10y + 4 = 0$

Two circles satisfy
given conditions.

21. $x^2 + (y - 3)^2 = 4$
    $(x - 0)^2 + (y - 3)^2 = 2^2$
    $h=0, \; k=3, \; r=2$
    Center $(0,3)$

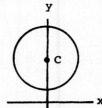

23. $4(x+1)^2 + 4(y-5)^2 = 81$

$(x+1)^2 + (y-5)^2 = \frac{81}{4} = (\frac{9}{2})^2$

$h = -1, \ k = 5, \ r = \frac{9}{2}$

Center $(-1,5)$

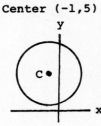

25. $x^2 + y^2 - 25 = 0$

$x^2 + y^2 = 25 = 5^2$

Center $(0,0), \ r = 5$

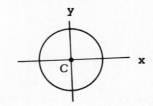

27. $x^2 + y^2 - 2x - 8 = 0$

$(x^2 - 2x) + y^2 = 8$

$(x^2 - 2x + 1) + y^2 = 8 + 1$

$(x - 1)^2 + y^2 = 9$

Center $(1,0), \ r = 3$

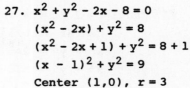

29. $x^2 + y^2 + 8x - 10y - 8 = 0$

$(x^2 + 8x) + (y^2 - 10y) = 8$

$(x^2 + 8x + 16) + (y^2 - 10y + 25) = 8 + 16 + 25$

$(x + 4)^2 + (y - 5)^2 = 49$

Center $(-4,5), \ r = 7$

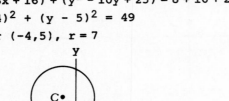

31. $2x^2 + 2y^2 - 4x - 8y - 1 = 0$

$x^2 + y^2 - 2x - 4y - \frac{1}{2} = 0$

$(x^2 - 2x) + (y^2 - 4y) = \frac{1}{2}$

$(x^2 - 2x + 1) + (y^2 - 4y + 4) = \frac{1}{2} + 1 + 4$

$(x - 1)^2 + (y - 2)^2 = \frac{11}{2}$

Center $(1,2), \ r = \sqrt{11/2} = \frac{1}{2}\sqrt{22}$

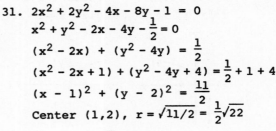

33. $x^2 + y^2 = 100$

$x^2 + (-y)^2 = 100 \qquad (-x)^2 + y^2 = 100 \qquad (-x)^2 + (-y)^2 = 100$

$x^2 + y^2 = 100 \qquad x^2 + y^2 = 100 \qquad x^2 + y^2 = 100$

(Sym. to x-axis)   (Sym. to y-axis)   (Sym. to origin)

35. $x^2 + y^2 + 8y - 9 = 0$

$x^2 + (-y)^2 + 8(-y) - 9 = 0 \qquad (-x)^2 + y^2 + 8y - 9 = 0 \qquad (-x)^2 + (-y)^2 + 8(-y) - 9 = 0$

$x^2 + y^2 \underset{\uparrow}{-} 8y - 9 = 0 \qquad x^2 + y^2 + 8y - 9 = 0 \qquad x^2 + y^2 \underset{\uparrow}{-} 8y - 9 = 0$

(Not sym. to x-axis)   (Sym. to y-axis)   (Not sym. to origin)

37. $x^2 - 6x + y^2 - 7 = 0$

crosses x-axis where y=0

$x^2 - 6x + 0 - 7 = 0$

$x^2 - 6x - 7 = 0$

$(x - 7)(x + 1) = 0$

$x = 7, \ -1$

Crosses at $(7,0)$ and $(-1,0)$

39. $\sqrt{(x - 2)^2 + (y - 4)^2} = 2(\sqrt{x^2 + y^2})$

$(x - 2)^2 + (y - 4)^2 = 4(x^2 + y^2)$

$x^2 - 4x + 4 + y^2 - 8y + 16 = 4x^2 + 4y^2$

$3x^2 + 3y^2 + 4x + 8y - 20 = 0$

Circle

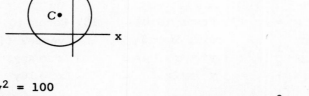

41. $R = \frac{mv}{Bq} = \frac{(1.67 \times 10^{-27} \text{ kg})(2.40 \times 10^8 \text{ m/s})}{(1.50 \text{ V} \cdot \text{s/m}^2)(1.60 \times 10^{-19} \text{ C})} = 1.67 \ \frac{\text{kg} \cdot \text{m/s}}{\text{V} \cdot \text{s} \cdot \text{C/m}^2} = 1.67 \ \frac{\text{kg} \cdot \text{m}^3}{\text{V} \cdot \text{s}^2 \cdot \text{C}}$

$= 1.67 \ \frac{\text{kg} \cdot \text{m}^3}{(\text{J/A} \cdot \text{s}) \cdot (\text{A} \cdot \text{s}) \cdot \text{s}^2} = 1.67 \ \frac{\text{kg} \cdot \text{m}^3}{\text{J} \cdot \text{s}^2} = 1.67 \ \frac{\text{kg} \cdot \text{m}^3}{\text{N} \cdot \text{m} \cdot \text{s}^2} = 1.67 \ \frac{\text{kg} \cdot \text{m}^2}{(\text{m} \cdot \text{kg/s}^2) \cdot \text{s}^2}$

$= 1.67 \text{ m}$ (see Appendix B) $x^2 + y^2 = (1.67)^2 = 2.79$ (center at origin)

43. Tank: $r = 2$ ft $= 24$ in.   Hole: $r = 3.00$ in.

      Center at origin   Center at $(0,-18)$

      $x^2 + y^2 = 24^2$ $x^2 + (y+18)^2 = 3^2$

      $x^2 + y^2 = 576$ $x^2 + y^2 + 36y + 324 = 9$

                              $x^2 + y^2 + 36y + 315 = 0$

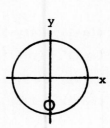

Exercises 1-6, p. 33

1. $y^2 = 4x$
$y^2 = 4px$
$4p = 4$, $p = 1$
focus $(1,0)$
dir. $x = -1$

3. $y^2 = -4x$
$y^2 = 4px$
$4p = -4$, $p = -1$
focus $(-1,0)$
dir. $x = 1$

5. $x^2 = 8y$
$x^2 = 4py$
$4p = 8$, $p = 2$
focus $(0,2)$
dir. $y = -2$

7. $x^2 = -4y$
$x^2 = 4py$
$4p = -4$, $p = -1$
focus $(0,-1)$
dir. $y = 1$

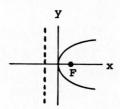 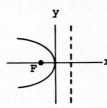 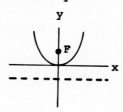 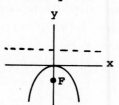

9. $y^2 = 2x$
$y^2 = 4px$
$4p = 2$, $p = \frac{1}{2}$
focus $(\frac{1}{2},0)$
dir. $x = -\frac{1}{2}$

11. $y = x^2$, $x^2 = y$
$x^2 = 4py$
$4p = 1$, $p = \frac{1}{4}$
focus $(0,\frac{1}{4})$
dir. $y = -\frac{1}{4}$

13. Focus $(3,0)$
dir. $x = -3$
$y^2 = 4px$, $p = 3$
$y^2 = 12x$

15. Focus $(0,4)$
vertex $(0,0)$
$x^2 = 4py$, $p = 4$
$x^2 = 16y$

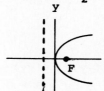 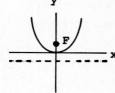

17. Vertex $(0,0)$
dir. $y = -1$
$x^2 = 4py$, $p = 1$
$x^2 = 4y$

19. Vertex $(0,0)$
Axis along y-axis
means $x^2 = 4py$.
$(-1,8)$ must satisfy
equation.
$(-1)^2 = 4p(8)$, $p = \frac{1}{32}$
$x^2 = 4(\frac{1}{32})y = \frac{1}{8}y$

21.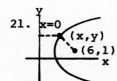
$\sqrt{(x-6)^2+(y-1)^2} = x$
$(x-6)^2 + (y-1)^2 = x^2$
$x^2 - 12x + 36 + y^2 - 2y + 1 = x^2$
$y^2 - 12x - 2y + 37 = 0$

23. Since focus is 2 units below
vertex, directrix is 2 units above.
$\sqrt{(x-1)^2+(y-1)^2} = 5 - y$
$(x-1)^2 + (y-1)^2 = (5-y)^2$
$x^2 - 2x + 1 + y^2 - 2y + 1 = 25 - 10y + y^2$
$x^2 - 2x + 8y - 23 = 0$

25. Incident ray: $m = \frac{4-0}{1-4} = -\frac{4}{3}$
$y - 4 = -\frac{4}{3}(x - 1)$ $y^2 = 16x$, $p = 4$
$3y - 12 = -4x + 4$
$4x + 3y - 16 = 0$
Reflected ray is
parallel to x-axis.
$y = 4$

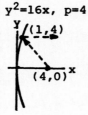 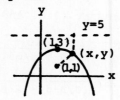

27. $y^2 = 4px$
   $(1.25)^2 = 4p(0.425)$
   $p = 0.919$ m

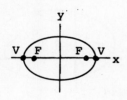

29. $x^2 = 4py$
   $10^2 = 4p(1)$
   $p = 25$
   $x^2 = 100y$

31. $T = 2\pi\sqrt{LC}$, $L = 1$ H
   $T = 2\pi\sqrt{C}$
   1 μF $= 10^{-6}$ F, 1 ms $= 10^{-3}$ s

C(μF)	T(ms)
1	6
50	44
100	63
150	77
200	89
250	99

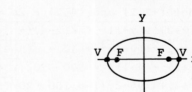

33. $x^2 = 4py$, $p = 2$
   $x^2 = 8y$
   (or $y^2 = 8x$ if shoreline
   is shown vertically.)

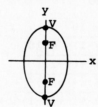

Exercises 1-7, p. 39

1. $\dfrac{x^2}{4} + \dfrac{y^2}{1} = 1$

   Major axis along x-axis
   $a^2 = 4$, $b^2 = 1$
   $a^2 = b^2 + c^2$, $4 = 1 + c^2$, $c^2 = 3$
   $a = 2$, $b = 1$, $c = \sqrt{3}$
   Vertices $(2,0)$, $(-2,0)$
   Foci $(\sqrt{3},0)$, $(-\sqrt{3},0)$

3. $\dfrac{x^2}{25} + \dfrac{y^2}{36} = 1$

   Major axis along y-axis
   $a^2 = 36$, $b^2 = 25$
   $a^2 = b^2 + c^2$, $36 = 25 + c^2$, $c^2 = 11$
   $a = 6$, $b = 5$, $c = \sqrt{11}$
   Vertices $(0,6)$, $(0,-6)$
   Foci $(0,\sqrt{11})$, $(0,-\sqrt{11})$

5. $4x^2 + 9y^2 = 36$

   $\dfrac{x^2}{9} + \dfrac{y^2}{4} = 1$

   Major axis along x-axis
   $a^2 = 9$, $b^2 = 4$
   $a^2 = b^2 + c^2$, $9 = 4 + c^2$, $c^2 = 5$
   $a = 3$, $b = 2$, $c = \sqrt{5}$
   Vertices $(3,0)$, $(-3,0)$
   Foci $(\sqrt{5},0)$, $(-\sqrt{5},0)$

7. $49x^2 + 4y^2 = 196$

   $\dfrac{x^2}{4} + \dfrac{y^2}{49} = 1$

   Major axis along y-axis
   $a^2 = 49$, $b^2 = 4$
   $a^2 = b^2 + c^2$, $49 = 4 + c^2$, $c^2 = 45$
   $a = 7$, $b = 2$, $c = \sqrt{45} = 3\sqrt{5}$
   Vertices $(0,7)$, $(0,-7)$
   Foci $(0,\sqrt{45})$, $(0,-\sqrt{45})$

9. $8x^2 + y^2 = 16$

   $\dfrac{x^2}{2} + \dfrac{y^2}{16} = 1$

   Major axis along y-axis
   $a^2 = 16$, $b^2 = 2$
   $a^2 = b^2 + c^2$, $16 = 2 + c^2$, $c^2 = 14$
   $a = 4$, $b = \sqrt{2}$, $c = \sqrt{14}$
   Vertices $(0,4)$, $(0,-4)$
   Foci $(0,\sqrt{14})$, $(0,-\sqrt{14})$

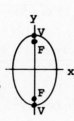
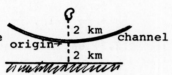

11. $4x^2 + 25y^2 = 25$

$\dfrac{x^2}{25/4} + \dfrac{y^2}{1} = 1$

Major axis along x-axis

$a^2 = \dfrac{25}{4}, \ b^2 = 1$

$a^2 = b^2 + c^2, \ \dfrac{25}{4} = 1 + c^2, \ c^2 = \dfrac{21}{4}$

$a = \dfrac{5}{2}, \ b = 1, \ c = \dfrac{1}{2}\sqrt{21}$

Vertices $(\dfrac{5}{2}, 0), (-\dfrac{5}{2}, 0)$

Foci $(\dfrac{1}{2}\sqrt{21}, -\dfrac{1}{2}\sqrt{21})$

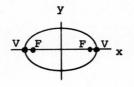

13. Vertex (15,0), focus (9,0)

Major axis along x-axis

$a = 15, \ c = 9$

$a^2 = b^2 + c^2, \ 225 = b^2 + 81$

$b^2 = 144$

$\dfrac{x^2}{225} + \dfrac{y^2}{144} = 1$

$144x^2 + 225y^2 = 32{,}400$

15. Focus (0,2)

Major axis along y-axis

$c = 2, \ a = 3$

$a^2 = b^2 + c^2$

$9 = b^2 + 4, \ b^2 = 5$

$\dfrac{x^2}{5} + \dfrac{y^2}{9} = 1$

$9x^2 + 5y^2 = 45$

17. Vertex (8,0), passes through (2,3)

Major axis along x-axis ; $a = 8$

$\dfrac{x^2}{a^2} + \dfrac{y^2}{b^2} = 1$, $\dfrac{x^2}{64} + \dfrac{y^2}{b^2} = 1$

(2,3) satifies equation

$\dfrac{4}{64} + \dfrac{9}{b^2} = 1, \ \dfrac{9}{b^2} = \dfrac{15}{16}, \ b^2 = \dfrac{48}{5}$

$\dfrac{x^2}{64} + \dfrac{y^2}{\frac{48}{5}} = 1, \ \dfrac{x^2}{64} + \dfrac{5y^2}{48} = 1, \ 3x^2 + 20y^2 = 192$

19. Passes through (2,2) and (1,4) ; points satisfy equation $\dfrac{x^2}{A} + \dfrac{y^2}{B} = 1$

Find A and B (a and b not used since direction of major axis not specified)

$\dfrac{4}{A} + \dfrac{4}{B} = 1 \qquad \dfrac{4}{A} + \dfrac{4}{B} = 1 \qquad\qquad \dfrac{4}{A} + \dfrac{4}{20} = 1, \ \dfrac{4}{A} = \dfrac{4}{5}, \ A = 5$

$\dfrac{1}{A} + \dfrac{16}{B} = 1 \qquad \dfrac{4}{A} + \dfrac{64}{B} = 4 \qquad\qquad \dfrac{x^2}{5} + \dfrac{y^2}{20} = 1, \quad 4x^2 + y^2 = 20$

$\qquad\qquad\qquad\qquad \dfrac{60}{B} = 3, \ B = 20 \qquad$ (Major axis along y-axis)

21. Foci (-2,1) and (4,1) : $2a = 10$

$\sqrt{(x+2)^2+(y-1)^2} + \sqrt{(x-4)^2+(y-1)^2} = 10$

$\sqrt{(x+2)^2+(y-1)^2} = 10 - \sqrt{(x-4)^2+(y-1)^2}$

$(x+2)^2+(y-1)^2 = 100 - 20\sqrt{(x-4)^2+(y-1)^2} + (x-4)^2 + (y-1)^2$

$x^2+4x+4+y^2-2y+1 = 100 - 20\sqrt{(x-4)^2+(y-1)^2} + x^2-8x+16+y^2-2y+1$

$20\sqrt{(x-4)^2+(y-1)^2} = 112 - 12x, \ 5\sqrt{(x-4)^2+(y-1)^2} = 28 - 3x$

$25[(x-4)^2 + (y-1)^2] = 784 - 168x + 9x^2$

$25x^2 - 200x + 400 + 25y^2 - 50y + 25 = 784 - 168x + 9x^2$

$16x^2 + 25y^2 - 32x - 50y - 359 = 0$

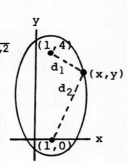

23. Vertices (1,5) and (1,-1) give $2a = 6$; Foci (1,4) and (1,0)

$\sqrt{(x-1)^2+(y-4)^2} + \sqrt{(x-1)^2+y^2} = 6 \ ; \ \sqrt{(x-1)^2+(y-4)^2} = 6 - \sqrt{(x-1)^2+y^2}$

$(x-1)^2 + (y-4)^2 = 36 - 12\sqrt{(x-1)^2+y^2} + (x-1)^2$

$x^2-2x+1+y^2-8y+16 = 36 - 12\sqrt{(x-1)^2+y^2} + x^2-2x+1+y^2$

$12\sqrt{(x-1)^2 + y^2} = 20 + 8y \ ; \ 3\sqrt{(x-1)^2 + y^2} = 5 + 2y$

$9[(x-1)^2 + y^2] = 25 + 20y + 4y^2$

$9x^2 - 18x + 9 + 9y^2 = 25 + 20y + 4y^2$

$9x^2 + 5y^2 - 18x - 20y - 16 = 0$

25. $2x^2 + 3y^2 - 8x - 4 = 0$

    $2x^2 + 3(-y)^2 - 8x - 4 = 0$

    $2x^2 + 3y^2 - 8x - 4 = 0$

    Equation is unchanged. Ellipse is symmetrical to the x-axis.

27. $2a = 100,\ b = 30$

$$\frac{x^2}{50^2} + \frac{y^2}{30^2} = 1$$

$$\frac{x^2}{2500} + \frac{y^2}{900} = 1$$

$$9x^2 + 25y^2 = 22,500$$

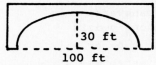

30 ft

100 ft

29. $2a = 2000+4000+4000+500$

      $= 10,500$

    $a = 5250$

    $c = 5250 - 4500 = 750$

    $a^2 = b^2 + c^2$

    $5250^2 = b^2 + 750^2$

    $b^2 = 2.70 \times 10^7$

    $a^2 = 2.76 \times 10^7$

$$\frac{x^2}{2.76\times10^7} + \frac{y^2}{2.70\times10^7} = 1$$

    $2.70x^2 + 2.76y^2 = 7.45\times10^7$

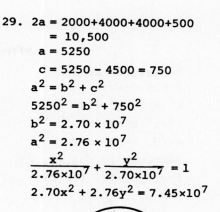

2000 mi   500 mi

4000 mi

31. **x is the major axis of the ellipse.**

    The minor axis is perpendicular to the page and equals the diameter of the pipe.

    $x^2 = 6.0^2 + 6.0^2 = 72$

    $x = \sqrt{72} = 8.5$ in.

    Minor axis = 6.0 in.

6.0 in.   45°   6.0 in.

33. **Base of each triangle is 6**

    perimeter = 14 in. , $2a = 14 - 6 = 8$ in.

    $a = 4,\ c = 3$

$a^2 = b^2 + c^2$

$16 = b^2 + 9$

$b^2 = 7$

$$\frac{x^2}{16} + \frac{y^2}{7} = 1$$

$7x^2 + 16y^2 = 112$

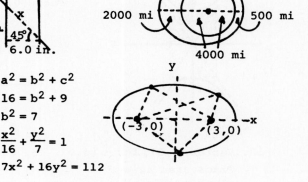

(-3,0)   (3,0)

Exercises 1-8, p. 45

1. $$\dfrac{x^2}{25} - \dfrac{y^2}{144} = 1$$

    Transverse axis along x-axis

    $a^2 = 25,\ b^2 = 144$

    $c^2 = a^2 + b^2 = 25+144 = 169$

    $a=5,\ b=12,\ c=13$

    Vertices (5,0), (-5,0)

    Foci (13,0), (-13,0)

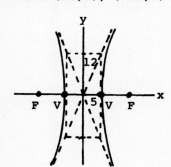

3. $$\dfrac{y^2}{9} - \dfrac{x^2}{1} = 1$$

    Transverse axis along y-axis

    $a^2 = 9,\ b^2 = 1$

    $c^2 = a^2 + b^2 = 9+1 = 10$

    $a=3,\ b=1,\ c=\sqrt{10}$

    Vertices (0,3), (0,-3)

    Foci $(0,\sqrt{10})$, $(0,-\sqrt{10})$

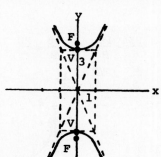

5. $2x^2 - y^2 = 4$

$$\dfrac{x^2}{2} - \dfrac{y^2}{4} = 1$$

    Transverse axis along x-axis

    $a^2 = 2,\ b^2 = 4$

    $c^2 = a^2 + b^2 = 2+4 = 6$

    $a=\sqrt{2},\ b=2,\ c=\sqrt{6}$

    Vertices $(\sqrt{2},0)$, $(-\sqrt{2},0)$

    Foci $(\sqrt{6},0)$, $(-\sqrt{6},0)$

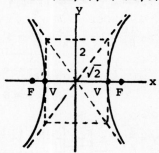

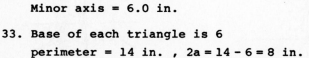

7. $2y^2 - 5x^2 = 10$

$\dfrac{y^2}{5} - \dfrac{x^2}{2} = 1$

Transverse axis along y-axis

$a^2 = 5$, $b^2 = 2$

$c^2 = a^2 + b^2 = 5 + 2 = 7$

$a = \sqrt{5}$, $b = \sqrt{2}$, $c = \sqrt{7}$

Vertices $(0,\sqrt{5}),(0,-\sqrt{5})$

Foci $(0,\sqrt{7})$, $(0,-\sqrt{7})$

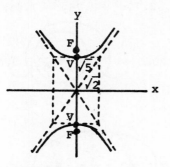

9. $4x^2 - y^2 + 4 = 0$

$4x^2 - y^2 = -4$

$-\dfrac{x^2}{1} + \dfrac{y^2}{4} = 1$

$\dfrac{y^2}{4} - \dfrac{x^2}{1} = 1$

Transverse axis along y-axis

$a^2 = 4$, $b^2 = 1$

$c^2 = a^2 + b^2 = 4 + 1 = 5$

$a = 2$, $b = 1$, $c = \sqrt{5}$

Vertices $(0,2)$, $(0,-2)$

Foci $(0,\sqrt{5})$, $(0,-\sqrt{5})$

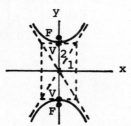

11. $4x^2 - 9y^2 = 16$

$\dfrac{x^2}{4} - \dfrac{y^2}{16/9} = 1$

Transverse axis along x-axis

$a^2 = 4$, $b^2 = \dfrac{16}{9}$

$c^2 = a^2 + b^2 = 4 + \dfrac{16}{9} = \dfrac{52}{9}$

$a = 2$, $b = \dfrac{4}{3}$, $c = \sqrt{\dfrac{52}{9}} = \dfrac{2}{3}\sqrt{13}$

Vertices $(2,0)$, $(-2,0)$

Foci $(\dfrac{2}{3}\sqrt{13},0)$, $(-\dfrac{2}{3}\sqrt{13},0)$

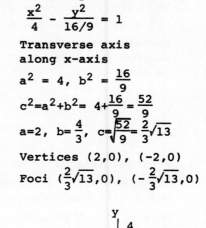

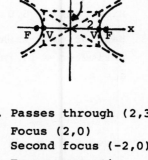

13. Vertex $(3,0)$, focus $(5,0)$

Transverse axis along x-axis

$a = 3$, $c = 5$

$c^2 = a^2 + b^2$, $25 = 9 + b^2$, $b^2 = 16$

$\dfrac{x^2}{9} - \dfrac{y^2}{16} = 1$

$16x^2 - 9y^2 = 144$

15. Conjugate axis $= 12$

Vertex $(0,10)$

Transverse axis along y-axis

$b = 6$, $a = 10$

$\dfrac{y^2}{100} - \dfrac{x^2}{36} = 1$

$9y^2 - 25x^2 = 900$

17. Passes through $(2,3)$

Focus $(2,0)$

Second focus $(-2,0)$

Transverse axis along x-axis, $c = 2$

$d_1 = \sqrt{(2+2)^2 + (3-0)^2} = 5$

$d_2 = \sqrt{(2-2)^2 + (3-0)^2} = 3$

$d_1 - d_2 = 5 - 3 = 2$

$2a = 2$, $a = 1$

$c^2 = a^2 + b^2$; $4 = 1 + b^2$, $b^2 = 3$

$\dfrac{x^2}{1} - \dfrac{y^2}{3} = 1$

$3x^2 - y^2 = 3$

19. Passes through $(5,4)$ and $(3,\dfrac{4}{5}\sqrt{5})$

Assume equation $\dfrac{x^2}{A} - \dfrac{y^2}{B} = 1$ (direction of transverse axis not specified)

Points satisfy equation

$\dfrac{25}{A} - \dfrac{16}{B} = 1 \qquad \dfrac{25}{A} - \dfrac{16}{B} = 1$

$\dfrac{9}{A} - \dfrac{16}{5B} = 1 \qquad \dfrac{45}{A} - \dfrac{16}{B} = 5$

$\qquad\qquad\qquad\qquad \overline{\qquad \dfrac{20}{A} = 4 \qquad}$

$\dfrac{25}{5} - \dfrac{16}{B} = 1$, $\dfrac{16}{B} = \dfrac{20}{5}$, $B = 4$

$\dfrac{x^2}{5} - \dfrac{y^2}{4} = 1$, $4x^2 - 5y^2 = 20$

$A = 5$

21. $xy = 2$

$y = \dfrac{2}{x}$

x	-4	-2	-1	$-\frac{1}{2}$	$\frac{1}{2}$	1	2	4
y	$-\frac{1}{2}$	-1	-2	-4	4	2	1	$\frac{1}{2}$

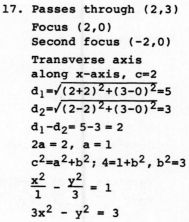

23. $xy = -2$

$y = -\dfrac{2}{x}$

x	-4	-2	-1	$-\frac{1}{2}$	$\frac{1}{2}$	1	2	4
y	$\frac{1}{2}$	1	2	4	-4	-2	-1	$-\frac{1}{2}$

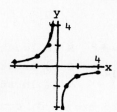

25. Foci $(1,2)$ and $(11,2)$, $2a = 8$

$\sqrt{(x-1)^2+(y-2)^2} - \sqrt{(x-11)^2+(y-2)^2} = 8$

$\sqrt{(x-1)^2+(y-2)^2} = 8 + \sqrt{(x-11)^2+(y-2)^2}$

$(x-1)^2+(y-2)^2 = 64 + 16\sqrt{(x-11)^2+(y-2)^2} + (x-11)^2+(y-2)^2$

$x^2-2x+1+y^2-4y+4 = 64+16\sqrt{(x-11)^2+(y-2)^2} + x^2-22x+121+y^2-4y+4$

$16\sqrt{(x-11)^2+(y-2)^2} = 20x-184$; $4\sqrt{(x-11)^2+(y-2)^2} = 5x-46$

$16[(x-11)^2+(y-2)^2] = 25x^2-460x+2116$

$16x^2 - 352x + 1936 + 16y^2 - 64y + 64 = 25x^2 - 460x + 2116$

$9x^2 - 16y^2 - 108x + 64y + 116 = 0$

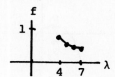

27. Center $(2,0)$ and vertex $(3,0)$ gives $a=1$, $2a=2$, transverse axis along x-axis
Conjugate axis of 6 gives $2b=6$, $b=3$; $c^2=a^2+b^2= 1+9 = 10$, $c = \sqrt{10}$
Foci at $(2+\sqrt{10},0)$ and $(2-\sqrt{10},0)$

$\sqrt{[x-(2+\sqrt{10})]^2+y^2} - \sqrt{[x-(2-\sqrt{10})]^2+y^2} = 2$

$\sqrt{[x-(2+\sqrt{10})]^2+y^2} = 2 + \sqrt{[x-(2-\sqrt{10})]^2+y^2}$

$[x-(2+\sqrt{10})]^2+y^2 = 4+4\sqrt{[x-(2-\sqrt{10})]^2+y^2} + [x-(2-\sqrt{10})]^2+y^2$

$x^2-2(2+\sqrt{10})x+(2+\sqrt{10})^2 = 4+4\sqrt{[x-(2-\sqrt{10})]^2+y^2} + x^2-2(2-\sqrt{10})x+(2-\sqrt{10})^2$

$4\sqrt{[x-(2-\sqrt{10})]^2+y^2} = -2(2+\sqrt{10})x+2(2-\sqrt{10})x+(2+\sqrt{10})^2-(2-\sqrt{10})^2-4$

$= -4x-2\sqrt{10}x+4x-2\sqrt{10}x+4+4\sqrt{10}+10-4+4\sqrt{10}-10-4$

$= -4\sqrt{10}x + 8\sqrt{10} - 4$

$\sqrt{[x-(2-\sqrt{10})]^2+y^2} = \sqrt{10}(2-x) - 1$

$x^2-2(2-\sqrt{10})x+(2-\sqrt{10})^2+y^2 = 10(4-4x+x^2)-2\sqrt{10}(2-x)+1$

$x^2-4x+2\sqrt{10}x+4-4\sqrt{10}+10+y^2 = 40-40x+10x^2-4\sqrt{10}+2\sqrt{10}x+1$

$9x^2 - y^2 -36x + 27 = 0$

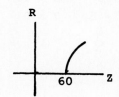

29. $z = \sqrt{(X_L - X_C)^2 + R^2}$

$X_L - X_C = 60\ \Omega$

$z = \sqrt{3600 + R^2}$

$z^2 = 3600 + R^2$

$z^2 - R^2 = 3600$

$\dfrac{z^2}{3600} - \dfrac{R^2}{3600} = 1$

$z \geq 0$ from original equation
$R \geq 0$ from physical meaning
$a=60$, $b=60$

31. $v = f\lambda$, $v = 3.0 \times 10^{10}$ cm/s

$f = \dfrac{3.0 \times 10^{10}}{\lambda}$

λ(cm) $(\times 10^{-5})$	f(1/s) $(\times 10^{15})$
4.0	0.75
5.0	0.60
6.0	0.50
7.0	0.43

33.

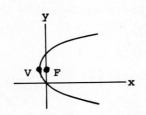

R is location of rifle

P is location of instrument

T is location of target

Time t_1 for sound to travel d_1 equals time t_3 for bullet to travel d_3 plus time t_2 for sound to travel d_2. Let v_s = velocity of sound and v_b = velocity of bullet. (distance = velocity × time)

$$t_1 = t_3 + t_2 \quad ; \quad \frac{d_1}{v_s} = \frac{d_3}{v_b} + \frac{d_2}{v_s} \quad ; \quad d_1 = \frac{v_s}{v_b}d_3 + d_2 \quad ; \quad d_1 - d_2 = \frac{v_s}{v_b}d_3$$

Since v_s, v_b and d_3 are all constants, $d_1 - d_2$ = constant and P lies on a hyperbola.

Exercises 1-9, p. 50

1. $(y-2)^2 = 4(x+1)$
    $x' = x+1, \; y' = y-2$
    $x' = x-h, \; y' = y-k$
    $h = -1, \; k = 2$
    $y'^2 = 4x'$
    Parabola, vertex $(-1,2)$
    Axis parallel to x-axis
    $4p = 4, \; p = 1$
    Focus $(-1+p, 2)$ or $(0,2)$

3. $\dfrac{(x-1)^2}{4} - \dfrac{(y-2)^2}{9} = 1$
    $x' = x-1, \; y' = y-2$
    $x' = x-h, \; y' = y-k$
    $h = 1, \; k = 2$
    $\dfrac{x'^2}{4} - \dfrac{y'^2}{9} = 1$
    Hyperbola, center $(1,2)$
    Transverse axis parallel to x-axis
    $a = 2, \; b = 3$

5. $\dfrac{(x+1)^2}{1} + \dfrac{y^2}{9} = 1$
    $x' = x+1, \; y' = y+0$
    $x' = x-h, \; y' = y-k$
    $h = -1, \; k = 0$
    $\dfrac{x'^2}{1} + \dfrac{y'^2}{9} = 1$
    Ellipse, center $(-1,0)$
    Major axis parallel to y-axis
    $a = 3, \; b = 1$

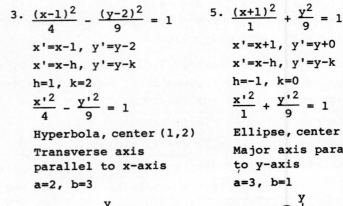

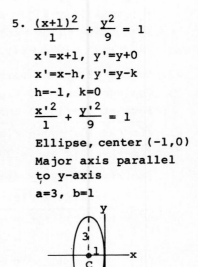

7. $(x+3)^2 = -12(y-1)$
    $x' = x+3, \; y' = y-1$
    $x' = x-h, \; y' = y-k$
    $h = -3, \; k = 1$
    $x'^2 = -12y'$
    Parabola, vertex $(-3,1)$
    Axis parallel to y-axis
    $4p = -12, \; p = -3$
    Focus $(-3, 1+p)$ or $(-3,-2)$

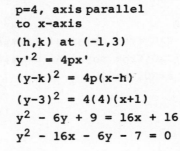

9. Parabola: vertex $(-1,3)$
    $p = 4$, axis parallel to x-axis
    (h,k) at $(-1,3)$
    $y'^2 = 4px'$
    $(y-k)^2 = 4p(x-h)$
    $(y-3)^2 = 4(4)(x+1)$
    $y^2 - 6y + 9 = 16x + 16$
    $y^2 - 16x - 6y - 7 = 0$

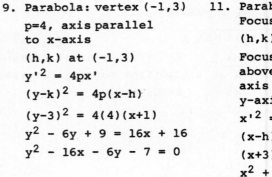

11. Parabola: vertex $(-3,2)$
    Focus $(-3,3)$
    (h,k) at $(-3,2)$
    Focus one unit above vertex means axis parallel to y-axis with $p=1$
    $x'^2 = 4py'$
    $(x-h)^2 = 4p(y-k)$
    $(x+3)^2 = 4(1)(y-2)$
    $x^2 + 6x + 9 = 4y - 8$
    $x^2 + 6x - 4y + 17 = 0$

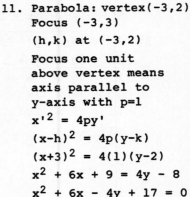

13. Ellipse: center $(-2,2)$
focus $(-5,2)$, vertex $(-7,2)$

(h,k) at $(-2,2)$

a=dist. center to vertex = 5
c=dist. center to focus = 3

focus to left of center means
major axis parallel to x-axis

$a^2 = b^2 + c^2$, $25 = b^2 + 9$, $b^2 = 16$

$$\frac{x'^2}{a^2} + \frac{y'^2}{b^2} = 1$$

$$\frac{(x-h)^2}{a^2} + \frac{(y-k)^2}{b^2} = 1$$

$$\frac{(x+2)^2}{25} + \frac{(y-2)^2}{16} = 1$$

$16(x+2)^2 + 25(y-2)^2 = 400$

$16x^2 + 25y^2 + 64x - 100y - 236 = 0$

15. Ellipse: vertices $(-2,-3)$ and $(-2,5)$
end of minor axis $(0,1)$

(h,k) midpoint between vertices
at $(-2,1)$

a= dist. center to vertex = 4
b= dist. center to end of minor axis = 2

one vertex above other means
major axis parallel to y-axis

$$\frac{y'^2}{a^2} + \frac{x'^2}{b^2} = 1$$

$$\frac{(y-k)^2}{a^2} + \frac{(x-h)^2}{b^2} = 1$$

$$\frac{(y-1)^2}{16} + \frac{(x+2)^2}{4} = 1$$

$(y-1)^2 + 4(x+2)^2 = 16$

$4x^2 + y^2 + 16x - 2y + 1 = 0$

17. Hyperbola: vertex $(-1,1)$
focus $(-1,4)$, center $(-1,2)$

(h,k) at $(-1,2)$

a=dist. center to vertex = 1
c=dist. center to focus = 2

focus above center means transverse axis parallel to y-axis

$c^2 = a^2 + b^2$, $4 = 1 + b^2$, $b^2 = 3$

$$\frac{y'^2}{a^2} - \frac{x'^2}{b^2} = 1$$

$$\frac{(y-k)^2}{a^2} - \frac{(x-h)^2}{b^2} = 1$$

$$\frac{(y-2)^2}{1} - \frac{(x+1)^2}{3} = 1$$

$3(y-2)^2 - (x+1)^2 = 3$

$x^2 - 3y^2 + 2x + 12y - 8 = 0$

19. Hyperbola: vertices $(2,1)$ and $(-4,1)$
focus $(-6,1)$

(h,k) midpoint between vertices at $(-1,1)$

dist. between vertices = $2a = 6$, a=3
c=dist. center to focus = 5

$c^2 = a^2 + b^2$, $25 = 9 + b^2$, $b^2 = 16$

one vertex to right of other means
transverse axis parallel to x-axis

$$\frac{x'^2}{a^2} - \frac{y'^2}{b^2} = 1$$

$$\frac{(x-h)^2}{a^2} - \frac{(y-k)^2}{b^2} = 1$$

$$\frac{(x+1)^2}{9} - \frac{(y-1)^2}{16} = 1$$

$16(x+1)^2 - 9(y-1)^2 = 144$

$16x^2 - 9y^2 + 32x + 18y - 137 = 0$

21. $x^2 + 2x - 4y - 3 = 0$

$x^2 + 2x = 4y + 3$

$x^2 + 2x + 1 = 4y + 3 + 1$

$(x + 1)^2 = 4(y + 1)$

$h = -1$, $k = -1$, $x'^2 = 4y'$

Parabola, vertex $(-1,-1)$

Axis parallel to y-axis
Focus $(-1,-1+p)$ or $(-1,0)$

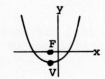

23. $4x^2 + 9y^2 + 24x = 0$

$4(x^2 + 6x) + 9y^2 = 0$

$4(x^2 + 6x + 9) + 9y^2 = 36$

$$\frac{(x+3)^2}{9} + \frac{y^2}{4} = 1 \; ; \; h=-3, \; k=0$$

$$\frac{x'^2}{9} + \frac{y'^2}{4} = 1$$

Ellipse: center $(-3,0)$, major axis
parallel to x-axis; a=3, b=2

25. $9x^2 - y^2 + 8y - 7 = 0$

$9x^2 - (y^2 - 8y) = 7$

$9x^2 - (y^2 - 8y + 16) = 7 - 16$

$9x^2 - (y-4)^2 = -9$

$\dfrac{(y-4)^2}{9} - \dfrac{x^2}{1} = 1$

h=0, k=4

$\dfrac{y'^2}{9} - \dfrac{x'^2}{1} = 1$

Hyperbola:center(0,4)
Transverse axis
parallel to y-axis
a=3, b=1

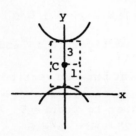

27. $2x^2 - 4x = 9y - 2$

$2(x^2 - 2x + 1) = 9y - 2 + 2$

$2(x - 1)^2 = 9y$

$(x - 1)^2 = \dfrac{9}{2}y$

h=1, k=0

$x'^2 = \dfrac{9}{2}y'$

Parabola:vertex(1,0)
Axis parallel to
y-axis

$4p = \dfrac{9}{2}, \quad p = \dfrac{9}{8}$

Focus(1,0+p) or $(1, \dfrac{9}{8})$

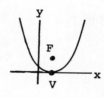

29. Hyperbola:vertex(3,-1)

Asymptotes x-y=-1 and
x+y=-3

Center at point of
intersection of
asymptotes

x-y=-1
x+y=-3
2x =-4, x=-2, y=-1

(h,k) at (-2,-1)

Vertex to right of center
means transverse axis
parallel to x-axis and
slopes of asymptotes
are ±b/a.

Slopes are ±1, or a=b

a=dist. center to vertex
or a=5

$\dfrac{x'^2}{a^2} - \dfrac{y'^2}{b^2} = 1$

$\dfrac{(x-h)^2}{a^2} - \dfrac{(y-k)^2}{b^2} = 1$

$\dfrac{(x+2)^2}{25} - \dfrac{(y+1)^2}{25} = 1$

$(x+2)^2 - (y+1)^2 = 25$

$x^2 - y^2 + 4x - 2y - 22 = 0$

31. $y^2 = 4x$, p = 1

Vertex (0,0), focus (1,0)

For first parabola

Vertex (1,0), focus (0,0)

Focus to left of vertex

means p=-1

$y'^2 = 4px'$

$(y-k)^2 = 4p(x-h)$

$(y-0)^2 = 4(-1)(x-1)$

$y^2 = -4x + 4$

$y^2 + 4x - 4 = 0$

33. $P = EI - rI^2$

E=6.0 V, r=0.3 Ω

$P = 6I - 0.3I^2$

$10P = 60I - 3I^2$

$3(I^2 - 20I) = -10P$

$3(I^2 - 20I + 100) = -10P + 300$

$(I - 10)^2 = -\dfrac{10}{3}(P - 30)$

Parabola:center(10,30)

$4p = -\dfrac{10}{3}, \quad p = -\dfrac{5}{6}$

35.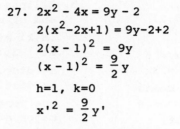

2a = 2.8+4.6 = 7.4, a=3.7

center is (3.7-2.8,0)
or (0.9,0); c=0.9

$a^2 = b^2 + c^2$, $3.7^2 = b^2 + 0.9^2$

$b^2 = 12.9$, b=3.6

$\dfrac{x'^2}{a^2} + \dfrac{y'^2}{b^2} = 1$

$\dfrac{(x-h)^2}{a^2} + \dfrac{(y-k)^2}{b^2} = 1$

$\dfrac{(x-0.9)^2}{3.7^2} + \dfrac{y^2}{3.6^2} = 1$

(all distances in
billions of miles)

Exercises 1-10, p. 55

1. $x^2 + 2y^2 - 2 = 0$
   A and C have
   same sign
   $A \neq C$, $B = 0$
   Ellipse

3. $2x^2 - y^2 - 1 = 0$
   A and C have
   different signs
   $B = 0$
   Hyperbola

5. $2x^2 + 2y^2 - 3y - 1 = 0$
   $A = C$, $B = 0$
   Circle

7. $2x^2 - x - y = 1$
   $2x^2 - x - y - 1 = 0$
   $B = 0$, $C = 0$
   Parabola

9. $x^2 = y^2 - 1$
   $x^2 - y^2 + 1 = 0$
   A and C have
   different signs
   $B = 0$
   Hyperbola

11. $x^2 = y - y^2$
    $x^2 + y^2 - y = 0$
    $A = C$, $B = 0$
    Circle

13. $x(y+3x) = x^2+xy-y^2+1$
    $xy+3x^2 = x^2+xy-y^2+1$
    $2x^2 + y^2 - 1 = 0$
    A and C have
    same sign
    $A \neq C$, $B = 0$
    Ellipse

15. $2xy + x - 3y = 6$
    $2xy + x - 3y - 6 = 0$
    $A = C = 0$, $B \neq 0$
    Hyperbola

17. $2x(x-y) = y(3-y-2x)$
    $2x^2 - 2xy = 3y - y^2 - 2xy$
    $2x^2 + y^2 - 3y = 0$
    A and C have same
    sign , $A \neq C$, $B = 0$
    Ellipse

19. $y(3 - 2y) = 2(x^2 - y^2)$
    $3y - 2y^2 = 2x^2 - 2y^2$
    $2x^2 - 3y = 0$
    $B = 0$, $C = 0$
    Parabola

21. $x^2 = 8(y - x - 2)$
    $x^2 = 8y - 8x - 16$
    $x^2 + 8x - 8y + 16 = 0$
    Parabola
    $x^2 + 8x = 8y - 16$
    $x^2+8x+16 = 8y-16+16$
    $(x + 4)^2 = 8y$
    Vertex $(-4,0)$

Axis parallel to y-axis
$4p=8$, $p=2$
Focus $(-4,0+p)$ or $(-4,2)$
Directrix $y = -2$

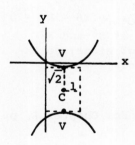

23. $y^2 = 2(x^2 - 2x - 2y)$
    $y^2 = 2x^2 - 4x - 4y$
    $2x^2 - y^2 - 4x - 4y = 0$
    Hyperbola
    $2(x^2 - 2x) - (y^2 + 4y) = 0$
    $2(x^2 - 2x + 1) - (y^2 + 4y + 4) = 2 - 4$
    $2(x - 1)^2 - (y + 2)^2 = -2$
    $\dfrac{(y + 2)^2}{2} - \dfrac{(x - 1)^2}{1} = 1$
    Center $(1, -2)$
    $a = \sqrt{2}$, $b = 1$
    Transverse axis parallel to y-axis
    Vertices $(1,-2+\sqrt{2})$, $(1,-2-\sqrt{2})$

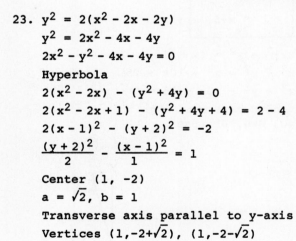

25. $y^2 + 42 = 2x(10 - x)$
$y^2 + 42 = 20x - 2x^2$
$2x^2 + y^2 - 20x + 42 = 0$
Ellipse
$2(x^2 - 10x) + y^2 = -42$
$2(x^2 - 10x + 25) + y^2 = -42 + 50$
$\dfrac{(x - 5)^2}{4} + \dfrac{y^2}{8} = 1$
Center $(5,0)$
$a = \sqrt{8} = 2\sqrt{2}$, $b = 2$
Major axis parallel to y-axis
Vertices $(5, 2\sqrt{2})$, $(5, -2\sqrt{2})$

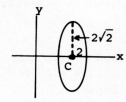

27. $4(y^2 - 4x - 2) = 5(4y - 5)$
$4y^2 - 16x - 8 = 20y - 25$
$4y^2 - 16x - 20y + 17 = 0$
Parabola
$4(y^2 - 5y) = 16x - 17$
$4(y^2 - 5y + \dfrac{25}{4}) = 16x - 17 + 25$
$4(y - \dfrac{5}{2})^2 = 16(x + \dfrac{1}{2})$
$(y - \dfrac{5}{2})^2 = 4(x + \dfrac{1}{2})$
Vertex $(-\dfrac{1}{2}, \dfrac{5}{2})$
Axis parallel to x-axis
$p = 1$, focus $(\dfrac{1}{2}, \dfrac{5}{2})$, directrix $x = -\dfrac{3}{2}$

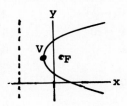

29. $x^2 + ky^2 = a^2$

(a) $k = 1$: $A = C$, $B = 0$, Circle

(b) $k < 0$: A and C have different signs, $B = 0$ Hyperbola

(c) $k > 0$ ($k \neq 1$) : A and C have same sign, $A \neq C$, $B = 0$, Ellipse

31. $Ax^2 + Bxy + Cy^2 + Dx + Ey + F = 0$
$A = C \neq 0$, $B = D = E = F = 0$
$Ax^2 + Ay^2 = 0$
Only point which satisfies equation is $(0,0)$.
(Equation can be written as $x^2 = -y^2$, true only for $y = 0$)

33. $z = \sqrt{(X_L - X_C)^2 + R^2}$
R and X_C are constant
$z^2 = (X_L - X_C)^2 + R^2$ $(z \geq 0)$
$z^2 - (X_L - X_C)^2 - R^2 = 0$
$z^2 - X_L^2 + 2X_C X_L - X_C^2 - R^2 = 0$
Coefficients af z^2 and X_L^2 have different signs; no ZX_L term; Hyperbola

35.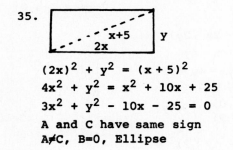

$(2x)^2 + y^2 = (x + 5)^2$
$4x^2 + y^2 = x^2 + 10x + 25$
$3x^2 + y^2 - 10x - 25 = 0$
A and C have same sign
$A \neq C$, $B = 0$, Ellipse

Review Exercises for Chapter 1, p. 56

1. Straight line
Passes through $(1, -7)$
Slope = 4
$y = mx + b$
$y = 4x + b$

$-7 = 4(1) + b$, $b = -11$
$y = 4x - 11$
$4x - y - 11 = 0$

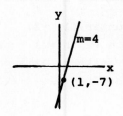

3. Straight line
   y-int. of −1, b=−1
   perpendicular to
   3x − 2y + 8 = 0
   2y = 3x + 8
   $y = \frac{3}{2}x + 4$, $m=\frac{3}{2}$
   Slope of required
   line $= -\frac{2}{3}$
   $y = mx + b$, $y = -\frac{2}{3}x - 1$
   2x + 3y + 3 = 0

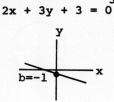

5. Circle: center(1,−2)
   Passes through (4,−3)
   $(x-h)^2 + (y-k)^2 = r^2$
   h=1, k=−2
   $(x - 1)^2 + (y+2)^2 = r^2$
   $(4-1)^2 + (-3+2)^2 = r^2$
   $r^2 = 10$
   $(x-1)^2 + (y+2)^2 = 10$
   $x^2 + y^2 - 2x + 4y - 5 = 0$

7. Parabola: focus (3,0)
   vertex (0,0)
   Axis along x-axis, p=3
   $y^2 = 4px$
   $y^2 = 12x$

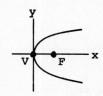

9. Ellipse: vertex (10,0)
   focus (8,0) , center (0,0)
   Major axis along x-axis
   a=10, c=8
   $a^2=b^2+c^2$, $100=b^2+64$, $b^2=36$
   $\frac{x^2}{a^2} + \frac{y^2}{b^2} = 1$, $\frac{x^2}{100} + \frac{y^2}{36} = 1$
   $9x^2 + 25y^2 = 900$

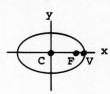

11. Hyperbola:vertex(0,13)
    Center(0,0)
    conjugate axis = 24
    Trans. axis along
    y-axis; a=13, b=12
    $\frac{y^2}{a^2} - \frac{x^2}{b^2} = 1$, $\frac{y^2}{169} - \frac{x^2}{144} = 1$
    $144y^2 - 169x^2 = 24,336$

13. $x^2 + y^2 + 6x - 7 = 0$
    $x^2 + 6x + y^2 = 7$
    $(x^2 +6x+9) + y^2 = 7+9$
    $(x + 3)^2 + y^2 = 16$
    Center (−3,0), r=4

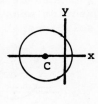

15. $x^2 = -20y$
    4p = −20, p = −5
    Vertex (0,0)
    focus (0,−5)
    directrix y = 5

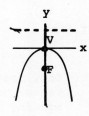

17. $16x^2 + y^2 = 16$
    $\frac{x^2}{1} + \frac{y^2}{16} = 1$
    Center (0,0), $a^2=16$, $b^2=1$
    $a^2=b^2+c^2$, $16=1+c^2$, $c^2=15$
    a=4, b=1, $c=\sqrt{15}$
    Vertices (0,4),(0,−4)
    Foci $(0,\sqrt{15})$, $(0,-\sqrt{15})$

19. $2x^2 - 5y^2 = 8$
    $\frac{x^2}{4} - \frac{y^2}{8/5} = 1$
    Center (0,0), $a^2=4$, $b^2=\frac{8}{5}$
    $c^2=a^2+b^2=4+\frac{8}{5}=\frac{28}{5}$
    a=2, $b=\frac{2}{5}\sqrt{10}$, $c=\frac{2}{5}\sqrt{35}$
    Vertices (2,0), (−2,0)
    Foci $(\frac{2}{5}\sqrt{35},0)$, $(-\frac{2}{5}\sqrt{35},0)$

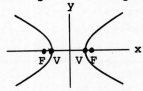

21. $x^2 - 8x - 4y - 16 = 0$
$x^2 - 8x = 4y + 16$
$x^2 - 8x + 16 = 4y + 16 + 16$
$(x-4)^2 = 4(y+8)$
Vertex $(4,-8)$
$4p=4$, $p=1$, focus$(4,-7)$

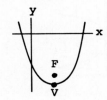

23. $4x^2 + y^2 - 16x + 2y + 13 = 0$
$4(x^2 - 4x) + (y^2 + 2y) = -13$
$4(x^2 - 4x + 4) + (y^2 + 2y + 1)$
$\qquad = -13 + 16 + 1$
$4(x-2)^2 + (y+1)^2 = 4$
$\dfrac{(x-2)^2}{1} + \dfrac{(y+1)^2}{4} = 1$
Center$(2,-1)$, $a=2$, $b=1$

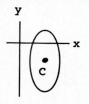

25. $x^2 + y^2 = 9$, circle
Center $(0,0)$, $r=3$
$4x^2 + y^2 = 16$
$\dfrac{x^2}{4} + \dfrac{y^2}{16} = 1$
Ellipse: center $(0,0)$
$a=4$, $b=2$
4 points of inter-
section mean 4 real
solutions

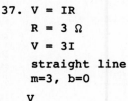

27. $x^2 + y^2 - 4y - 5 = 0$
$x^2 + (y^2 - 4y + 4) = 5+4$
$x^2 + (y-2)^2 = 9$
Circle, center$(0,2)$, $r=3$
$y^2 - 4x^2 - 4 = 0$
$\dfrac{y^2}{4} - \dfrac{x^2}{1} = 1$
Hyperbola
Center$(0,0)$, $a=2$, $b=1$
2 points of inter-
section mean 2 real
solutions

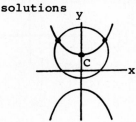

29. $25x^2 + 4y^2 = 100$
$\underline{4x^2 + 9y^2 = 36}$
$225x^2 + 36y^2 = 900$
$\underline{16x^2 + 36y^2 = 144}$
$209x^2 \qquad = 756$
$x = \pm\sqrt{\dfrac{756}{209}} = \pm1.90$

$100x^2 + 16y^2 = 400$
$\underline{100x^2 + 225y^2 = 900}$
$209y^2 = 500$
$y = \pm\sqrt{\dfrac{500}{209}} = \pm1.55$

$(1.90, 1.55)$, $(1.90, -1.55)$, $(-1.90, 1.55)$, $(-1.90, -1.55)$

31. $A(-3,11)$, $B(2,-1)$, $C(14,4)$

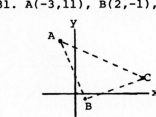

$m_{AB} = \dfrac{-1-11}{2+3} = -\dfrac{12}{5}$, $m_{BC} = \dfrac{4+1}{14-2} = \dfrac{5}{12}$
$m_{AB} = -1/m_{BC}$, $AB \perp BC$
Triangle ABC is right triangle
$d^2_{AB} = (2+3)^2 + (-1-11)^2 = 25+144 = 169$
$d^2_{BC} = (14-2)^2 + (4+1)^2 = 144+25 = 169$
$d^2_{AC} = (14+3)^2 + (4-11)^2 = 289+49 = 338$
$d^2_{AC} = d^2_{AB} + d^2_{BC}$; Triangle ABC is right triangle

33.

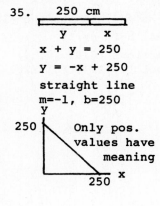

$\sqrt{(x-3)^2 + (y-1)^2} = y+3$
$(x-3)^2 + (y-1)^2 = (y+3)^2$
$x^2 - 6x + 9 + y^2 - 2y + 1 = y^2 + 6y + 9$
$x^2 - 6x - 8y + 1 = 0$
Vertex $(3,-1)$, $p=2$
$(x-3)^2 = 4(2)(y+1)^2$
$x^2 - 6x - 8y + 1 = 0$

35. 250 cm

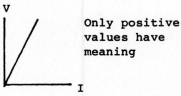

$x + y = 250$
$y = -x + 250$
straight line
$m=-1$, $b=250$
Only pos.
values have
meaning

37. $V = IR$
$R = 3\ \Omega$
$V = 3I$
straight line
$m=3$, $b=0$
Only positive
values have
meaning

39. $\alpha = \dfrac{R_1}{R_1 + R_2}$, $\alpha = \dfrac{1}{2}$

$\dfrac{1}{2} = \dfrac{R_1}{R_1 + R_2}$

$R_1 + R_2 = 2R_1 \quad R_1 = R_2$

st. line, m=1, b=0

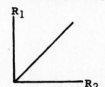

41.

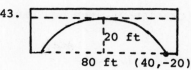

$2\pi r = 4.50$

$r = 0.716, \; r^2 = 0.513$

$x^2 + y^2 = 0.513$

43.

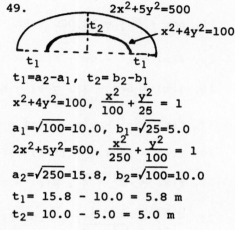

80 ft   (40,-20)

$x^2 = 4py$

$40^2 = 4p(-20)$

$p = -20$

$x^2 = -80y$

45.

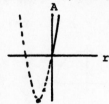

$A = 2\pi r^2 + 2\pi r(10)$

$\quad = 2\pi(r^2 + 10r)$

$A + 50\pi = 2\pi(r^2 + 10r + 25)$

$(r+5)^2 = \dfrac{1}{2\pi}(A + 50\pi)$

Parabola
vertex (-5,-50π)

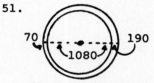

47. $y = 120t - 16t^2, \; x = 60t$

$t = x/60$

$y = 120\left(\dfrac{x}{60}\right) - 16\left(\dfrac{x}{60}\right)^2$

$\quad = 2x - \dfrac{x^2}{225}$

$225y = 450x - x^2$

$x^2 - 450x = -225y$

$x^2 - 450x + 50625$
$\quad = 50625 - 225y$
$\quad = 225(225 - y)$

$(x-225)^2 = -225(y-225)$

Parabola: V(225,225)

Will hit ground
when y=0

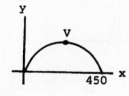

49.

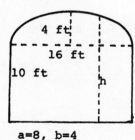

$2x^2 + 5y^2 = 500$

$x^2 + 4y^2 = 100$

$t_1 = a_2 - a_1, \quad t_2 = b_2 - b_1$

$x^2 + 4y^2 = 100, \quad \dfrac{x^2}{100} + \dfrac{y^2}{25} = 1$

$a_1 = \sqrt{100} = 10.0, \quad b_1 = \sqrt{25} = 5.0$

$2x^2 + 5y^2 = 500, \quad \dfrac{x^2}{250} + \dfrac{y^2}{100} = 1$

$a_2 = \sqrt{250} = 15.8, \quad b_2 = \sqrt{100} = 10.0$

$t_1 = 15.8 - 10.0 = 5.8 \; m$

$t_2 = 10.0 - 5.0 = 5.0 \; m$

51.

All measurements in miles

$2a = 70 + 1080 + 1080 + 190 = 2420$

$a = 1210, \quad c = a - (1080 + 70) = 60$

$a^2 = b^2 + c^2, \quad 1210^2 = b^2 + 60^2$

$b^2 = 1.461 \times 10^6$

$\dfrac{x^2}{1.464 \times 10^6} + \dfrac{y^2}{1.461 \times 10^6} = 1$

53. $W = k(T_2^2 - T_1^2)$

W is constant

$T_2^2 - T_1^2 = \dfrac{W}{k}$

$\dfrac{T_2^2}{\frac{W}{k}} - \dfrac{T_1^2}{\frac{W}{k}} = 1$

Hyperbola

$a = b = \sqrt{W/k}$

55.

a=8, b=4

$\dfrac{x^2}{64} + \dfrac{y^2}{16} = 1$

Find y+10 for x=4 ft

$\dfrac{16}{64} + \dfrac{y^2}{16} = 1$

$16 + 4y^2 = 64$

$4y^2 = 48, \quad y^2 = 12$

$y = \sqrt{12} = 3.5 \; ft$

$h = 10 + y = 13.5 \; ft$

CHAPTER 2   **THE DERIVATIVE**

Exercises 2-1, p. 65

1. From geometry: $A = s^2$ 3. From geometry: $V = \frac{1}{3}\pi r^2 h$; $V = \frac{1}{3}\pi r^2(8) = \frac{8}{3}\pi r^2$

5. $d = rt$; $d = 55t$ 7. $p = 100c - 100(3) = 100(c - 3)$

9. $I = Prt$; $I = 200(0.12)t = 24t$

11. $6n_1 + 9n_2 = 12,000$; $9n_2 = 12,000 - 6n_1$; $n_2 = -\frac{2}{3}n_1 + \frac{4000}{3}$

13. $f(x) = x^2 - 9x$; $f(3) = 3^2 - 9(3) = -18$; $f(-5) = (-5)^2 - 9(-5) = 70$

15. $g(x) = 6\sqrt{x} - 3$; $g(9) = 6\sqrt{9} - 3 = 15$; $g(\frac{1}{4}) = 6\sqrt{\frac{1}{4}} - 3 = 0$

17. $f(x) = ax^2 - a^2x$; $f(a) = a(a^2) - a^2(a) = 0$; $f(2a) = a(2a)^2 - a^2(2a) = 2a^3$

19. $T(t) = 5t + 7$; $T(-t) = 5(-t) + 7 = -5t + 7$; $T[T(t)] = 5(5t + 7) + 7 = 25t + 42$

21. $F(x) = \sqrt{x^2 + 4}$, $G(x) = x - 1$; $F[G(x)] = \sqrt{(x - 1)^2 + 4} = \sqrt{x^2 - 2x + 5}$

23. $\frac{G(x)}{F(x)} = \frac{x - 1}{\sqrt{x^2 + 4}}$ 25. $f(x) = 6 - 3x$ 27. $f(x) = \sqrt{9 - x^2}$; $y = \sqrt{9 - x^2}$

                                              $y = -3x + 6$ $y^2 = 9 - x^2$; $x^2 + y^2 = 9$

                                 st. line: m=-3, b=6   circle ($y \geq 0$), r=3

29. $\sqrt{\sqrt[3]{x^6 + 1}} = {}^{2(3)}\!\sqrt{x^6 + 1} = \sqrt[6]{x^6 + 1}$ 31. $\frac{(4x - 5)^4}{(4x - 5)^{1/2}} = (4x - 5)^{4 - 1/2} = (4x - 5)^{7/2}$

33. $(2x + 1)^{1/2} + (x + 3)(2x + 1)^{-1/2}$ 35. $\frac{(x^2 + 1)^{1/2} - x^2(x^2 + 1)^{-1/2}}{x^2 + 1}$

    $= (2x + 1)^{1/2} + \frac{x + 3}{(2x + 1)^{1/2}}$ $= \frac{(x^2 + 1)^{1/2}[(x^2 + 1)^{1/2} - x^2(x^2 + 1)^{-1/2}]}{(x^2 + 1)^{1/2}(x^2 + 1)}$

    $= \frac{(2x + 1)^{1/2}(2x + 1)^{1/2} + x + 3}{(2x + 1)^{1/2}}$ $= \frac{x^2 + 1 - x^2}{(x^2 + 1)^{3/2}} = \frac{1}{(x^2 + 1)^{3/2}}$

    $= \frac{2x + 1 + x + 3}{(2x + 1)^{1/2}} = \frac{3x + 4}{(2x + 1)^{1/2}}$

37.   No. Two values of   39. (a) Yes. For each x there is one y.
               y for each x>0.   (b) No. Two values of x for y = 5.

41. $Y(y) = \dfrac{y + 1}{y - 1}$ not defined for y = 1. Division by zero. Therefore, $y \neq 1$.

43. $F(y) = \sqrt{y - 1}$ not defined for y < 1. Imaginary values. Therefore, $y \geq 1$.

45. $f(x) = \begin{array}{l} x + 1 \quad \text{for } x < 1 \\ \sqrt{x + 3} \quad \text{for } x \geq 1 \end{array}$; f(-2) = -2 + 1 = -1 ; $f(2) = \sqrt{2 + 3} = \sqrt{5}$

47.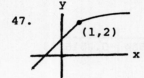

49. $p(T) = \dfrac{273}{T - 273}$; $p(308) = \dfrac{273}{308 - 273} = \dfrac{273}{35} = 7.8$

51. $s(t) = 16t^2$; $s(t + 2) = 16(t + 2)^2 = 16t^2 + 64t + 64$ feet

53. $p = 4s$; $s = \dfrac{p}{4}$; $c = 60 - p$; $2\pi r = 60 - p$; $r = \dfrac{60 - p}{2\pi}$

$A = s^2 + \pi r^2 = (\dfrac{p}{4})^2 + \pi(\dfrac{60 - p}{2\pi})^2 = \dfrac{1}{16}p^2 + \dfrac{(60 - p)^2}{4\pi}$

$= \dfrac{\pi p^2 + 4(3600 - 120p + p^2)}{16\pi} = \dfrac{(4 + \pi)p^2 - 480p + 14,400}{16\pi}$

55. $A = xy$; $y = \dfrac{A}{x}$; $p = 2y + 2x = 2(\dfrac{A}{x}) + 2x = 2(\dfrac{A}{x} + x)$

Exercises 2-2, p. 75

1. f(x) = 3x - 2 is continuous for all real x. f(x) is defined for all x. Small change is x produces a small change in f(x).

3. $f(x) = \dfrac{1}{x + 3}$ is not continuous for x = -3. f(x) is not defined for x = -3, division by zero. f(x) is continuous for all other real values of x.

5. $f(x) = \dfrac{1}{\sqrt{x}}$ is continuous for x > 0. Imaginary values for x < 0 and division by zero for x = 0.

7. Graph of f(x) has no breaks. Continuous for all x. f(x) is defined for all x and a small change in x produces a small change in f(x).

9. f(x) is not continuous for x = 1. f(x) is defined for x = 1, but a small change in x can produce a change of at least 3 in f(x).

11. f(x) is continuous for x < 2. f(x) is not defined for x > 2. f(x) is defined for x = 2, but a small change is not possible if it makes x > 2. (Considering only values of $x \leq 2$, we may state that a small change in x produces a small change in f(x). See Exercise 48.)

13. f(x) = 3x - 2

x	4.000	3.500	3.100	3.010	3.001	2.999	2.990	2.900	2.500	2.000
f(x)	10.000	8.500	7.300	7.030	7.003	6.997	6.970	6.700	5.500	4.000

$\lim\limits_{x \to 3} f(x) = 7$

15. $f(x) = \dfrac{x^3 - x}{x - 1}$

x	0.900	0.990	0.999	1.001	1.010	1.100
f(x)	1.71	1.9701	1.997001	2.003001	2.0301	2.31

$\lim\limits_{x \to 1} f(x) = 2$

17. $f(x) = \dfrac{2 - \sqrt{x + 2}}{x - 2}$

x	2.1	2.01	2.001	1.999	1.99	1.9
f(x)	-0.2484567	-0.2498440	-0.2499850	-0.2500155	-0.2501564	-0.2515823

$\lim\limits_{x \to 2} f(x) = -0.25$

19. $f(x) = \dfrac{2x + 1}{5x - 3}$

x	10	100	1000
f(x)	0.4468085	0.4044266	0.4004403

$\lim\limits_{x \to \infty} f(x) = 0.4$

21. $f(x) = x + 4$ is continuous at $x = 3$. Therefore, $\lim\limits_{x \to 3} x + 4 = 3 + 4 = 7$

23. $f(x) = \dfrac{x^2 - 1}{x + 1}$ is continuous at $x = 2$. Therefore, $\lim\limits_{x \to 2} \dfrac{x^2 - 1}{x + 1} = \dfrac{4 - 1}{2 + 1} = 1$

25. $\lim\limits_{x \to 0} \dfrac{x^2 + x}{x} = \lim\limits_{x \to 0} x + 1 = 1$

27. $\lim\limits_{x \to -1} \dfrac{x^2 - 1}{x + 1} = \lim\limits_{x \to -1} \dfrac{(x - 1)(x + 1)}{x + 1} = \lim\limits_{x \to -1} x - 1 = -2$

29. $\lim\limits_{x \to 1} \dfrac{x^3 - x}{x - 1} = \lim\limits_{x \to 1} \dfrac{x(x - 1)(x + 1)}{x - 1} = \lim\limits_{x \to 1} x(x + 1) = 2$

31. $\lim\limits_{x \to 3} \dfrac{x^2 - 2x - 3}{3 - x} = \lim\limits_{x \to 3} \dfrac{(x - 3)(x + 1)}{3 - x} = \lim\limits_{x \to 3} -(x + 1) = -4$

33. $\lim\limits_{x \to -1} \sqrt{x}(x + 1)$ does not exist. $f(-1) = 0$, but is not defined for other $x < 0$. Thus, a small change in x cannot be made.

35. $\lim\limits_{x \to \infty} \dfrac{x^2}{1 - 2x} = 0$ since the denominator increases without bound.

37. $\lim\limits_{x \to \infty} \dfrac{3x^2 + 5}{x^2 - 2} = \lim\limits_{x \to \infty} \dfrac{3 + \dfrac{5}{x^2}}{1 - \dfrac{2}{x^2}} = \dfrac{3}{1} = 3$ 39. $\lim\limits_{x \to \infty} \dfrac{2x - 6}{x^2 - 9} = \lim\limits_{x \to \infty} \dfrac{\dfrac{2}{x} - \dfrac{6}{x^2}}{1 - \dfrac{9}{x^2}} = \dfrac{0}{1} = 0$

41. $f(x) = \dfrac{x^2 - 3x}{x}$ $\lim\limits_{x \to 0} f(x) = -3$

x	0.1	0.01	0.001	-0.001	-0.01	-0.1
f(x)	-2.9	-2.99	-2.999	-3.001	-3.01	-3.1

$\lim\limits_{x \to 0} \dfrac{x^2 - 3x}{x} = \lim\limits_{x \to 0} x - 3 = -3$

43. $f(x) = \dfrac{2x^2 + x}{x^2 - 3}$ $\lim\limits_{x \to \infty} f(x) = 2$ $\lim\limits_{x \to \infty} \dfrac{2x^2 + x}{x^2 - 3} = \lim\limits_{x \to \infty} \dfrac{2 + \dfrac{1}{x}}{1 - \dfrac{3}{x^2}} = \dfrac{2}{1} = 2$

x	10	100	1000
f(x)	2.1649485	2.0106032	2.001006

45. $B = \dfrac{ki}{r} = \dfrac{6.30 \times 10^{-7} i}{0.100} = 6.30 \times 10^{-6} i$

i	0.200	0.290	0.299	0.301	0.310	0.400
B	1.26	1.83	1.884	1.896	1.95	2.52

$(\times 10^{-6})$ $\lim\limits_{i \to 0.3} B = 1.89 \times 10^{-6}$ T

47. If the temperature decreases 10% each minute, at the end of a minute it is 90% of the temperature at the beginning of the minute. Thus, T = 100°C for t = 0, T = 90°C for t = 1 min, T = 81°C for t = 2 min, or $T = 100(0.9)^t$. Thus, $T = 100(0.9)^{10} = 34.9°C$ for t = 10 min. Also, $\lim\limits_{t \to \infty} T = 0°C$.

Exercises 2-3, p. 82

1. $y = x^2$; P is (2,4)

	Q_1	Q_2	Q_3	Q_4	P
x_2	1.5	1.9	1.99	1.999	2
y_2	2.25	3.61	3.9601	3.996001	4
$y_2 - 4$	-1.75	-0.39	-0.0399	-0.003999	
$x_2 - 2$	-0.5	-0.1	-0.01	-0.001	
$m = \dfrac{y_2 - 4}{x_2 - 2}$	3.5	3.9	3.99	3.999	

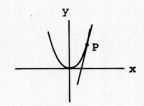

$m_{tan} = 4$

3. $y = x^2 + 2x$; P is (-3,3)

	Q_1	Q_2	Q_3	Q_4	P
x_2	-2.5	-2.9	-2.99	-2.999	-3
y_2	1.25	2.61	2.9601	2.996001	3
$y_2 - 3$	-1.75	-0.39	-0.0399	-0.003999	
$x_2 + 3$	0.5	0.1	0.01	0.001	
$m = \dfrac{y_2 - 3}{x_2 + 3}$	-3.5	-3.9	-3.99	-3.999	

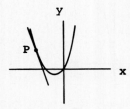

$m_{tan} = -4$

5. $y = x^2$
$4 + \Delta y = (2 + \Delta x)^2$
$\qquad = 4 + 4\Delta x + (\Delta x)^2$
$\Delta y = 4\Delta x + (\Delta x)^2$
$m_{PQ} = \dfrac{\Delta y}{\Delta x} = 4 + \Delta x$
As $\Delta x \to 0$, $m_{PQ} \to 4$
$m_{tan} = \lim\limits_{\Delta x \to 0} m_{PQ} = 4$

7. $y = x^2 + 2x$
$3 + \Delta y = (-3 + \Delta x)^2 + 2(-3 + \Delta x)$
$\qquad = 9 - 6\Delta x + (\Delta x)^2 - 6 + 2\Delta x$
$\Delta y = -4\Delta x + (\Delta x)^2$
$m_{PQ} = \dfrac{\Delta y}{\Delta x} = -4 + \Delta x$
As $\Delta x \to 0$, $m_{PQ} \to -4$
$m_{tan} = \lim\limits_{\Delta x \to 0} m_{PQ} = -4$

9. $y = x^2$; $x = 2$, $x = -1$

$y_1 + \Delta y = (x_1 + \Delta x)^2$

           $= x_1^2 + 2x_1\Delta x + (\Delta x)^2$

    $\Delta y = 2x_1\Delta x + (\Delta x)^2$

    $\dfrac{\Delta y}{\Delta x} = 2x_1 + \Delta x$

As $\Delta x \to 0$, $\dfrac{\Delta y}{\Delta x} \to 2x_1$

    $m_{tan} = 2x_1$

For $x_1 = 2$, $m_{tan} = 2(2) = 4$

For $x_1 = -1$, $m_{tan} = 2(-1) = -2$

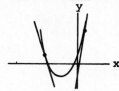

11. $y = x^2 + 2x$; $x = -3$, $x = 1$

$y_1 + \Delta y = (x_1 + \Delta x)^2 + 2(x_1 + \Delta x)$

         $= x_1^2 + 2x_1\Delta x + (\Delta x)^2 + 2x_1 + 2\Delta x$

    $\Delta y = 2x_1\Delta x + (\Delta x)^2 + 2\Delta x$

    $\dfrac{\Delta y}{\Delta x} = 2x_1 + \Delta x + 2$

As $\Delta x \to 0$, $\dfrac{\Delta y}{\Delta x} \to 2x_1 + 2$

    $m_{tan} = 2x_1 + 2$

For $x_1 = -3$, $m_{tan} = 2(-3) + 2 = -4$

For $x_1 = 1$, $m_{tan} = 2(1) + 2 = 4$

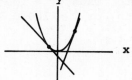

13. $y = x^2 + 4x + 5$; $x = -3$, $x = 2$

$y_1 + \Delta y = (x_1 + \Delta x)^2 + 4(x_1 + \Delta x) + 5$

        $= x_1^2 + 2x_1\Delta x + (\Delta x)^2 + 4x_1 + 4\Delta x + 5$

    $\Delta y = 2x_1\Delta x + (\Delta x)^2 + 4\Delta x$

    $\dfrac{\Delta y}{\Delta x} = 2x_1 + \Delta x + 4$

As $\Delta x \to 0$, $\dfrac{\Delta y}{\Delta x} \to 2x_1 + 4$

    $m_{tan} = 2x_1 + 4$

For $x_1 = -3$, $m_{tan} = 2(-3) + 4 = -2$

For $x_1 = 2$, $m_{tan} = 2(2) + 4 = 8$

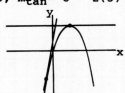

15. $y = 6x - x^2$; $x = -2$, $x = 3$

$y_1 + \Delta y = 6(x_1 + \Delta x) - (x_1 + \Delta x)^2$

        $= 6x_1 + 6\Delta x - x_1^2 - 2x_1\Delta x - (\Delta x)^2$

    $\Delta y = 6\Delta x - 2x_1\Delta x - (\Delta x)^2$

    $\dfrac{\Delta y}{\Delta x} = 6 - 2x_1 - \Delta x$

As $\Delta x \to 0$, $\dfrac{\Delta y}{\Delta x} \to 6 - 2x_1$

    $m_{tan} = 6 - 2x_1$

For $x_1 = -2$, $m_{tan} = 6 - 2(-2) = 10$

For $x_1 = 3$, $m_{tan} = 6 - 2(3) = 0$

17. $y = x^3 - 2x$; $x = -1$, $x = 0$, $x = 1$

$y_1 + \Delta y = (x_1 + \Delta x)^3 - 2(x_1 + \Delta x)$

        $= x_1^3 + 3x_1^2\Delta x + 3x_1(\Delta x)^2 + (\Delta x)^3 - 2x_1 - 2\Delta x$

    $\Delta y = 3x_1^2\Delta x + 3x_1(\Delta x)^2 + (\Delta x)^3 - 2\Delta x$

    $\dfrac{\Delta y}{\Delta x} = 3x_1^2 + 3x_1\Delta x + (\Delta x)^2 - 2$

As $\Delta x \to 0$, $\dfrac{\Delta y}{\Delta x} \to 3x_1^2 - 2$, $m_{tan} = 3x_1^2 - 2$

For $x_1 = -1$, $m_{tan} = 3(-1)^2 - 2 = 1$

For $x_1 = 0$, $m_{tan} = 3(0)^2 - 2 = -2$

For $x_1 = 1$, $m_{tan} = 3(1)^2 - 2 = 1$

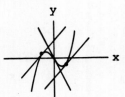

19. $y = x^4$; $x = 0$, $x = 0.5$, $x = 1$

$y_1 + \Delta y = (x_1 + \Delta x)^4$

$= x_1^4 + 4x_1^3 \Delta x + 6x_1^2 (\Delta x)^2 + 4x_1(\Delta x)^3 + (\Delta x)^4$

$\Delta y = 4x_1^3 \Delta x + 6x_1^2 (\Delta x)^2 + 4x_1(\Delta x)^3 + (\Delta x)^4$

$\dfrac{\Delta y}{\Delta x} = 4x_1^3 + 6x_1^2 \Delta x + 4x_1(\Delta x)^2 + (\Delta x)^3$

As $\Delta x \to 0$, $\dfrac{\Delta y}{\Delta x} \to 4x_1^3$, $m_{tan} = 4x_1^3$

For $x_1 = 0$, $m_{tan} = 4(0)^3 = 0$

For $x_1 = 0.5$, $m_{tan} = 4(0.5)^3 = 0.5$

For $x_1 = 1$, $m_{tan} = 4(1)^3 = 4$

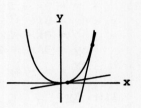

21. $y = x^2 + 2$; $P(2,6)$, $Q(2.1,6.41)$

$y_1 + \Delta y = (x_1 + \Delta x)^2 + 2$

$= x_1^2 + 2x_1\Delta x + (\Delta x)^2 + 2$

$\Delta y = 2x_1\Delta x + (\Delta x)^2$

$\dfrac{\Delta y}{\Delta x} = 2x_1 + \Delta x$

As $\Delta x \to 0$, $\dfrac{\Delta y}{\Delta x} \to 2x_1$; $m_{tan} = 2x_1$

For $x=2$, $m_{tan} = 2(2) = 4$ (inst. rate of change)

$\dfrac{\Delta y}{\Delta x}\bigg|_{PQ} = \dfrac{6.41 - 6}{2.1 - 2} = \dfrac{0.41}{0.1} = 4.1$ (av. rate of change)

23. $y = 9 - x^3$; $P(2,1)$, $Q(2.1,-0.261)$

$y_1 + \Delta y = 9 - (x_1 + \Delta x)^3$

$= 9 - x_1^3 - 3x_1^2 \Delta x - 3x_1(\Delta x)^2 - (\Delta x)^3$

$\Delta y = -3x_1^2 \Delta x - 3x_1(\Delta x)^2 - (\Delta x)^3$

$\dfrac{\Delta y}{\Delta x} = -3x_1^2 - 3x_1\Delta x - (\Delta x)^2$

As $\Delta x \to 0$, $\dfrac{\Delta y}{\Delta x} \to -3x_1^2$; $m_{tan} = -3x_1^2$

For $x = 2$, $m_{tan} = -3(2)^2 = -12$ (inst. rate of change)

$\dfrac{\Delta y}{\Delta x}\bigg|_{PQ} = \dfrac{-0.261 - 1}{2.1 - 2} = \dfrac{-1.261}{0.1} = -12.61$ (av. rate of change)

Exercises 2-4, p. 87

1. $y = 3x - 1$

$y + \Delta y = 3(x + \Delta x) - 1$

$= 3x + 3\Delta x - 1$

$\Delta y = 3\Delta x$

$\dfrac{\Delta y}{\Delta x} = 3$

$\lim\limits_{\Delta x \to 0} \dfrac{\Delta y}{\Delta x} = 3$

3. $y = 1 - 2x$

$y + \Delta y = 1 - 2(x + \Delta x)$

$= 1 - 2x - 2\Delta x$

$\Delta y = -2\Delta x$

$\dfrac{\Delta y}{\Delta x} = -2$

$\lim\limits_{\Delta x \to 0} \dfrac{\Delta y}{\Delta x} = -2$

5. $y = x^2 - 1$

$y + \Delta y = (x + \Delta x)^2 - 1$

$= x^2 + 2x\Delta x + (\Delta x)^2 - 1$

$\Delta y = 2x\Delta x + (\Delta x)^2$

$\dfrac{\Delta y}{\Delta x} = 2x + \Delta x$

$\lim\limits_{\Delta x \to 0} \dfrac{\Delta y}{\Delta x} = 2x$

7. $y = 5x^2$

$y + \Delta y = 5(x + \Delta x)^2$

$= 5x^2 + 10x\Delta x + 5(\Delta x)^2$

$\Delta y = 10x\Delta x + 5(\Delta x)^2$

$\dfrac{\Delta y}{\Delta x} = 10x + 5\Delta x$

$\lim\limits_{\Delta x \to 0} \dfrac{\Delta y}{\Delta x} = 10x$

9. $y = x^2 - 7x$

$y + \Delta y = (x + \Delta x)^2 - 7(x + \Delta x)$

$= x^2 + 2x\Delta x + (\Delta x)^2 - 7x - 7\Delta x$

$\Delta y = 2x\Delta x + (\Delta x)^2 - 7\Delta x$

$\dfrac{\Delta y}{\Delta x} = 2x + \Delta x - 7$

$\lim\limits_{\Delta x \to 0} \dfrac{\Delta y}{\Delta x} = 2x - 7$

11. $y = 8x - 2x^2$
$y + \Delta y = 8(x + \Delta x) - 2(x + \Delta x)^2$
$= 8x + 8\Delta x - 2x^2 - 4x\Delta x - 2(\Delta x)^2$
$\Delta y = 8\Delta x - 4x\Delta x - 2(\Delta x)^2$
$\dfrac{\Delta y}{\Delta x} = 8 - 4x - 2\Delta x$
$\lim\limits_{\Delta x \to 0} \dfrac{\Delta y}{\Delta x} = 8 - 4x$

13. $y = x^3 + 4x - 6$
$y + \Delta y = (x + \Delta x)^3 + 4(x + \Delta x) - 6$
$= x^3 + 3x^2\Delta x + 3x(\Delta x)^2 + (\Delta x)^3$
$\quad + 4x + 4\Delta x - 6$
$\Delta y = 3x^2\Delta x + 3x(\Delta x)^2 + (\Delta x)^3 + 4\Delta x$
$\dfrac{\Delta y}{\Delta x} = 3x^2 + 3x\Delta x + (\Delta x)^2 + 4$
$\lim\limits_{\Delta x \to 0} \dfrac{\Delta y}{\Delta x} = 3x^2 + 4$

15. $y = \dfrac{1}{x + 2}$
$y + \Delta y = \dfrac{1}{x + \Delta x + 2}$
$\Delta y = \dfrac{1}{x + \Delta x + 2} - \dfrac{1}{x + 2}$
$= \dfrac{x + 2 - (x + \Delta x + 2)}{(x + \Delta x + 2)(x + 2)}$
$= \dfrac{-\Delta x}{(x + \Delta x + 2)(x + 2)}$
$\dfrac{\Delta y}{\Delta x} = \dfrac{-1}{(x + \Delta x + 2)(x + 2)}$
$\lim\limits_{\Delta x \to 0} \dfrac{\Delta y}{\Delta x} = \dfrac{-1}{(x + 2)^2}$

17. $y = x + \dfrac{1}{x}$
$y + \Delta y = x + \Delta x + \dfrac{1}{x + \Delta x}$
$\Delta y = \Delta x + \dfrac{1}{x + \Delta x} - \dfrac{1}{x}$
$= \Delta x + \dfrac{x - (x + \Delta x)}{(x + \Delta x)x}$
$= \Delta x - \dfrac{\Delta x}{x(x + \Delta x)}$
$\dfrac{\Delta y}{\Delta x} = 1 - \dfrac{1}{x(x + \Delta x)}$
$\lim\limits_{\Delta x \to 0} \dfrac{\Delta y}{\Delta x} = 1 - \dfrac{1}{x^2}$

19. $y = \dfrac{2}{x^2}$
$y + \Delta y = \dfrac{2}{(x + \Delta x)^2}$
$\Delta y = \dfrac{2}{(x + \Delta x)^2} - \dfrac{2}{x^2}$
$= \dfrac{2x^2 - 2(x + \Delta x)^2}{(x + \Delta x)^2 x^2}$
$= \dfrac{-4x\Delta x - 2(\Delta x)^2}{(x + \Delta x)^2 x^2}$
$\dfrac{\Delta y}{\Delta x} = \dfrac{-4x - 2\Delta x}{(x + \Delta x)^2 x^2}$
$\lim\limits_{\Delta x \to 0} \dfrac{\Delta y}{\Delta x} = \dfrac{-4x}{x^4} = -\dfrac{4}{x^3}$

21. $y = x^4 + x^3 + x^2 + x$
$y + \Delta y = (x + \Delta x)^4 + (x + \Delta x)^3 + (x + \Delta x)^2 + (x + \Delta x)$
$= x^4 + 4x^3\Delta x + 6x^2(\Delta x)^2 + 4x(\Delta x)^3 + (\Delta x)^4 + x^3 + 3x^2\Delta x + 3x(\Delta x)^2 + (\Delta x)^3 + x^2 + 2x\Delta x + (\Delta x)^2 + x + \Delta x$
$\Delta y = 4x^3\Delta x + 6x^2(\Delta x)^2 + 4x(\Delta x)^3 + (\Delta x)^4 + 3x^2\Delta x + 3x(\Delta x)^2 + (\Delta x)^3 + 2x\Delta x + (\Delta x)^2 + \Delta x$
$\dfrac{\Delta y}{\Delta x} = 4x^3 + 6x^2\Delta x + 4x(\Delta x)^2 + (\Delta x)^3 + 3x^2 + 3x\Delta x + (\Delta x)^2 + 2x + \Delta x + 1$
$\lim\limits_{\Delta x \to 0} \dfrac{\Delta y}{\Delta x} = 4x^3 + 3x^2 + 2x + 1$

23. $y = x^4 - \dfrac{2}{x}$
$y + \Delta y = (x + \Delta x)^4 - \dfrac{2}{x + \Delta x} = x^4 + 4x^3\Delta x + 6x^2(\Delta x)^2 + 4x(\Delta x)^3 + (\Delta x)^4 - \dfrac{2}{x + \Delta x}$
$\Delta y = 4x^3\Delta x + 6x^2(\Delta x)^2 + 4x(\Delta x)^3 + (\Delta x)^4 - \dfrac{2}{x + \Delta x} + \dfrac{2}{x}$
$= 4x^3\Delta x + 6x^2(\Delta x)^2 + 4x(\Delta x)^3 + (\Delta x)^4 + \dfrac{2\Delta x}{x(x + \Delta x)}$
$\dfrac{\Delta y}{\Delta x} = 4x^3 + 6x^2\Delta x + 4x(\Delta x)^2 + (\Delta x)^3 + \dfrac{2}{x(x + \Delta x)}$
$\lim\limits_{\Delta x \to 0} \dfrac{\Delta y}{\Delta x} = 4x^3 + \dfrac{2}{x^2}$

25. $y = 3x^2 - 2x$
$y + \Delta y = 3(x + \Delta x)^2 - 2(x + \Delta x) = 3x^2 + 6x\Delta x + 3(\Delta x)^2 - 2x - 2\Delta x$
$\Delta y = 6x\Delta x + 3(\Delta x)^2 - 2\Delta x$
$\dfrac{\Delta y}{\Delta x} = 6x + 3\Delta x - 2$ $\lim\limits_{\Delta x \to 0} \dfrac{\Delta y}{\Delta x} = 6x - 2$ $\dfrac{dy}{dx}\Big|_{x=-1} = 6(-1) - 2 = -8$

27.
$$y = \frac{6}{x + 3}$$
$$y + \Delta y = \frac{6}{x + \Delta x + 3}$$
$$\Delta y = \frac{6}{x+\Delta x+3} - \frac{6}{x+3}$$
$$= \frac{6(x+3)-6(x+\Delta x+3)}{(x+\Delta x+3)(x+3)}$$
$$= \frac{-6\Delta x}{(x+\Delta x+3)(x+3)}$$
$$\frac{\Delta y}{\Delta x} = \frac{-6}{(x+\Delta x+3)(x+3)}$$
$$\lim_{\Delta x \to 0} \frac{\Delta y}{\Delta x} = \frac{-6}{(x+3)^2}$$
$$\frac{dy}{dx}\Big|_{x=3} = \frac{-6}{6^2} = -\frac{1}{6}$$

29.
$$y = \sqrt{x+1}$$
$$y^2 = x + 1 \quad (y \geq 0)$$
$$(y+\Delta y)^2 = x + \Delta x + 1$$
$$y^2 + 2y\Delta y + (\Delta y)^2 = x + \Delta x + 1$$
$$2y\Delta y + (\Delta y)^2 = \Delta x$$
$$\Delta y(2y + \Delta y) = \Delta x$$
$$\Delta y = \frac{\Delta x}{2y + \Delta y}$$
$$\frac{\Delta y}{\Delta x} = \frac{1}{2y + \Delta y}$$
$$\lim_{\Delta x \to 0} \frac{\Delta y}{\Delta x} = \frac{1}{2y} = \frac{1}{2\sqrt{x+1}}$$
$$(\Delta y \to 0 \text{ also})$$

31.
$$y = \sqrt{1-3x}$$
$$y^2 = 1 - 3x \quad (y \geq 0)$$
$$(y+\Delta y)^2 = 1 - 3(x + \Delta x)$$
$$y^2 + 2y\Delta y + (\Delta y)^2 = 1 - 3x - 3\Delta x$$
$$2y\Delta y + (\Delta y)^2 = -3\Delta x$$
$$\Delta y(2y + \Delta y) = -3\Delta x$$
$$\Delta y = \frac{-3\Delta x}{2y + \Delta y}$$
$$\frac{\Delta y}{\Delta x} = \frac{-3}{2y + \Delta y}$$
$$\lim_{\Delta x \to 0} \frac{\Delta y}{\Delta x} = \frac{-3}{2y} = -\frac{3}{2\sqrt{1-3x}}$$
$$(\Delta y \to 0 \text{ also})$$

Exercises 2-5, p. 91

1. $s = 4t + 10$; $t = 3$ s

t(seconds)	2.00	2.50	2.90	2.99	2.999
s(feet)	18.00	20.00	21.60	21.96	21.996
$\Delta s = 22-s$	4.00	2.00	0.40	0.04	0.004
$\Delta t = 3-t$	1.00	0.50	0.10	0.01	0.001
$v = \Delta s/\Delta t$	4.00	4.00	4.00	4.00	4.00

Inst. velocity for t = 3 s is
$$\lim_{\Delta t \to 0} \frac{\Delta s}{\Delta t} = 4.00 \text{ ft/s}$$

3. $s = 1 - 2t^2$; $t = 4$ s

t(seconds)	3.00	3.50	3.90	3.99	3.999
s(feet)	-17.00	-23.50	-29.42	-30.8402	-30.984002
$\Delta s = -31-s$	-14.00	-7.50	-1.58	-0.1598	-0.015998
$\Delta t = 4-t$	1.00	0.50	0.10	0.01	0.001
$v = \Delta s/\Delta t$	-14.0	-15.0	-15.8	-15.98	-15.998

Inst. velocity for t = 4 s is
$$\lim_{\Delta t \to 0} \frac{\Delta s}{\Delta t} = -16.0 \text{ ft/s}$$

5.
$$s = 4t + 10$$
$$s + \Delta s = 4(t + \Delta t) + 10$$
$$\Delta s = 4\Delta t$$
$$\frac{\Delta s}{\Delta t} = 4$$
$$\lim_{\Delta t \to 0} \frac{\Delta s}{\Delta t} = v = 4$$
$$v\big|_{t=3} = 4.00 \text{ ft/s}$$

7.
$$s = 1 - 2t^2$$
$$s + \Delta s = 1 - 2(t + \Delta t)^2$$
$$\Delta s = -4t\Delta t - 2(\Delta t)^2$$
$$\frac{\Delta s}{\Delta t} = -4t - 2\Delta t$$
$$\lim_{\Delta t \to 0} \frac{\Delta s}{\Delta t} = v = -4t$$
$$v\big|_{t=4} = -16.0 \text{ ft/s}$$

9.
$$s = 3t - \frac{2}{t}$$
$$s + \Delta s = 3(t+\Delta t) - \frac{2}{t+\Delta t}$$
$$\Delta s = 3\Delta t - \frac{2}{t+\Delta t} + \frac{2}{t}$$
$$= 3\Delta t + \frac{2\Delta t}{(t+\Delta t)t}$$
$$\frac{\Delta s}{\Delta t} = 3 + \frac{2}{(t+\Delta t)t}$$
$$\lim_{\Delta t \to 0} \frac{\Delta s}{\Delta t} = v = 3 + \frac{2}{t^2}$$

11.
$$s = t^3 - 6t + 2$$
$$s + \Delta s = (t+\Delta t)^3 - 6(t+\Delta t) + 2$$
$$\Delta s = 3t^2\Delta t + 3t(\Delta t)^2 + (\Delta t)^3 - 6\Delta t$$
$$\frac{\Delta s}{\Delta t} = 3t^2 + 3t\Delta t + (\Delta t)^2 - 6$$
$$\lim_{\Delta t \to 0} \frac{\Delta s}{\Delta t} = v = 3t^2 - 6$$

13.
$$v = 6t^2 - 4t + 2$$
$$v + \Delta v = 6(t+\Delta t)^2 - 4(t+\Delta t) + 2$$
$$\Delta v = 12t\Delta t + 6(\Delta t)^2 - 4\Delta t$$
$$\frac{\Delta v}{\Delta t} = 12t + 6\Delta t - 4$$
$$\lim_{\Delta t \to 0} \frac{\Delta v}{\Delta t} = a = 12t - 4$$

15.

$$s = 1 - 2t^2$$
$$s + \Delta s = 1 - 2(t + \Delta t)^2$$
$$\Delta s = -4t\Delta t - 2(\Delta t)^2$$
$$\frac{\Delta s}{\Delta t} = -4t - 2\Delta t$$
$$\lim_{\Delta t \to 0} \frac{\Delta s}{\Delta t} = v = -4t$$
$$v + \Delta v = -4(t + \Delta t)$$
$$\Delta v = -4\Delta t \; ; \; \frac{\Delta v}{\Delta t} = -4$$
$$\lim_{\Delta t \to 0} \frac{\Delta v}{\Delta t} = a = -4$$

17.

$$q = 30 - 2t$$
$$q + \Delta q = 30 - 2(t + \Delta t)$$
$$\Delta q = -2\Delta t$$
$$\frac{\Delta q}{\Delta t} = -2$$
$$\lim_{\Delta t \to 0} \frac{\Delta q}{\Delta t} = i = -2$$

19.

$$A = \pi r^2$$
$$A + \Delta A = \pi(r + \Delta r)^2$$
$$\Delta A = 2\pi r \Delta r + \pi(\Delta r)^2$$
$$\frac{\Delta A}{\Delta r} = 2\pi r + \pi \Delta r$$
$$\lim_{\Delta r \to 0} \frac{\Delta A}{\Delta r} = 2\pi r$$
$$\frac{dA}{dr}\bigg|_{r=3} = 2\pi(3) = 6\pi$$

21.

$$R = 50(1 + 0.0053T + 0.00001T^2)$$
$$= 50 + 0.265T + 0.0005T^2$$
$$R + \Delta R = 50 + 0.265(T+\Delta T) + 0.0005(T+\Delta T)^2$$
$$\Delta R = 0.265\Delta T + 0.001T\Delta T + 0.0005(\Delta T)^2$$
$$\frac{\Delta R}{\Delta T} = 0.265 + 0.001T + 0.0005\Delta T$$
$$\lim_{\Delta T \to 0} \frac{\Delta R}{\Delta T} = 0.265 + 0.001T$$
$$\frac{dR}{dT}\bigg|_{T=20°C} = 0.265 + 0.001(20) = 0.285 \; \Omega/°C$$

23.

$$H = \frac{5000}{t^2 + 10}$$
$$H + \Delta H = \frac{5000}{(t + \Delta t)^2 + 10}$$
$$\Delta H = \frac{5000}{(t+\Delta t)^2 + 10} - \frac{5000}{t^2 + 10}$$
$$= \frac{-10000t\Delta t - 5000(\Delta t)^2}{[(t+\Delta t)^2+10](t^2+ 10)}$$
$$\frac{\Delta H}{\Delta t} = \frac{-10000t - 5000\Delta t}{[(t+\Delta t)^2+10](t^2+ 10)}$$
$$\lim_{\Delta t \to 0} \frac{\Delta H}{\Delta t} = \frac{-10000t}{(t^2+10)^2}$$
$$\frac{dH}{dt}\bigg|_{t=3} = \frac{-10000(3)}{(3^2+10)^2} = -83.1 \; W/m^2 \cdot h$$

25.

$$\frac{r}{d} = \frac{4}{4} \; ; \; r = d$$

Let V = vol. of oil

$$V = \frac{1}{3}\pi r^2 d = \frac{1}{3}\pi d^2 d = \frac{1}{3}\pi d^3$$
$$V + \Delta V = \frac{1}{3}\pi(d + \Delta d)^3 \qquad \Delta V = \frac{1}{3}\pi[3d^2\Delta d + 3d(\Delta d)^2 + (\Delta d)^3]$$
$$\frac{\Delta V}{\Delta d} = \frac{1}{3}\pi[3d^2 + 3d\Delta d + (\Delta d)^2] \qquad \lim_{\Delta d \to 0} \frac{\Delta V}{\Delta d} = \frac{1}{3}\pi(3d^2) = \pi d^2$$

27. Let r = distance from source

$$I = \frac{k}{r^2} \; ; \; 25 = \frac{k}{(0.20)^2} \; ; \; k = 1 \; lx \cdot m^2 \; ; \; I = \frac{1}{r^2}$$
$$I = \frac{1}{r^2} \qquad I + \Delta I = \frac{1}{(r + \Delta r)^2} \qquad \Delta I = \frac{1}{(r + \Delta r)^2} - \frac{1}{r^2} = \frac{-2r\Delta r - (\Delta r)^2}{(r + \Delta r)^2 r^2}$$
$$\frac{\Delta I}{\Delta r} = \frac{-2r - \Delta r}{(r+\Delta r)^2 r^2} \qquad \lim_{\Delta r \to 0} \frac{\Delta I}{\Delta r} = \frac{-2r}{r^4} = -\frac{2}{r^3}$$
$$\frac{dI}{dr}\bigg|_{r=0.01 \; m} = -\frac{2}{(0.01)^3} = -2 \times 10^6 \; lx/m$$
$$\frac{dI}{dr}\bigg|_{r=0.10 \; m} = -\frac{2}{(0.10)^3} = -2 \times 10^3 \; lx/m$$

Exercises 2-6, p. 97

1. $y = x^5$

$\dfrac{dy}{dx} = 5x^{5-1} = 5x^4$

Eq.(2-23)

3. $y = 4x^9$

$\dfrac{dy}{dx} = 4(9x^8) = 36x^8$

Eq.(2-24)

5. $y = x^4 - 6$

$\dfrac{dy}{dx} = 4x^3 - 0 = 4x^3$

Eqs.(2-23,22,25)

7. $y = x^2 + 2x$

$\dfrac{dy}{dx} = 2x + 2(1)$

$= 2x + 2$

9. $y = 5x^3 - x - 1$

$\dfrac{dy}{dx} = 5(3x^2) - 1 - 0$

$= 15x^2 - 1$

11. $y = x^8 - 4x^7 - x$

$\dfrac{dy}{dx} = 8x^7 - 4(7x^6) - 1$

$= 8x^7 - 28x^6 - 1$

13. $y = 6x^7 - 5x^3 + 2$

$\dfrac{dy}{dx} = 6(7x^6) - 5(3x^2)$

$= 42x^6 - 15x^2$

15. $y = \frac{1}{3}x^3 + \frac{1}{2}x^2$

$\dfrac{dy}{dx} = \frac{1}{3}(3x^2) + \frac{1}{2}(2x)$

$= x^2 + x$

17. $y = 6x^2 - 8x + 1$

$\dfrac{dy}{dx} = 12x - 8$

$\dfrac{dy}{dx}\Big|_{x=2} = 12(2) - 8 = 16$

19. $y = 2x^3 + 9x - 7$

$\dfrac{dy}{dx} = 6x^2 + 9$

$\dfrac{dy}{dx}\Big|_{x=-2} = 6(-2)^2 + 9 = 33$

21. $y = 2x^6 - 6x^2$

$\dfrac{dy}{dx} = 12x^5 - 12x$

$\dfrac{dy}{dx}\Big|_{x=2} = 12(2)^5 - 12(2) = 360$

$= m_{tan}$ at $x=2$

23. $y = 6x - 2x^4$

$\dfrac{dy}{dx} = 6 - 8x^3$

$\dfrac{dy}{dx}\Big|_{x=-1} = 6 - 8(-1)^3 = 14$

$= m_{tan}$ at $x=-1$

25. $s = 6t^5 - 5t + 2$

$\dfrac{ds}{dt} = 30t^4 - 5$

$= v$

27. $s = 120t - 16t^2$

$\dfrac{ds}{dt} = 120 - 32t$

$= v$

29. $s = 2t^3 - 4t^2$

$\dfrac{ds}{dt} = 6t^2 - 8t$

$v_{t=4} = 6(4)^2 - 8(4)$

$= 64$

31. $s = 80t - 4.9t^2$

$\dfrac{ds}{dt} = 80 - 9.8t$

$v_{t=3} = 80 - 9.8(3)$

$= 50.6$

33. $y = 3x^2 - 6x$

$\dfrac{dy}{dx} = 6x - 6$

Tangent parallel to x-axis where slope is zero. $6x - 6 = 0$, $x=1$

35. $V = \pi r^2 h$

$r = h$

$V = \pi r^3$

$\dfrac{dV}{dr} = 3\pi r^2$

37. $V = 4.0T + 0.005T^2$

$\dfrac{dV}{dt} = 4.0 + 0.01T$

$\dfrac{dV}{dt}\Big|_{T=200°C} = 4.0 + 0.01(200)$

$= 6.0$ V/°C

39. $h = 30x^2 - x^3$

$\dfrac{dh}{dx} = 60x - 3x^2$

$\dfrac{dh}{dx}\Big|_{x=10\text{ km}} = 60(10) - 3(10)^2$

$= 300$ m/km

41. $F = x^4 - 12x^3 + 46x^2 - 60x + 25$

$\dfrac{dF}{dx} = 4x^3 - 36x^2 + 92x - 60$

$\dfrac{dF}{dx}\Big|_{x=4\text{ cm}} = 4(4)^3 - 36(4)^2 + 92(4) - 60 = -12$ N/cm

Exercises 2-7, p. 102

1. $y = x^2(3x + 2)$

$u = x^2 \quad v = 3x + 2$

$\dfrac{du}{dx} = 2x \quad \dfrac{dv}{dx} = 3$

$\dfrac{dy}{dx} = x^2(3) + (3x + 2)(2x)$

$= 9x^2 + 4x$

3. $y = 6x(3x^2 - 5x)$

$u = 6x \quad v = 3x^2 - 5x$

$\dfrac{du}{dx} = 6 \quad \dfrac{dv}{dx} = 6x - 5$

$\dfrac{dy}{dx} = 6x(6x-5) + (3x^2-5x)(6)$

$= 54x^2 - 60x$

5. $y = (x + 2)(2x - 5)$

$u = x + 2 \quad v = 2x - 5$

$\dfrac{du}{dx} = 1 \quad \dfrac{dv}{dx} = 2$

$\dfrac{dy}{dx} = (x+2)(2) + (2x-5)(1)$

$= 4x - 1$

7. $y = (x^4 - 3x^2 + 3)(1 - 2x^3)$

   $u = x^4 - 3x^2 + 3 \quad v = 1 - 2x^3$

   $\dfrac{du}{dx} = 4x^3 - 6x \qquad \dfrac{dv}{dx} = -6x^2$

   $\dfrac{dy}{dx} = (x^4 - 3x^2 + 3)(-6x^2) + (1 - 2x^3)(4x^3 - 6x)$

   $\quad = -6x^6 + 18x^4 - 18x^2 + 4x^3 - 6x - 8x^6 + 12x^4$

   $\quad = -14x^6 + 30x^4 + 4x^3 - 18x^2 - 6x$

9. $y = (2x - 7)(5 - 2x)$

   $\dfrac{dy}{dx} = (2x - 7)(-2) + (5 - 2x)(2)$

   $\quad = -8x + 24$

   $y = -4x^2 + 24x - 35$

   $\dfrac{dy}{dx} = -8x + 24$

11. $y = (x^3 - 1)(2x^2 - x - 1)$

   $\dfrac{dy}{dx} = (x^3 - 1)(4x - 1) + (2x^2 - x - 1)(3x^2)$

   $\quad = 4x^4 - x^3 - 4x + 1 + 6x^4 - 3x^3 - 3x^2$

   $\quad = 10x^4 - 4x^3 - 3x^2 - 4x + 1$

   $y = 2x^5 - x^4 - x^3 - 2x^2 + x + 1$

   $\dfrac{dy}{dx} = 10x^4 - 4x^3 - 3x^2 - 4x + 1$

13. $y = \dfrac{x}{2x + 3}$

   $u = x \qquad v = 2x + 3$

   $\dfrac{du}{dx} = 1 \qquad \dfrac{dv}{dx} = 2$

   $\dfrac{dy}{dx} = \dfrac{(2x + 3)(1) - x(2)}{(2x + 3)^2}$

   $\quad = \dfrac{3}{(2x + 3)^2}$

15. $y = \dfrac{1}{x^2 + 1}$

   $u = 1 \quad v = x^2 + 1$

   $\dfrac{du}{dx} = 0 \quad \dfrac{dv}{dx} = 2x$

   $\dfrac{dy}{dx} = \dfrac{(x^2 + 1)(0) - 1(2x)}{(x^2 + 1)^2}$

   $\quad = \dfrac{-2x}{(x^2 + 1)^2}$

17. $y = \dfrac{x^2}{3 - 2x}$

   $u = x^2 \quad v = 3 - 2x$

   $\dfrac{du}{dx} = 2x \quad \dfrac{dv}{dx} = -2$

   $\dfrac{dy}{dx} = \dfrac{(3-2x)(2x) - x^2(-2)}{(3 - 2x)^2}$

   $\quad = \dfrac{6x - 2x^2}{(3 - 2x)^2}$

19. $y = \dfrac{2x - 1}{3x^2 + 2}$

   $u = 2x - 1 \quad v = 3x^2 + 2$

   $\dfrac{du}{dx} = 2 \qquad \dfrac{dv}{dx} = 6x$

   $\dfrac{dy}{dx} = \dfrac{(3x^2+2)(2) - (2x-1)(6x)}{(3x^2 + 2)^2}$

   $\quad = \dfrac{-6x^2 + 6x + 4}{(3x + 2)^2}$

21. $y = \dfrac{x + 8}{x^2 + x + 2}$

   $u = x + 8 \quad v = x^2 + x + 2$

   $\dfrac{du}{dx} = 1 \qquad \dfrac{dv}{dx} = 2x + 1$

   $\dfrac{dy}{dx} = \dfrac{(x^2+x+2)(1) - (x+8)(2x+1)}{(x^2 + x + 2)^2}$

   $\quad = \dfrac{-x^2 - 16x - 6}{(x^2 + x + 2)^2}$

23. $y = \dfrac{2x^2 - x - 1}{x^3 + 2x^2}$

   $u = 2x^2 - x - 1 \quad v = x^3 + 2x^2$

   $\dfrac{du}{dx} = 4x - 1 \qquad \dfrac{dv}{dx} = 3x^2 + 4x$

   $\dfrac{dy}{dx} = \dfrac{(x^3 + 2x^2)(4x - 1) - (2x^2 - x - 1)(3x^2 + 4x)}{(x^3 + 2x^2)^2}$

   $\quad = \dfrac{4x^4 - x^3 + 8x^3 - 2x^2 - 6x^4 - 8x^3 + 3x^3 + 4x^2 + 3x^2 + 4x}{(x^3 + 2x^2)^2}$

   $\quad = \dfrac{-2x^4 + 2x^3 + 5x^2 + 4x}{(x^3 + 2x^2)^2}$

25. $y = (3x - 1)(4 - 7x) \; ; \; x = 3$

   $\dfrac{dy}{dx} = (3x - 1)(-7) + (4 - 7x)(3)$

   $\quad = -42x + 19$

   $\dfrac{dy}{dx}\bigg|_{x=3} = -42(3) + 19 = -107$

27. $y = \dfrac{3x - 5}{2x + 3} \; ; \; x = -2$

   $\dfrac{dy}{dx} = \dfrac{(2x + 3)(3) - (3x - 5)(2)}{(2x + 3)^2}$

   $\quad = \dfrac{19}{(2x + 3)^2}$

   $\dfrac{dy}{dx}\bigg|_{x=-2} = \dfrac{19}{[2(-2) + 3]^2} = 19$

29. $y = \dfrac{x^2(1 - 2x)}{3x - 7}$

   $\dfrac{dy}{dx} = \dfrac{(3x-7)[x^2(-2)+(1-2x)(2x)] - x^2(1-2x)(3)}{(3x - 7)^2}$

   $\quad = \dfrac{(3x-7)(-6x^2+2x) - 3x^2 + 6x^3}{(3x - 7)^2} = \dfrac{-12x^3+45x^2-14x}{(3x - 7)^2}$

31. $y = (4x + 1)(x^4 - 1)$

   $\dfrac{dy}{dx} = (4x+1)(4x^3) + (x^4-1)(4)$

   $\quad = 20x^4 + 4x^3 - 4$

   $\dfrac{dy}{dx}\bigg|_{x=-1} = 20(-1)^4 + 4(-1)^3 - 4 = 12$

33. $y = \dfrac{x}{x^2 + 1}$

$\dfrac{dy}{dx} = \dfrac{(x^2 + 1)(1) - x(2x)}{(x^2 + 1)^2}$

$= \dfrac{1 - x^2}{(x^2 + 1)^2}$

$\dfrac{1 - x^2}{(x^2+1)^2} = 0$; $1 - x^2 = 0$; $x = -1, 1$

35. $s = \dfrac{2}{t^2}$

$v = \dfrac{ds}{dt} = \dfrac{t^2(0) - 2(2t)}{(t^2)^2}$

$= -\dfrac{4}{t^3}$

$v|_{t=4\ s} = -\dfrac{4}{4^3} = -\dfrac{1}{16}$ ft/s

37. $V = IR$

$= (5.00 + 0.01t^2)(15.00 - 0.10t)$

$\dfrac{dV}{dt} = (5.00 + 0.01t^2)(-0.10)$
$+ (15.00 - 0.10t)(0.02t)$

$= -0.003t^2 + 0.30t - 0.50$

$\dfrac{dV}{dt}\Big|_{t=5\ s} = -0.003(5)^2 + 0.30(5) - 0.50$
$= 0.925$ V/s

39. $P = \dfrac{E^2 r}{R^2 + 2Rr + r^2}$

$\dfrac{dP}{dr} = \dfrac{(R^2 + 2Rr + r^2)(E^2) - E^2 r(2R + 2r)}{(R^2 + 2Rr + r^2)^2}$

$= \dfrac{E^2(R^2 - r^2)}{[(R + r)^2]^2} = \dfrac{E^2(R + r)(R - r)}{(R + r)^4}$

$= \dfrac{E^2(R - r)}{(R + r)^3}$

Exercises 2-8, p. 108

1. $y = \sqrt{x} = x^{1/2}$

$\dfrac{dy}{dx} = \dfrac{1}{2}x^{1/2 - 1} = \dfrac{1}{2}x^{-1/2}$

$= \dfrac{1}{2x^{1/2}}$

3. $y = \dfrac{1}{x^2} = x^{-2}$

$\dfrac{dy}{dx} = -2x^{-2-1} = -2x^{-3}$

$= -\dfrac{2}{x^3}$

5. $y = \dfrac{3}{\sqrt[3]{x}} = 3x^{-1/3}$

$\dfrac{dy}{dx} = 3(-\dfrac{1}{3})x^{-4/3}$

$= -\dfrac{1}{x^{4/3}}$

7. $y = x\sqrt{x} - \dfrac{1}{x} = x^{3/2} - x^{-1}$

$\dfrac{dy}{dx} = \dfrac{3}{2}x^{1/2} - (-x^{-2})$

$= \dfrac{3}{2}x^{1/2} + \dfrac{1}{x^2}$

9. $y = (x^2 + 1)^5$

$u = x^2 + 1,\ \dfrac{du}{dx} = 2x,\ n = 5$

$\dfrac{dy}{dx} = 5(x^2 + 1)^4(2x)$

$= 10x(x^2 + 1)^4$

11. $y = 2(7 - 4x^3)^8$

$u = 7 - 4x^3,\ \dfrac{du}{dx} = -12x^2,\ n = 8$

$\dfrac{dy}{dx} = 2(8)(7 - 4x^3)^7(-12x^2)$

$= -192x^2(7 - 4x^3)^7$

13. $y = (2x^3 - 3)^{1/3}$

$u = 2x^3 - 3,\ \dfrac{du}{dx} = 6x^2, n = \dfrac{1}{3}$

$\dfrac{dy}{dx} = \dfrac{1}{3}(2x^3 - 3)^{-2/3}(6x^2)$

$= \dfrac{2x^2}{(2x^3 - 3)^{2/3}}$

15. $y = \dfrac{1}{(1 - x^2)^4} = (1 - x^2)^{-4}$

$u = 1 - x^2,\ \dfrac{du}{dx} = -2x, n = -4$

$\dfrac{dy}{dx} = -4(1 - x^2)^{-5}(-2x)$

$= \dfrac{8x}{(1 - x^2)^5}$

17. $y = 4(2x^4 - 5)^{3/4}$

$u = 2x^4 - 5,\ \dfrac{du}{dx} = 8x^3, n = \dfrac{3}{4}$

$\dfrac{dy}{dx} = 4(\dfrac{3}{4})(2x^4 - 5)^{-1/4}(8x^3)$

$= \dfrac{24x^3}{(2x^4 - 5)^{1/4}}$

19. $y = \sqrt[4]{1 - 8x^2} = (1 - 8x^2)^{1/4}$

$u = 1 - 8x^2,\ \dfrac{du}{dx} = -16x,\ n = \dfrac{1}{4}$

$\dfrac{dy}{dx} = \dfrac{1}{4}(1 - 8x^2)^{-3/4}(-16x)$

$= \dfrac{-4x}{(1 - 8x^2)^{3/4}}$

21. $y = x\sqrt{x - 1} = x(x - 1)^{1/2}$

$\dfrac{dy}{dx} = x(\dfrac{1}{2})(x - 1)^{-1/2}(1) + (x - 1)^{1/2}(1)$

$= \dfrac{x}{2(x-1)^{1/2}} + (x - 1)^{1/2}$

$= \dfrac{x + 2(x-1)^{1/2}(x-1)^{1/2}}{2(x-1)^{1/2}}$

$= \dfrac{x + 2(x - 1)}{2(x - 1)^{1/2}} = \dfrac{3x - 2}{2(x - 1)^{1/2}}$

23. $y = \dfrac{\sqrt{4x+3}}{8x+1} = \dfrac{(4x+3)^{1/2}}{8x+1}$

$\dfrac{dy}{dx} = \dfrac{(8x+1)(1/2)(4x+3)^{-1/2}(4) - (4x+3)^{1/2}(8)}{(8x+1)^2} = \dfrac{\dfrac{2(8x+1)}{(4x+3)^{1/2}} - 8(4x+3)^{1/2}}{(8x+1)^2}$

$= \dfrac{\dfrac{2(8x+1)}{(4x+3)^{1/2}} - \dfrac{8(4x+3)^{1/2}(4x+3)^{1/2}}{(4x+3)^{1/2}}}{(8x+1)^2} = \dfrac{2(8x+1) - 8(4x+3)}{(4x+3)^{1/2}(8x+1)^2}$

$= \dfrac{-16x - 22}{(4x+3)^{1/2}(8x+1)^2}$

25. $y = \sqrt{3x+4} = (3x+4)^{1/2}$; $x = 7$

$\dfrac{dy}{dx} = \dfrac{1}{2}(3x+4)^{-1/2}(3) = \dfrac{3}{2(3x+4)^{1/2}}$

$\left.\dfrac{dy}{dx}\right|_{x=7} = \dfrac{3}{2[3(7)+4]^{1/2}} = \dfrac{3}{10}$

27. $y = \dfrac{\sqrt{x}}{1-x} = \dfrac{x^{1/2}}{1-x}$; $x = 4$

$\dfrac{dy}{dx} = \dfrac{(1-x)(1/2)(x^{-1/2}) - x^{1/2}(-1)}{(1-x)^2}$

$= \dfrac{\dfrac{1-x}{2x^{1/2}} + x^{1/2}}{(1-x)^2} = \dfrac{1-x+2x}{2x^{1/2}(1-x)^2}$

$= \dfrac{1+x}{2x^{1/2}(1-x)^2}$

$\left.\dfrac{dy}{dx}\right|_{x=4} = \dfrac{1+4}{2(4)^{1/2}(1-4)^2} = \dfrac{5}{36}$

29. (a) $y = \dfrac{1}{x^3}$

$\dfrac{dy}{dx} = \dfrac{x^3(0) - 1(3x^2)}{(x^3)^2}$

$= \dfrac{-3x^2}{x^6} = -\dfrac{3}{x^4}$

(b) $y = \dfrac{1}{x^3} = x^{-3}$

$\dfrac{dy}{dx} = -3x^{-4}$

$= -\dfrac{3}{x^4}$

31. $y = \dfrac{x^2}{\sqrt{x^2+1}} = \dfrac{x^2}{(x^2+1)^{1/2}}$

$\dfrac{dy}{dx} = \dfrac{(x^2+1)^{1/2}(2x) - x^2(1/2)(x^2+1)^{-1/2}(2x)}{x^2+1}$

$= \dfrac{2x(x^2+1)^{1/2} - \dfrac{x^3}{(x^2+1)^{1/2}}}{x^2+1}$

$= \dfrac{2x(x^2+1) - x^3}{(x^2+1)^{3/2}} = \dfrac{x^3+2x}{(x^2+1)^{3/2}}$

$\dfrac{x^3+2x}{(x^2+1)^{3/2}} = 0$; $x^3+2x = 0$; $x(x^2+2) = 0$; $x = 0$

33. $y^2 = 4x$; $(1,2)$

$y = 2x^{1/2}$

$\dfrac{dy}{dx} = 2\left(\dfrac{1}{2}\right)x^{-1/2} = \dfrac{1}{x^{1/2}}$

$\left.\dfrac{dy}{dx}\right|_{x=1} = \dfrac{1}{1^{1/2}} = 1$

35. $s = (t^3 - t)^{4/3}$

$v = \dfrac{ds}{dt} = \dfrac{4}{3}(t^3 - t)^{1/3}(3t^2 - 1)$

$v|_{t=4} = \dfrac{4}{3}(4^3 - 4)^{1/3}[3(4)^2 - 1]$

$= \dfrac{4}{3}(60^{1/3})(47) = \dfrac{188}{3}\sqrt[3]{60}$

37. $P = \dfrac{k}{v^{3/2}}$; $300 = \dfrac{k}{100^{3/2}}$; $k = 3 \times 10^5$

$P = \dfrac{3 \times 10^5}{v^{3/2}} = 3 \times 10^5 v^{-3/2}$

$\dfrac{dP}{dv} = 3 \times 10^5\left(-\dfrac{3}{2}\right)v^{-5/2} = -\dfrac{4.5 \times 10^5}{v^{5/2}}$

$\left.\dfrac{dP}{dv}\right|_{v=100 \text{ cm}^3} = \dfrac{-4.5 \times 10^5}{100^{5/2}} = -4.5 \text{ kPa/cm}^3$

39. $H = \dfrac{4000}{\sqrt{t^6+100}} = 4000(t^6+100)^{-1/2}$

$\dfrac{dH}{dt} = 4000\left(-\dfrac{1}{2}\right)(t^6+100)^{-3/2}(6t^5) = -\dfrac{12000t^5}{(t^6+100)^{3/2}}$

$\left.\dfrac{dH}{dt}\right|_{t=4 \text{ h}} = -\dfrac{12000(4)^5}{(4^6+100)^{3/2}} = -45.2 \text{ W/m}^2\cdot\text{h}$

41. $V = \dfrac{kq}{\sqrt{x^2 + b^2}} = kq(x^2 + b^2)^{-1/2}$

$\dfrac{dV}{dx} = kq(-\dfrac{1}{2})(x^2 + b^2)^{-3/2}(2x)$

$= -\dfrac{kqx}{(x^2 + b^2)^{3/2}}$

43. $H = \dfrac{k}{[r^2 + (\ell/2)^2]^{3/2}} = k[r^2 + (\ell/2)^2]^{-3/2}$

$\dfrac{dH}{dr} = k(-\dfrac{3}{2})[r^2 + (\ell/2)^2]^{-5/2}(2r)$

$= -\dfrac{3kr}{[r^2 + (\ell/2)^2]^{5/2}}$

Exercises 2-9, p. 112

1. $3x + 2y = 5$

$3(1) + 2\dfrac{dy}{dx} = 0$

$\dfrac{dy}{dx} = -\dfrac{3}{2}$

3. $4y - 3x^2 = x$

$4\dfrac{dy}{dx} - 6x = 1$

$\dfrac{dy}{dx} = \dfrac{6x + 1}{4}$

5. $x^2 - y^2 - 9 = 0$

$2x - 2y\dfrac{dy}{dx} - 0 = 0$

$\dfrac{dy}{dx} = \dfrac{x}{y}$

7. $y^5 = x^2 - 1$

$5y^4\dfrac{dy}{dx} = 2x - 0$

$\dfrac{dy}{dx} = \dfrac{2x}{5y^4}$

9. $y^2 + y = x^2 - 4$

$2y\dfrac{dy}{dx} + \dfrac{dy}{dx} = 2x - 0$

$\dfrac{dy}{dx}(2y + 1) = 2x$

$\dfrac{dy}{dx} = \dfrac{2x}{2y + 1}$

11. $y + 3xy - 4 = 0$

$\dfrac{dy}{dx} + 3[x\dfrac{dy}{dx} + y(1)] - 0 = 0$

$\dfrac{dy}{dx}(3x + 1) = -3y$

$\dfrac{dy}{dx} = -\dfrac{3y}{3x + 1}$

13. $xy^3 + 3y + x^2 = 9$

$[x(3y^2\dfrac{dy}{dx}) + y^3(1)] + 3\dfrac{dy}{dx} + 2x = 0$

$\dfrac{dy}{dx}(3xy^2 + 3) = -2x - y^3$

$\dfrac{dy}{dx} = \dfrac{-2x - y^3}{3xy^2 + 3}$

15. $3x^2y^2 + y = 3x - 1$

$3[x^2(2y\dfrac{dy}{dx}) + y^2(2x)] + \dfrac{dy}{dx} = 3 - 0$

$\dfrac{dy}{dx}(6x^2y + 1) = 3 - 6xy^2$

$\dfrac{dy}{dx} = \dfrac{3 - 6xy^2}{6x^2y + 1}$

17. $(2y - x)^4 + x^2 = y + 3$

$4(2y - x)^3(2\dfrac{dy}{dx} - 1) + 2x = \dfrac{dy}{dx} + 0$

$\dfrac{dy}{dx}[8(2y - x)^3 - 1] = 4(2y - x)^3 - 2x$

$\dfrac{dy}{dx} = \dfrac{4(2y - x)^3 - 2x}{8(2y - x)^3 - 1}$

19. $2(x^2 + 1)^3 + (y^2 + 1)^2 = 17$

$6(x^2 + 1)^2(2x) + 2(y^2 + 1)(2y\dfrac{dy}{dx}) = 0$

$\dfrac{dy}{dx} = \dfrac{-3x(x^2 + 1)^2}{y(y^2 + 1)}$

21. $3x^3y^2 - 2y^3 = -4$; $(1,2)$

$3[x^3(2y\dfrac{dy}{dx}) + y^2(3x^2)] - 6y^2\dfrac{dy}{dx} = 0$

$\dfrac{dy}{dx}(6x^3y - 6y^2) = -9x^2y^2$

$\dfrac{dy}{dx} = \dfrac{-3x^2y}{2x^3 - 2y}$

$\dfrac{dy}{dx}\bigg|_{(1,2)} = \dfrac{-3(1)^2(2)}{2(1)^3 - 2(2)} = 3$

23. $5y^4 + 7 = x^4 - 3y$; $(3,-2)$

$20y^3\dfrac{dy}{dx} + 0 = 4x^3 - 3\dfrac{dy}{dx}$

$\dfrac{dy}{dx}(20y^3 + 3) = 4x^3$ $\dfrac{dy}{dx} = \dfrac{4x^3}{20y^3 + 3}$

$\dfrac{dy}{dx}\bigg|_{(3,-2)} = \dfrac{4(3)^3}{20(-2)^3 + 3} = -\dfrac{108}{157}$

25. $xy + y^2 + 2 = 0$; $(-3,1)$

$[x\dfrac{dy}{dx} + y(1)] + 2y\dfrac{dy}{dx} + 0 = 0$

$\dfrac{dy}{dx} = \dfrac{-y}{x + 2y}$

$\dfrac{dy}{dx}\bigg|_{(-3,1)} = \dfrac{-1}{-3 + 2(1)} = 1$

$= m_{\tan}$ at $(-3,1)$

27. $x^2 + y^2 = a^2$

$2x + 2y\dfrac{dy}{dx} = 0$

$\dfrac{dy}{dx} = -\dfrac{x}{y}$

Exercises 2-10, p. 116

1. $y = x^3 + x^2$

$\dfrac{dy}{dx} = 3x^2 + 2x$

$\dfrac{d^2y}{dx^2} = 6x + 2$

$\dfrac{d^3y}{dx^3} = 6 \qquad \dfrac{d^4y}{dx^4} = 0$

$\dfrac{d^ny}{dx^n} = 0 \quad (n \ge 4)$

3. $f(x) = x^3 - 6x^4$

$f'(x) = 3x^2 - 24x^3$

$f''(x) = 6x - 72x^2$

$f'''(x) = 6 - 144x$

$f^{(4)}(x) = -144$

$f^{(5)}(x) = 0$

$f^{(n)}(x) = 0 \quad (n \ge 5)$

5. $y = (1 - 2x)^4$

$y' = 4(1 - 2x)^3(-2) = -8(1 - 2x)^3$

$y'' = -24(1 - 2x)^2(-2) = 48(1 - 2x)^2$

$y''' = 96(1 - 2x)(-2) = -192(1 - 2x)$

$y^{(4)} = -192(-2) = 384$

$y^{(5)} = 0$

$y^{(n)} = 0 \quad (n \ge 5)$

7. $f(x) = x(2x + 1)^3$

$f'(x) = x[3(2x+1)^2(2)] + (2x+1)^3(1) = (2x+1)^2[6x+(2x+1)] = (8x+1)(2x+1)^2$

$f''(x) = (8x+1)[2(2x+1)(2)] + (2x+1)^2(8) = (2x+1)[(32x+4)+(16x+8)] = 12(2x+1)(4x+1)$

$f'''(x) = 12[(2x+1)(4) + (4x+1)(2)] = 24(8x + 3)$

$f^{(4)}(x) = 192 \qquad f^{(n)}(x) = 0 \quad (n \ge 5)$

9. $y = 2x^7 - x^6 - 3x$

$y' = 14x^6 - 6x^5 - 3$

$y'' = 84x^5 - 30x^4$

11. $y = 2x + \sqrt{x} = 2x + x^{1/2}$

$y' = 2 + \dfrac{1}{2}x^{-1/2}$

$y'' = \dfrac{1}{2}(-\dfrac{1}{2})x^{-3/2}$

$= -\dfrac{1}{4x^{3/2}}$

13. $f(x) = \sqrt[4]{8x - 3} = (8x - 3)^{1/4}$

$f'(x) = \dfrac{1}{4}(8x - 3)^{-3/4}(8)$

$= 2(8x - 3)^{-3/4}$

$f''(x) = 2(-\dfrac{3}{4})(8x - 3)^{-7/4}(8)$

$= -\dfrac{12}{(8x - 3)^{7/4}}$

15. $f(x) = \dfrac{4}{\sqrt{1 - 2x}} = 4(1 - 2x)^{-1/2}$

$f'(x) = 4(-\dfrac{1}{2})(1-2x)^{-3/2}(-2) = 4(1-2x)^{-3/2}$

$f''(x) = 4(-\dfrac{3}{2})(1-2x)^{-5/2}(-2) = \dfrac{12}{(1 - 2x)^{5/2}}$

17. $y = 2(2 - 5x)^4$

$y' = 8(2-5x)^3(-5) = -40(2 - 5x)^3$

$y'' = -120(2 - 5x)^2(-5)$

$= 600(2 - 5x)^2$

19. $y = (3x^2 - 1)^5$

$y' = 5(3x^2 - 1)^4(6x) = 30x(3x^2 - 1)^4$

$y'' = 30x[4(3x^2 - 1)^3(6x)] + (3x^2 - 1)^4(30)$

$= (3x^2 - 1)^3[720x^2 + 30(3x^2 - 1)]$

$= 30(27x^2 - 1)(3x^2 - 1)^3$

21. $f(x) = \dfrac{2x}{1 - x}$

$f'(x) = \dfrac{(1 - x)(2) - 2x(-1)}{(1 - x)^2}$

$= \dfrac{2}{(1 - x)^2} = 2(1 - x)^{-2}$

$f''(x) = -4(1 - x)^{-3}(-1) = \dfrac{4}{(1 - x)^3}$

23. $y = \dfrac{x^2}{x + 1}$

$y' = \dfrac{(x + 1)(2x) - x^2(1)}{(x + 1)^2} = \dfrac{x^2 + 2x}{(x + 1)^2}$

$y'' = \dfrac{(x + 1)^2(2x + 2) - (x^2 + 2x)(2)(x + 1)(1)}{(x + 1)^4}$

$= \dfrac{2(x + 1)[(x + 1)^2 - (x^2 + 2x)]}{(x + 1)^4}$

$= \dfrac{2}{(x + 1)^3}$

25. $x^2 - y^2 = 9$

$2x - 2yy' = 0$

$x - yy' = 0 \qquad y' = \dfrac{x}{y}$

$1 - [yy'' + (y')^2] = 0$

$yy'' = 1 - (y')^2$

$y'' = \dfrac{1 - (y')^2}{y} = \dfrac{1 - (x/y)^2}{y}$

$= \dfrac{y^2 - x^2}{y^3} = -\dfrac{9}{y^3}$

27. $x^2 - xy = 1 - y^2$

$2x - [xy' + y(1)] = 0 - 2yy'$

$2x - xy' - y + 2yy' = 0$ $\qquad y' = \dfrac{y - 2x}{2y - x}$

$2 - [xy'' + y'(1)] - y' + 2[yy'' + (y')^2] = 0$

$y''(2y - x) = 2y' - 2 - 2(y')^2$

$y'' = \dfrac{2y' - 2 - 2(y')^2}{2y - x} = \dfrac{2[(\frac{y - 2x}{2y - x}) - 1 - (\frac{y - 2x}{2y - x})^2]}{2y - x}$

$\quad = \dfrac{2[(y - 2x)(2y - x) - (2y - x)^2 - (y - 2x)^2]}{(2y - x)^3} = -\dfrac{6(x^2 - xy + y^2)}{(2y - x)^3}$

29. $f(x) = \sqrt{x^2 + 9} = (x^2 + 9)^{1/2}$

$f'(x) = \frac{1}{2}(x^2 + 9)^{-1/2}(2x) = \dfrac{x}{(x^2 + 9)^{1/2}}$

$f''(x) = \dfrac{(x^2 + 9)^{1/2}(1) - x(1/2)(x^2 + 9)^{-1/2}(2x)}{x^2 + 9}$

$\quad = \dfrac{9}{(x^2 + 9)^{3/2}}$

$f''(4) = \dfrac{9}{(4^2 + 9)^{3/2}} = \dfrac{9}{125}$

31. $y = 3x^{2/3} - \dfrac{2}{x} = 3x^{2/3} - 2x^{-1}$

$y' = 3(\frac{2}{3})x^{-1/3} - 2(-x^{-2})$

$\quad = 2x^{-1/3} + 2x^{-2}$

$y'' = 2(-\frac{1}{3})x^{-4/3} + 2(-2)x^{-3}$

$\quad = -\dfrac{2}{3x^{4/3}} - \dfrac{4}{x^3}$

$y''|_{x=-8} = -\dfrac{2}{3(-8)^{4/3}} - \dfrac{4}{(-8)^3}$

$\quad = -\dfrac{1}{24} + \dfrac{1}{128} = \dfrac{-13}{384}$

33. $s = 2250t - 16.1t^2$

$v = \dfrac{ds}{dt} = 2250 - 32.2t$

$a = \dfrac{dv}{dt} = -32.2 \text{ ft/s}^2$

35. $y = 0.0001(x^5 - 25x^2)$

$\dfrac{dy}{dx} = 0.0001(5x^4 - 50x)$

$\dfrac{d^2y}{dx^2} = 0.0001(20x^3 - 50)$

$\quad = 0.001(2x^3 - 5)$

$\dfrac{d^2y}{dx^2}\Big|_{x=3} \text{ m} = 0.001[2(3)^3 - 5]$

$\quad = 0.049/\text{m}$

37. $q = \sqrt{2t + 1} - 1$

$\dfrac{dq}{dt} = \frac{1}{2}(2t + 1)^{-1/2}(2)$

$\quad = (2t + 1)^{-1/2}$

$\dfrac{d^2q}{dt^2} = -\frac{1}{2}(2t + 1)^{-3/2}(2)$

$\quad = \dfrac{-1}{(2t + 1)^{3/2}}$

$V = L\dfrac{d^2q}{dt^2} = -\dfrac{1.60}{(2t + 1)^{3/2}}$

Review Exercises for Chapter 2, p. 118

1. $f(x) = 8 - 3x$ is continuous at $x = 4$. Therefore, $\lim\limits_{x \to 4} (8 - 3x) = 8 - 3(4) = -4$.

3. $f(x) = \dfrac{2x + 5}{x - 1}$ is continuous at $x = -3$. Therefore, $\lim\limits_{x \to -3} \dfrac{2x + 5}{x - 1} = \dfrac{2(-3)+5}{-3-1} = \dfrac{1}{4}$

5. $\lim\limits_{x \to 2} \dfrac{4x - 8}{x^2 - 4} = \lim\limits_{x \to 2} \dfrac{4(x - 2)}{(x+2)(x-2)} = \lim\limits_{x \to 2} \dfrac{4}{x + 2} = 1$

7. $\lim\limits_{x \to 2} \dfrac{x^2 + 3x - 10}{x^2 - x - 2} = \lim\limits_{x \to 2} \dfrac{(x + 5)(x - 2)}{(x + 1)(x - 2)} = \lim\limits_{x \to 2} \dfrac{x + 5}{x + 1} = \dfrac{7}{3}$

9. $\lim\limits_{x \to \infty} \dfrac{2 + \frac{1}{x + 4}}{3 - \frac{1}{x^2}} = \dfrac{2}{3}$ since $\dfrac{1}{x + 4}$ and $\dfrac{1}{x^2}$ approach zero as $x \to \infty$.

11. $\lim\limits_{x \to \infty} \dfrac{x - 2x^3}{1 + x^3} = \lim\limits_{x \to \infty} \dfrac{\frac{1}{x^2} - 2}{\frac{1}{x^3} + 1} = -2$

13. $y = 7 + 5x$ 15. $y = 6 - 2x^2$ 17. $y = \dfrac{2}{x^2}$

$\qquad y + \Delta y = 7 + 5(x + \Delta x)$ $\qquad y + \Delta y = 6 - 2(x + \Delta x)^2$ $y + \Delta y = \dfrac{2}{(x + \Delta x)^2}$

$\qquad\qquad \Delta y = 5\Delta x$ $\qquad\qquad \Delta y = -4x\Delta x - 2(\Delta x)^2$ $\quad \Delta y = \dfrac{2}{(x + \Delta x)^2} - \dfrac{2}{x^2}$

$\qquad\qquad \dfrac{\Delta y}{\Delta x} = 5$ $\qquad\qquad \dfrac{\Delta y}{\Delta x} = -4x - 2\Delta x$ $\qquad = \dfrac{-4x\Delta x - 2(\Delta x)^2}{(x + \Delta x)^2 x^2}$

$\qquad \lim\limits_{\Delta x \to 0} \dfrac{\Delta y}{\Delta x} = 5$ $\qquad \lim\limits_{\Delta x \to 0} \dfrac{\Delta y}{\Delta x} = -4x$

$\qquad\qquad\qquad\qquad\qquad\qquad\qquad\qquad\qquad \dfrac{\Delta y}{\Delta x} = \dfrac{-4x - 2\Delta x}{(x + \Delta x)^2 x^2}$

19. $y = \sqrt{x + 5}$

$\qquad\qquad y^2 = x + 5 \quad (y \geq 0)$ $\qquad\qquad \lim\limits_{\Delta x \to 0} \dfrac{\Delta y}{\Delta x} = \dfrac{-4x}{x^4} = -\dfrac{4}{x^3}$

$\qquad (y + \Delta y)^2 = x + \Delta x + 5$

$\qquad y^2 + 2y\Delta y + (\Delta y)^2 = x + \Delta x + 5$

$\qquad\qquad 2y\Delta y + (\Delta y)^2 = \Delta x$ 21. $y = 2x^7 - 3x^2 + 5$ 23. $y = 4\sqrt{x} - \dfrac{1}{x} = 4x^{1/2} - x^{-1}$

$\qquad\qquad \Delta y(2y + \Delta y) = \Delta x$ $\qquad \dfrac{dy}{dx} = 14x^6 - 6x$ $\qquad \dfrac{dy}{dx} = 2x^{-1/2} + x^{-2} = \dfrac{2}{\sqrt{x}} + \dfrac{1}{x^2}$

$\qquad\qquad \dfrac{\Delta y}{\Delta x} = \dfrac{1}{2y + \Delta y}$

$\qquad\qquad\qquad\qquad\qquad$ 25. $y = \dfrac{x}{1 - x}$

$\qquad \lim\limits_{\Delta x \to 0} \dfrac{\Delta y}{\Delta x} = \dfrac{1}{2y} = \dfrac{1}{2\sqrt{x + 5}}$ $\qquad \dfrac{dy}{dx} = \dfrac{(1 - x)(1) - x(-1)}{(1 - x)^2} = \dfrac{1}{(1 - x)^2}$

27. $y = (2 - 3x)^4$ 29. $y = \dfrac{3}{(5 - 2x^2)^{3/4}} = 3(5 - 2x^2)^{-3/4}$

$\qquad \dfrac{dy}{dx} = 4(2 - 3x)^3(-3) = -12(2 - 3x)^3$ $\qquad \dfrac{dy}{dx} = 3\left(-\dfrac{3}{4}\right)(5 - 2x^2)^{-7/4}(-4x) = \dfrac{9x}{(5 - 2x^2)^{7/4}}$

31. $y = x^2\sqrt{1 - 6x} = x^2(1 - 6x)^{1/2}$ 33. $y = \dfrac{\sqrt{4x + 3}}{2x} = \dfrac{(4x + 3)^{1/2}}{2x}$

$\qquad \dfrac{dy}{dx} = x^2\left(\dfrac{1}{2}\right)(1 - 6x)^{-1/2}(-6) + (1 - 6x)^{1/2}(2x)$ $\quad \dfrac{dy}{dx} = \dfrac{2x(1/2)(4x + 3)^{-1/2}(4) - (4x + 3)^{1/2}(2)}{(2x)^2}$

$\qquad\quad = \dfrac{-3x^2}{(1 - 6x)^{1/2}} + 2x(1 - 6x)^{1/2}$ $\qquad\quad = \dfrac{4x - 2(4x + 3)}{4x^2(4x + 3)^{1/2}} = \dfrac{-2x - 3}{2x^2(4x + 3)^{1/2}}$

$\qquad\quad = \dfrac{-3x^2 + 2x(1 - 6x)}{(1 - 6x)^{1/2}} = \dfrac{-15x^2 + 2x}{(1 - 6x)^{1/2}}$

35. $(2x - 3y)^3 = x^2 - y$ 37. $y = \dfrac{4}{x} + 2\sqrt[3]{x} = 4x^{-1} + 2x^{1/3} \,;\; x = 8$

$\qquad 3(2x - 3y)^2\left(2 - 3\dfrac{dy}{dx}\right) = 2x - \dfrac{dy}{dx}$

$\qquad\qquad\qquad\qquad\qquad\qquad\qquad \dfrac{dy}{dx} = -4x^{-2} + \dfrac{2}{3}x^{-2/3} = -\dfrac{4}{x^2} + \dfrac{2}{3x^{2/3}}$

$\qquad \dfrac{dy}{dx}[1 - 9(2x - 3y)^2] = 2x - 6(2x - 3y)^2$

$\qquad\qquad\qquad\qquad\qquad\qquad\qquad \dfrac{dy}{dx}\Big|_{x=8} = -\dfrac{4}{8^2} + \dfrac{2}{3(8)^{2/3}} = -\dfrac{1}{16} + \dfrac{1}{6} = \dfrac{5}{48}$

$\qquad \dfrac{dy}{dx} = \dfrac{2x - 6(2x - 3y)^2}{1 - 9(2x - 3y)^2}$

39. $y = 2x\sqrt{4x + 1} = 2x(4x + 1)^{1/2} \,;\; x = 6$ 41. $y = 3x^4 - \dfrac{1}{x} = 3x^4 - x^{-1}$

$\qquad \dfrac{dy}{dx} = 2x\left(\dfrac{1}{2}\right)(4x + 1)^{-1/2}(4) + (4x + 1)^{1/2}(2)$ $\qquad y' = 12x^3 + x^{-2}$

$\qquad\quad = \dfrac{4x}{(4x + 1)^{1/2}} + 2(4x + 1)^{1/2} = \dfrac{4x + 2(4x + 1)}{(4x + 1)^{1/2}}$ $\qquad y'' = 36x^2 - 2x^{-3}$

$\qquad\quad = \dfrac{12x + 2}{(4x + 1)^{1/2}}$ $\qquad\qquad = 36x^2 - \dfrac{2}{x^3}$

$\qquad \dfrac{dy}{dx}\Big|_{x=6} = \dfrac{12(6) + 2}{[4(6) + 1]^{1/2}} = \dfrac{74}{5}$

43. $y = \dfrac{1 - 3x}{1 + 4x}$

$y' = \dfrac{(1 + 4x)(-3) - (1 - 3x)(4)}{(1 + 4x)^2}$

$= \dfrac{-7}{(1 + 4x)^2} = -7(1 + 4x)^{-2}$

$y'' = 14(1 + 4x)^{-3}(4) = \dfrac{56}{(1 + 4x)^3}$

45. $\displaystyle\lim_{c_1 \to c_2} \dfrac{c_1 c_2}{c_1 + c_2} = \dfrac{c_2 c_2}{c_2 + c_2} = \dfrac{1}{2} c_2$

$\displaystyle\lim_{c_1 \to 0} \dfrac{c_1 c_2}{c_1 + c_2} = 0$

47. $y = 7x^4 - x^3$; $(-1, 8)$

$\dfrac{dy}{dx} = 28x^3 - 3x^2$

$\dfrac{dy}{dx}\bigg|_{x=-1} = 28(-1)^3 - 3(-1)^2$

$= -31 = m_{tan}$ at $(-1, 8)$

49. $p = 2(50 + w)$

$= 100 + 2w$

$\dfrac{dp}{dw} = 2$

51. $s = \dfrac{t}{3t + 1}$

$v = \dfrac{ds}{dt} = \dfrac{(3t+1)(1) - t(3)}{(3t + 1)^2}$

$= \dfrac{1}{(3t + 1)^2}$

$v\big|_{t=3} s = \dfrac{1}{[3(3)+1]^2} = 0.01$ m/s

53. $I = 0.1t + 5.0$

$\dfrac{dI}{dt} = 0.1$

$E = L\dfrac{dI}{dt} = 0.4(0.1)$

$= 0.04$ V

55. $E = \dfrac{k}{r^2} = kr^{-2}$

$\dfrac{dE}{dr} = -2kr^{-3}$

$= -\dfrac{2k}{r^3}$

57. $V = \pi r^2 h$; $100 = \pi r^2 h$; $h = \dfrac{100}{\pi r^2}$

$A = 2\pi r^2 + 2\pi rh = 2\pi r^2 + 2\pi r(\dfrac{100}{\pi r^2})$

$= 2\pi r^2 + \dfrac{200}{r} = 2\pi r^2 + 200r^{-1}$

$\dfrac{dA}{dr} = 4\pi r - 200r^{-2} = 4\pi r - \dfrac{200}{r^2}$

59. $v = \sqrt{v_o^2 + 2as} = (v_o^2 + 2as)^{1/2}$

$\dfrac{dv}{ds} = \dfrac{1}{2}(v_o^2 + 2as)^{-1/2}(2a)$

$= \dfrac{a}{(v_o^2 + 2as)^{1/2}}$

61. $e = 100(1 - \dfrac{1}{(V_1/V_2)^{0.4}})$

$= 100(1 - \dfrac{v_2^{.4}}{v_1^{.4}})$

$\dfrac{de}{dV_1} = -100(v_2^{0.4})(-0.4)(v_1^{-1.4})$

$= \dfrac{40v_2^{0.4}}{v_1^{1.4}}$

63. See Fig. 2-23.

$A = xy = x(4 - x^2) = 4x - x^3$

$\dfrac{dA}{dx} = 4 - 3x^2$

65. $V = \dfrac{1,500,000}{2t + 10} = 1,500,000(2t + 10)^{-1}$

$\dfrac{dV}{dt} = -1,500,000(2t + 10)^{-2}(2) = -3,000,000(2t + 10)^{-2} = -\dfrac{3,000,000}{(2t + 10)^2}$

$\dfrac{d^2V}{dt^2} = 6,000,000(2t + 10)^{-3}(2) = \dfrac{12,000,000}{(2t + 10)^3}$

$\dfrac{dV}{dt}\bigg|_{t=5} = \dfrac{-3,000,000}{[2(5)+10]^2} = -\$7500$/year (rate of appreciation - it is decreasing)

$\dfrac{d^2V}{dt^2}\bigg|_{t=5} = \dfrac{12,000,000}{[2(5)+10]^3} = \$1500$/year^2 (rate at which appreciation changes - it is increasing)

(Machinery is depreciating, but the depreciation is lessening.)

CHAPTER 3   APPLICATIONS OF THE DERIVATIVE

Exercises 3-1, p. 124

1. $y = x^2 + 2$ at $(2,6)$
   $y' = 2x$, $y'\big|_{x=2} = 2(2) = 4$
   $m = 4$, $(x_1, y_1) = (2,6)$
   $y - y_1 = m(x - x_1)$
   $y - 6 = 4(x - 2)$
   $4x - y - 2 = 0$

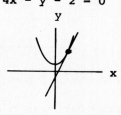

3. $y = \dfrac{1}{x^2+1} = (x^2+1)^{-1}$ at $(1,\tfrac{1}{2})$
   $y' = -(x^2+1)^{-2}(2x) = \dfrac{-2x}{(x^2+1)^2}$
   $y'\big|_{x=1} = \dfrac{-2}{(1+1)^2} = -\dfrac{1}{2}$, $m = -\dfrac{1}{2}$
   $y - y_1 = m(x - x_1)$
   $y - \dfrac{1}{2} = -\dfrac{1}{2}(x - 1)$
   $2y + x - 2 = 0$

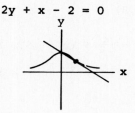

5. $y = 6x - x^2$ at $(1,5)$
   $y' = 6 - 2x$
   $y'\big|_{x=1} = 6 - 2 = 4$, $m = -\dfrac{1}{4}$
   $y - y_1 = m(x - x_1)$
   $y - 5 = -\dfrac{1}{4}(x - 1)$
   $x + 4y - 21 = 0$

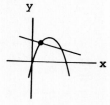

7. $y = \dfrac{1}{(x^2+1)^2}$ at $(1,\tfrac{1}{4})$
   $y = (x^2+1)^{-2}$
   $y' = -2(x^2+1)^{-3}(2x)$
   $\quad = \dfrac{-4x}{(x^2+1)^3}$
   $y'\big|_{x=1} = \dfrac{-4}{(1+1)^3} = -\dfrac{1}{2}$
   $m = 2$, $y - y_1 = m(x - x_1)$
   $y - \dfrac{1}{4} = 2(x - 1)$
   $8x - 4y - 7 = 0$

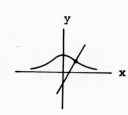

9. $y = \dfrac{1}{\sqrt{x^2 + 1}}$ where $x = \sqrt{3}$
   $y = (x^2+1)^{-1/2}$
   $y' = -\dfrac{1}{2}(x^2+1)^{-3/2}(2x)$
   $\quad = \dfrac{-x}{(x^2+1)^{3/2}}$
   $y'\big|_{x=\sqrt{3}} = \dfrac{-\sqrt{3}}{(3+1)^{3/2}} = -\dfrac{\sqrt{3}}{8}$
   Tangent line: $m = -\dfrac{1}{8}\sqrt{3}$
   for $x = \sqrt{3}$, $y = \dfrac{1}{\sqrt{3+1}} = \dfrac{1}{2}$
   $y - \dfrac{1}{2} = -\dfrac{1}{8}\sqrt{3}(x - \sqrt{3})$
   $8y - 4 = -\sqrt{3}x + 3$
   $\sqrt{3}x + 8y - 7 = 0$
   Normal line: $m = 8/\sqrt{3}$
   $y - \dfrac{1}{2} = \dfrac{8}{\sqrt{3}}(x - \sqrt{3})$
   $2\sqrt{3}y - \sqrt{3} = 16x - 16\sqrt{3}$
   $16x - 2\sqrt{3}y - 15\sqrt{3} = 0$

11. Parabola: vertex$(0,3)$
    focus $(0,0)$, where $x = -1$
    $(x-h)^2 = 4p(y-k)$
    $(h,k) = (0,3)$, $p = -3$
    $(x-0)^2 = 4(-3)(y-3)$
    $x^2 = -12(y-3)$, $y = -\dfrac{1}{12}x^2 + 3$
    $y' = -\dfrac{1}{6}x$, $y'\big|_{x=-1} = \dfrac{1}{6}$
    Tangent line: $m = \dfrac{1}{6}$
    for $x = -1$, $y = -\dfrac{1}{12}(1)^2 + 3 = \dfrac{35}{12}$
    $y - \dfrac{35}{12} = \dfrac{1}{6}(x + 1)$
    $12y - 35 = 2x + 2$
    $2x - 12y + 37 = 0$
    Normal line: $m = -6$
    $y - \dfrac{35}{12} = -6(x + 1)$
    $12y - 35 = -72x - 72$
    $72x + 12y + 37 = 0$

13. $y = x^2 - 2x$, $m_{\tan} = 2$
    $y' = 2x - 2$
    $2 = 2x - 2$, $x = 2$
    for $x = 2$, $y = 4 - 4 = 0$
    $y - y_1 = m(x - x_1)$
    $y - 0 = 2(x - 2)$
    $y = 2x - 4$

15. $y = (2x - 1)^3$, $m_{norm} = -\dfrac{1}{24}$ $(x>0)$
    $y' = 3(2x-1)^2(2) = 6(2x-1)^2$
    $m_{\tan} = 24$, $24 = 6(2x-1)^2$, $(2x-1)^2 = 4$, $2x - 1 = \pm 2$
    $x = -\dfrac{1}{2}, \dfrac{3}{2}$ (use $x = \dfrac{3}{2}$ since $x>0$); For $x = \dfrac{3}{2}$, $y = (3-1)^3 = 8$
    $y - y_1 = m(x - x_1)$, $y - 8 = -\dfrac{1}{24}(x - \dfrac{3}{2})$
    $48y - 384 = -2x + 3$, $2x + 48y - 387 = 0$

17. $y = 4x^2 - 8x$ at $(-1,12)$

$y' = 8x - 8$

$y'|_{x=-1} = -8 - 8 = -16$

$m = -16$, $(x_1,y_1) = (-1,12)$

$y - 12 = -16(x + 1)$

at x-int., $y = 0$

$-12 = -16(x + 1)$

$3 = 4(x + 1)$, $x = -\frac{1}{4}$

21.

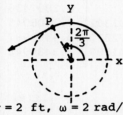

$r = 2$ ft, $\omega = 2$ rad/s, $t = \frac{\pi}{3}$ s

$\theta = \omega t = (2\frac{rad}{s})(\frac{\pi}{3} s) = \frac{2\pi}{3}$ rad

Point P: $x = r\cos\theta = 2\cos\frac{2\pi}{3} = -1$

$y = r\sin\theta = 2\sin\frac{2\pi}{3} = \sqrt{3}$

(If necessary, trig. review in Section 6-1.)

Circle: $x^2 + y^2 = 4$

$2x + 2yy' = 0$, $y' = -\frac{x}{y}$

$(x_1,y_1) = (-1,\sqrt{3})$, $m = -\frac{-1}{\sqrt{3}} = \frac{1}{\sqrt{3}}$

$y - \sqrt{3} = \frac{1}{\sqrt{3}}(x + 1)$

$\sqrt{3}y - 3 = x + 1$, $x - \sqrt{3}y + 4 = 0$

19. Parabola: vertex$(0,0)$
point on curve $(100,30)$

$x^2 = 4py$, $100^2 = 4p(30)$

$p = \frac{250}{3}$, $x^2 = 4(\frac{250}{3})y$

$y = \frac{3}{1000}x^2$, $y' = \frac{3}{500}x$

$y'|_{x=100} = \frac{300}{500} = \frac{3}{5}$

$(x_1,y_1) = (100,30)$, $m = \frac{3}{5}$

$y - 30 = \frac{3}{5}(x - 100)$, $3x - 5y - 150 = 0$

23. $y = \frac{4}{x^2 + 1} = 4(x^2 + 1)^{-1}$

$y' = -4(x^2+1)^{-2}(2x) = \frac{-8x}{(x^2+1)^2}$

at $x=-1$: $y = \frac{4}{1+1} = 2$, $y' = \frac{-8(-1)}{(1+1)^2} = 2$

$(x_1,y_1) = (-1,2)$, $m = -\frac{1}{2}$

$y - 2 = -\frac{1}{2}(x+1)$, $x + 2y - 3 = 0$

at $x=0$: $y = \frac{4}{0+1} = 4$, $y' = \frac{-8(0)}{(0+1)^2} = 0$

m_{norm} is undefined, line vertical, $x = 0$

at $x=1$: $y = \frac{4}{1+1} = 2$, $y' = \frac{-8(1)}{(1+1)^2} = -2$

$(x_1,y_1) = (1,2)$, $m = \frac{1}{2}$

$y - 2 = \frac{1}{2}(x-1)$, $x - 2y + 3 = 0$

Exercises 3-2, p. 128

1. $x^2 - 2x - 5 = 0$

$f(x) = x^2 - 2x - 5$

$f'(x) = 2x - 2$

$f(3) = 3^2 - 2(3) - 5 = -2$

$f(4) = 4^2 - 2(4) - 5 = 3$

Let $x_1 = 3.5$

n	x_n	$f(x_n)$	$f'(x_n)$	$x_n - \dfrac{f(x_n)}{f'(x_n)}$
1	3.5	0.25	5.0	3.45
2	3.45	0.0025	4.9	3.4494898
3	3.4494898			

Using quadratic formula: $x = \frac{-(-2)\pm\sqrt{(-2)^2 - 4(1)(-5)}}{2(1)} = \frac{2\pm\sqrt{24}}{2} = 1\pm\sqrt{6}$

Between 3 and 4, $x = 1 + \sqrt{6} = 3.4494897$; Results agree to 6 decimal places

3. $3x^2 - 5x - 1 = 0$

$f(x) = 3x^2 - 5x - 1$

$f'(x) = 6x - 5$

$f(-1) = 7$, $f(0) = -1$

n	x_n	$f(x_n)$	$f'(x_n)$	$x_n - \dfrac{f(x_n)}{f'(x_n)}$
1	-0.2	0.12	-6.2	-0.1806452
2	-0.1806452	0.0011238	-6.0838710	-0.1804604
3	-0.1804604			

Let $x_1 = -0.2$

Using quadratic formula: $x = \frac{-(-5)\pm\sqrt{(-5)^2 - 4(3)(-1)}}{2(3)} = \frac{5\pm\sqrt{37}}{6}$; Use $x = \frac{5-\sqrt{37}}{6}$

Between -1 and 0, $x = -0.1804604$; Results agree to 7 decimal places

5. $x^3 - 6x^2 + 10x - 4 = 0$

$f(x) = x^3 - 6x^2 + 10x - 4$

$f'(x) = 3x^2 - 12x + 10$

$f(0) = -4$, $f(1) = 1$

Let $x_1 = 0.8$

n	x_n	$f(x_n)$	$f'(x_n)$	$x_n - \dfrac{f(x_n)}{f'(x_n)}$
1	0.8	0.672	2.32	0.5103448
2	0.5103448	-0.3263425	4.6572176	0.5804173
3	0.5804173	-0.0215992	4.0456455	0.5857561
4	0.5857561	-0.0001212	4.0002572	0.5857864
5	0.5857864			

$x_4 = x_5 = 0.5858$ (to 4 decimal places)

7. $x^3 + 5x^2 + x - 1 = 0$

$f(x) = x^3 + 5x^2 + x - 1$

$f'(x) = 3x^2 + 10x + 1$

$f(0) = -1$, $f(1) = 6$

Let $x_1 = 0.2$

n	x_n	$f(x_n)$	$f'(x_n)$	$x_n - \dfrac{f(x_n)}{f'(x_n)}$
1	0.2	-0.592	3.12	0.3897436
2	0.3897436	0.2084460	5.3531361	0.3508045
3	0.3508045	0.0092950	4.8772370	0.3488987
4	0.3488987	0.0000220	4.8541785	0.3488942
5	0.3488942			

$x_4 = x_5 = 0.3489$ (to 4 decimal places)

9. $x^4 - x^3 - 3x^2 - x - 4 = 0$

$f(x) = x^4 - x^3 - 3x^2 - x - 4$

$f'(x) = 4x^3 - 3x^2 - 6x - 1$

$f(2) = -10$, $f(3) = 20$

Let $x_1 = 2.3$

n	x_n	$f(x_n)$	$f'(x_n)$	$x_n - \dfrac{f(x_n)}{f'(x_n)}$
1	2.3	-6.3529000	17.998000	2.6529781
2	2.6529781	3.0972726	36.657000	2.5684848
3	2.5684848	0.2174993	31.576096	2.5615967
4	2.5615967	0.0013671	31.179597	2.5615528
5	2.5615528			

$x_4 = x_5 = 2.5616$ (to 4 decimal places)

11. $x^4 - 2x^3 - 8x - 16 = 0$

$f(x) = x^4 - 2x^3 - 8x - 16$

$f'(x) = 4x^3 - 6x^2 - 8$

$f(0) = -16$, $f(-1) = -5$,

$f(-2) = 32$; Let $x_1 = -1.2$

n	x_n	$f(x_n)$	$f'(x_n)$	$x_n - \dfrac{f(x_n)}{f'(x_n)}$
1	-1.2	-0.8704000	-23.552	-1.2369565
2	-1.2369565	0.0219791	-24.750847	-1.2360685
3	-1.2360685	0.0000131	-24.721377	-1.2360680
4	-1.2360680			

$x_3 = x_4 = -1.23607$ (to 5 decimal places)

13. $2x^2 = \sqrt{2x+1}$

$f(x) = 2x^2 - \sqrt{2x+1}$

$f'(x) = 4x - \dfrac{1}{2}(2x+1)^{-1/2}(2)$

$\qquad = 4x - \dfrac{1}{\sqrt{2x+1}}$

$f(0) = -1$, $f(1) = 2 - \sqrt{3} = 0.27$

Let $x_1 = 0.7$

n	x_n	$f(x_n)$	$f'(x_n)$	$x_n - \dfrac{f(x_n)}{f'(x_n)}$
1	0.7	-0.5691933	2.1545028	0.9641878
2	0.9641878	0.1480665	3.2723829	0.9189405
3	0.9189405	0.0043021	3.0821497	0.9175447
4	0.9175447	0.0000041	3.0762742	0.9175433
5	0.9175433			

$x_4 = x_5 = 0.91754$ (to 5 decimal places)

15. $x = \dfrac{1}{\sqrt{x+2}}$, $f(x) = x - \dfrac{1}{\sqrt{x+2}}$

$f'(x) = 1 - \left(-\dfrac{1}{2}\right)(x+2)^{-3/2}$

$\qquad = 1 + \dfrac{1}{2(x+2)^{3/2}}$

$x > 0$ since $x = 1/\sqrt{x+2}$ and $1/\sqrt{x+2} > 0$.

$f(0.1) = -0.6$, $f(1) = 0.4$, Let $x_1 = 0.5$

n	x_n	$f(x_n)$	$f'(x_n)$	$x_n - \dfrac{f(x_n)}{f'(x_n)}$
1	0.5	-0.1324555	1.1264911	0.6175824
2	0.6175824	-0.0005049	1.1180645	0.6180340
3	0.6180340	-6.9×10^{-9}	1.1180340	0.6180340
4	0.6180340			

$x_4 = x_3 = 0.6180340$ (to 7 decimal places)

17. $x^3 - 2x^2 - 5x + 4 = 0$
$f(x) = x^3 - 2x^2 - 5x + 4$
$f'(x) = 3x^2 - 4x - 5$

x	f(x)
-2	-2 ← Root 1
-1	6
0	4 ← Root 2
1	-2
2	-6
3	-2 ← Root 3
4	16

n	x_n	$f(x_n)$	$f'(x_n)$	$x_n - \dfrac{f(x_n)}{f'(x_n)}$
1	-1.7	1.807	10.47	-1.8725884
2	-1.8725884	-0.2166267	13.010115	-1.8559377
3	-1.8559377	-0.0021074	12.757265	-1.8557725
4	-1.8557725			
1	0.7	-0.137	-6.33	0.6783570
2	0.6783570	0.0000367	-6.3329233	0.6783628
3	0.6783628			
1	3.1	-0.929	11.43	3.1812773
2	3.1812773	0.0487608	12.636467	3.1774186
3	3.1774186	0.0001123	12.578293	3.1774097
4	3.1774097			

Root 1: Let $x_1 = -1.7$, $x_4 = -1.8557725$, Root 2: Let $x_1 = 0.7$, $x_3 = 0.6783628$
Root 3: Let $x_1 = 3.1$, $x_4 = 3.1774097$

19. $V = \dfrac{4}{3}\pi r^3$

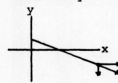

$d = 10.0000$ m
$r = 5.0000$ m
$500.0 = \dfrac{4}{3}\pi(5.0000 - t)^3$

n	t_n	$f(t_n)$	$f'(t_n)$	$t_n - \dfrac{f(t_n)}{f'(t_n)}$
1	0.1	1.7172	72.03	0.0761599
2	0.0761599	-0.0083683	72.732603	0.0762750
3	0.0762750	-0.0000002	72.729204	0.0762750
4	0.0762750			

$(5-t)^3 = \dfrac{375}{\pi} = 119.3662$ (additional significant digits not shown on left)
$125 - 75t + 15t^2 - t^3 = 119.3662$, $f(t) = t^3 - 15t^2 + 75t - 5.6338$
Let $t_1 = 0.1$ $f'(t) = 3t^2 - 30t + 75$
$t = 0.0763$ m (to four decimal places)

Exercises 3-3, p. 135

1. $x = 3t$, $y = 1 - t$, $t = 4$
$v_x = \dfrac{dx}{dt} = 3$, $v_x\big|_{t=4} = 3$
$v_y = \dfrac{dy}{dt} = -1$, $v_y\big|_{t=4} = -1$
$v\big|_{t=4} = \sqrt{3^2 + (-1)^2} = 3.16$
$\tan\theta = \dfrac{-1}{3} = -0.3333$
$\theta = -18.4°$ or $341.6°$
$(v_x > 0, v_y < 0)$

3. $x = \dfrac{10}{2t+3} = 10(2t+3)^{-1}$, $y = \dfrac{1}{t^2} = t^{-2}$, $t = 2$
$v_x = \dfrac{dx}{dt} = -10(2t+3)^{-2}(2) = \dfrac{-20}{(2t+3)^2}$
$v_y = \dfrac{dy}{dt} = -2t^{-3} = \dfrac{-2}{t^3}$
$v_x\big|_{t=2} = \dfrac{-20}{(4+3)^2} = -0.408$, $v_y\big|_{t=2} = \dfrac{-2}{2^3} = -0.250$
$v\big|_{t=2} = \sqrt{(-0.408)^2 + (-0.250)^2} = 0.479$
$\tan\theta = \dfrac{-0.250}{-0.408} = 0.6127$, $\theta = 211.5°$ $(v_x < 0, v_y < 0)$

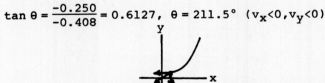

5. $x = 3t$, $y = 1 - t$, $t = 4$
$v_x = \dfrac{dx}{dt} = 3$, $a_x = \dfrac{dv_x}{dt} = \dfrac{d^2x}{dt^2} = 0$; $v_y = \dfrac{dy}{dt} = -1$, $a_y = \dfrac{dv_y}{dt} = \dfrac{d^2y}{dt^2} = 0$
$a = \sqrt{0^2 + 0^2} = 0$ (not accelerating for all values of t)

7. $x = \dfrac{10}{2t + 3}$, $y = \dfrac{1}{t^2}$, t=2 (see #3)

$v_x = \dfrac{dx}{dt} = \dfrac{-20}{(2t + 3)^2}$

$a_x = \dfrac{dv_x}{dt} = \dfrac{d^2x}{dt^2} = 40(2t + 3)^{-3}(2) = \dfrac{80}{(2t + 3)^3}$

$v_y = \dfrac{dy}{dt} = \dfrac{-2}{t^3}$, $a_y = \dfrac{dv_y}{dt} = \dfrac{d^2y}{dt^2} = \dfrac{6}{t^4}$

$a_x|_{t=2} = \dfrac{80}{(4+3)^3} = 0.233$

$a_y|_{t=2} = \dfrac{6}{2^4} = 0.375$

$a|_{t=1} = \sqrt{0.233^2 + 0.375^2} = 0.442$

$\tan \theta = \dfrac{0.375}{0.233} = 1.609$, $\theta = 58.1°$

$\qquad\qquad\qquad (a_x>0, a_y>0)$

11. $y^2 = 2x + 1$, $v_x = \sqrt{2x + 3}$, $x = 4$

$2y\dfrac{dy}{dt} = 2\dfrac{dx}{dt}$; $y = \sqrt{2x+1}$ (upper half)

$\dfrac{dy}{dt} = v_y = \dfrac{1}{y}v_x$

when x=4, $v_x = \sqrt{2(4) + 3} = \sqrt{11}$

$y = \sqrt{2(4)+1} = 3$, $v_y = \dfrac{\sqrt{11}}{3} = 1.106$

15. $x = 120t$, $y = 160t - 16t^2$ (see #13)

$v_x = \dfrac{dx}{dt} = 120$, $a_x = \dfrac{dv_x}{dt} = \dfrac{d^2x}{dt^2} = 0$ (all t)

$v_y = \dfrac{dy}{dt} = 160 - 32t$, $a_y = \dfrac{dv_y}{dt} = \dfrac{d^2y}{dt^2} = -32$ ft/s^2

for t=3 s and t=6 s, a = 32 ft/s^2, θ = 270.0°

17. $x = 10(\sqrt{1 + t^4} - 1)$, $y = 40t^{3/2}$

$v_x = 10(\dfrac{1}{2})(1 + t^4)^{-1/2}(4t^3) = \dfrac{20t^3}{(1 + t^4)^{1/2}}$

$v_y = 40(\dfrac{3}{2})t^{1/2} = 60t^{1/2}$

for t = 10 s: $v_x = \dfrac{20(10)^3}{[1+10^4]^{1/2}} = 200$ m/s

$\qquad v_y = 60(10)^{1/2} = 190$ m/s

$\qquad v = \sqrt{200^2 + 190^2} = 276$ m/s

$\qquad \tan \theta = \dfrac{190}{200} = 0.95$, $\theta = 43.5°$

for t = 100 s

$\qquad v_x = \dfrac{20(100)^3}{[1+100^4]^{1/2}} = 2000$ m/s

$\qquad v_y = 60(100)^{1/2} = 600$ m/s

$\qquad v = \sqrt{2000^2 + 600^2} = 2090$ m/s

$\qquad \tan \theta = \dfrac{600}{2000} = 0.300$, $\theta = 16.7°$

9. $y = 5x^3$ at (1,5), $v_x = 20$

$\dfrac{dy}{dt} = 15x^2\dfrac{dx}{dt}$

$v_y = 15x^2 v_x$

at (1,5) $v_y = 15(1)^2(20) = 300$

$v = \sqrt{20^2 + 300^2} = 301$

$\tan \theta = \dfrac{300}{20} = 15.0$, $\theta = 86.2°$

$(v_x > 0, v_y > 0)$

13. $x = 120t$, $y = 160t - 16t^2$

$v_x = \dfrac{dx}{dt} = 120$ ft/s (for all t)

$v_y = \dfrac{dy}{dt} = 160 - 32t$

$v_y|_{t=3} = 160 - 32(3) = 64$ ft/s

$v_y|_{t=3} = 160 - 32(6) = -32$ ft/s

$v|_{t=3} = \sqrt{120^2 + 64^2} = 136$ ft/s

$\tan \theta = \dfrac{64}{120} = 0.5333$, $\theta = 28.1°$

$v|_{t=6} = \sqrt{120^2 + (-32)^2} = 124$ ft/s

$\tan \theta = \dfrac{-32}{120} = -0.2666$, $\theta = -14.9°$ or 345.1°

$\qquad\qquad (v_x>0, v_y<0)$

19. $y = x - \dfrac{1}{90}x^3$, $v_x = x$

$\dfrac{dy}{dt} = \dfrac{dx}{dt} - \dfrac{1}{30}x^2\dfrac{dx}{dt}$

$v_y = v_x - \dfrac{1}{30}x^2 v_x = x - \dfrac{1}{30}x^2(x)$

$\qquad = x - \dfrac{1}{30}x^3$

Rocket hits ground when y=0.

$x - \dfrac{1}{90}x^3 = x(1 - \dfrac{1}{90}x^2) = 0$

x = 0 when it starts

When it hits $1 - \dfrac{1}{90}x^2 = 0$

$x^2 = 90$, x = 9.487 mi

For x = 9.487 mi, $v_x = x = 9.487$ mi/min

$\qquad v_y = x - \dfrac{1}{30}x^3 = 9.487 - \dfrac{9.487^3}{30}$

$\qquad\qquad = -18.98$ mi/min

$\qquad v = \sqrt{9.487^2 + (-18.98)^2} = 21.2$ mi/min

$\qquad \tan \theta = -18.98/9.487 = 2.001$

$\qquad \theta = -63.4°$ or 296.6°

Exercises 3-4, p. 139

1. $R = 4.000 + 0.003T^2$, $\dfrac{dT}{dt} = 0.100°C/s$

$\dfrac{dR}{dt} = 2(0.003T)\dfrac{dT}{dt} = 0.006T\dfrac{dT}{dt}$

$\dfrac{dR}{dt}\Big|_{T=150°C} = 0.006(150)(0.100)$

$= 0.0900\ \Omega/s$

3. $p = 30\sqrt{10x - x^2} - 50)$, $\dfrac{dx}{dt} = 0.200$ tons/week^2

$\dfrac{dp}{dt} = 30(\tfrac{1}{2})(10x - x^2)^{-1/2}(10 - 2x)\dfrac{dx}{dt}$

$= \dfrac{30(5 - x)dx/dt}{(10x - x^2)^{1/2}}$

$\dfrac{dp}{dt}\Big|_{x=4} = \dfrac{30(5 - 4)(0.200)}{(40 - 16)^{1/2}} = 1.22$ \$/week

5. $T = \pi\sqrt{\dfrac{L}{96}}$, $\dfrac{dL}{dt} = -0.100$ in./s

$\dfrac{dT}{dt} = \pi(\tfrac{1}{2})(\dfrac{L}{96})^{-1/2}(\dfrac{1}{96}\dfrac{dL}{dt}) = \dfrac{\pi}{2\sqrt{96L}}\dfrac{dL}{dt}$

$\dfrac{dT}{dt}\Big|_{L=16.0\ in.} = \dfrac{\pi}{2\sqrt{96(16)}}(-0.100)$

$= -0.00401$

7. $V = \dfrac{0.03I}{r^2}$, $\dfrac{dI}{dt} = 0.020$ A/s, $r = 0.040$ in. (constant)

$\dfrac{dV}{dt} = \dfrac{0.03}{r^2}\dfrac{dI}{dt}$, $\dfrac{dV}{dt}\Big|_{all\ t} = \dfrac{0.03}{(0.04)^2}(0.020)$

$= 0.375$ V/s

9. Let s = length of side of square
$A = s^2$, $\dfrac{ds}{dt} = 5.00$ ft/min

$\dfrac{dA}{dt} = 2s\dfrac{ds}{dt}$, $\dfrac{dA}{dt}\Big|_{s=10.0\ ft} = 2(10.0)(5.00)$

$= 100$ ft^2/min

11. $V = \pi r^2 h$, $r = 3.00$ in. $= 0.250$ ft (constant)

$V = \pi(0.25^2)h = 0.0625\pi h$

$\dfrac{dV}{dt} = 2.00$ ft^3/s

$\dfrac{dV}{dt} = 0.0625\pi\dfrac{dh}{dt}$

$2.00 = 0.0625\pi\dfrac{dh}{dt}$, $\dfrac{dh}{dt} = 10.2$ ft/s

13. $P = \dfrac{k}{V}$, $\dfrac{dV}{dt} = 20.0$ cm^3/min

$V = 600$ cm^3 when $P = 200$ kPa

$200 = \dfrac{k}{600}$, $k = 1.20 \times 10^5$ kPa·cm^3

$P = \dfrac{1.20\times10^5}{V}$, $\dfrac{dP}{dt} = \dfrac{-1.20\times10^5}{V^2}\dfrac{dV}{dt}$

$\dfrac{dP}{dt}\Big|_{V=800\ cm^3} = \dfrac{-1.20\times10^5}{800^2}(20.0)$

$= -3.75$ kPa/min

15. $V = \dfrac{4}{3}\pi r^3$, $\dfrac{dr}{dt} = 5.00$ mm/s

$\dfrac{dV}{dt} = \dfrac{4\pi}{3}(3r^2\dfrac{dr}{dt}) = 4\pi r^2\dfrac{dr}{dt}$

$\dfrac{dV}{dt}\Big|_{r=200\ mm} = 4\pi(200)^2(5.00)$

$= 2.51 \times 10^6$ mm^3/s

17. $\dfrac{dy}{dt} = 80.0$ km/h

$\dfrac{dx}{dt} = 100$ km/h

$z^2 = x^2 + y^2$, $2z\dfrac{dz}{dt} = 2x\dfrac{dx}{dt} + 2y\dfrac{dy}{dt}$

When t=0.0500 h, x=100(0.0500)=5.00 km

y=80.0(0.0500)=4.00 km

$z = \sqrt{5.00^2 + 4.00^2} = 6.40$ km

$z\dfrac{dz}{dt} = x\dfrac{dx}{dt} + y\dfrac{dy}{dt}$, $6.40\dfrac{dz}{dt}=5.00(100)+4.00(80.0)$

$\dfrac{dz}{dt} = \dfrac{500 + 320}{6.40} = 128$ km/h

19. See Fig. 3-20

$\dfrac{dy}{dt} = -10.0$ ft/s

$y^2 = 20.0^2 + x^2$

$2y\dfrac{dy}{dt} = 2x\dfrac{dx}{dt}$, $\dfrac{dx}{dt} = \dfrac{y}{x}\dfrac{dy}{dt}$

When y=36.0 ft

$x = \sqrt{36.0^2 - 20.0^2} = 29.9$ ft

$\dfrac{dx}{dt}\Big|_{x=36.0\ ft} = \dfrac{36.0}{29.9}(-10.0)$

$= -12.0$ ft/s

(x decreasing, boat approaching wharf)

21.

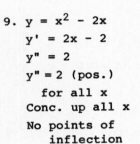

$\dfrac{dx}{dt} = 1600$ mi/h, $\dfrac{dy}{dt} = 1800$ mi/h, $z^2 = x^2 + y^2$

$2z\dfrac{dz}{dt} = 2x\dfrac{dx}{dt} + 2y\dfrac{dy}{dt}$, $z\dfrac{dz}{dt} = x\dfrac{dx}{dt} + y\dfrac{dy}{dt}$

After 0.5 h, $x = 800$ mi, $z = 900$ mi, $y = \sqrt{900^2 - 800^2} = 412$ mi

$900(1800) = 800(1600) + 412\dfrac{dy}{dt}$, $\dfrac{dy}{dt} = \dfrac{340000}{412} = 825$ mi/h

Exercises 3-5, p. 146

1. $y = x^2 - 2x$
   $y' = 2x - 2$
   $2x-2 = 0$ for $x=1$
   $x<1$, $y'<0$, y dec.
   $x>1$, $y'>0$, y inc.

3. $y = 12x - x^3$
   $y' = 12 - 3x^2 = 3(4 - x^2)$
   $= 3(2 + x)(2 - x)$
   $3(2+x)(2-x) = 0$ for $x=-2,2$
   $x<-2$, $y'<0$, y dec.
   $-2<x<2$, $y'>0$, y inc.
   $x>2$, $y'<0$, y dec.

5. $y = x^2 - 2x$, $y' = 2x - 2$, $y" = 2$
   $y'=2x-2=0$ for $x=1$
   $y" = 2$ (pos.) for all x
   When $x=1$, $y=-1$, $(1,-1)$ rel.min.

   See Exer. 1 : For $x<1$, y dec.,
                       $x>1$, y inc.
   $(1,-1)$ is rel. min.

7. $y = 12x - x^3$
   $y' = 12 - 3x^2$, $y" = -6x$
   $y'=12-3x^2=0$ for $x=-2,2$
   For $x=-2$, $y"=12$(pos.)
    $y=-16$, $(-2,-16)$ rel.min.
   For $x=2$, $y"=-12$(neg.)
    $y=16$, $(2,16)$ rel. max.

   See Exer. 3:
   For $x<-2$, y dec.
    $-2<x<2$, y inc.
      $x>2$, y dec.
   $(-2,-16)$ rel. min.
   $(2,16)$ rel max.

9. $y = x^2 - 2x$
   $y' = 2x - 2$
   $y" = 2$
   $y" = 2$ (pos.)
    for all x
   Conc. up all x
   No points of
   inflection

11. $y = 12x - x^3$, $y' = 12 -3x^2$, $y"=-6x$
    $y" = 0$ for $x=0$
    $x<0$, $y">0$ (pos.), y conc. up
    $x>0$, $y"<0$ (neg.), y conc. down
    When $x=0$, $y=0$
    $(0,0)$ is point of inflection

13. $y = x^2 - 2x$
    $x<1$, y dec. (Exer. 1)
    $x>1$, y inc.
    $(1,-1)$ rel. min. (Exer. 5)
    Conc. up all x (Exer. 9)

15. $y = 12x - x^3$
    $x<-2$, y dec. (Exer. 3)
    $-2<x<2$, y inc.
    $x>2$, y dec.
    $(-2,-16)$ rel. min. (Exer. 7)
    $(2,16)$ rel max.
    $x<0$, conc. up (Exer. 11)
    $x>0$, conc. down, $(0,0)$ pt. of infl.

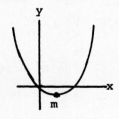

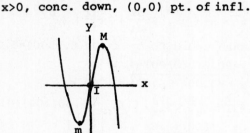

17. $y = 12x - 2x^2$

    $y' = 12 - 4x$, $y'' = -4$

    $y' = 12 - 4x = 0$, $x = 3$

    $x<3$, $y'>0$, y inc.

    $x>3$, $y'<0$, y dec.

    For $x=3$, $y=18$

    $(3,18)$ is rel. max.

    $y''=-4$ (neg.) all x

    conc. down all x

    no points of infl.

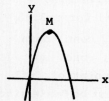

19. $y = 2x^3 + 6x^2$

    $y'=6x^2+12x$, $y''=12x+12$

    $y'=6x^2+12x=6x(x+2)$

      $=0$ for $x=0,-2$

    $x<-2$, $y'>0$, y inc.

    $-2<x<0$, $y'<0$, y dec.

    $x>0$, $y'>0$, y inc.

    For $x=-2$, $y=8$

    $(-2,8)$ is rel. max.

    For $x=0$, $y=0$

    $(0,0)$ is rel. min.

    $y''=12x+12=12(x+1)$

      $=0$ for $x=-1$

    For $x=-1$, $y=4$

    $x<-1$, $y''<0$, conc. down

    $x>-1$, $y''>0$, conc. up

    $(-1,4)$ is point of infl.

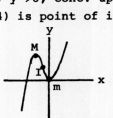

21. $y = x^5 - 5x$

    $y'=5x^4-5$, $y'' = 20x^3$

    $y'=5x^4-5=5(x^2+1)(x+1)(x-1)$

      $= 0$ for $x=-1, 1$

    $x<-1$, $y'>0$, y inc.

    $-1<x<1$, $y'<0$, y dec.

    $x>1$, $y'>0$, y inc.

    For $x=-1$, $y=4$

    $(-1,4)$ is rel. max.

    For $x=1$, $y=-4$

    $(1,-4)$ is rel min.

    $y'' = 20x^3 = 0$, $x = 0$

    $x<0$, $y''<0$, conc. down

    $x>0$, $y''>0$, conc. up

    For $x=0$, $y=0$

    $(0,0)$ is point of infl.

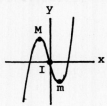

23. $y = 4x^3 - 3x^4$

    $y' = 12x^2 - 12x^3$

    $y'' = 24x - 36x^2$

    $y' = 12x^2 - 12x^3 = 12x^2(1 - x)$

      $= 0$ for $x=0,1$

    $x<0$, $y'>0$, y inc.

    $0<x<1$, $y'>0$, y inc.

    $x>1$, $y'<0$, y dec.

    For $x=1$, $y=1$, $(1,1)$ is rel. max.

    $y''=24x-36x^2 = 12x(2-3x) = 0$

      for $x =0, \frac{2}{3}$

    $x<0$, $y''<0$, conc. down

    $0<x<\frac{2}{3}$, $y''>0$, conc. up

    $x>\frac{2}{3}$, $y''<0$, conc. down

    For $x=0$, $y=0$; for $x=\frac{2}{3}$, $y=\frac{16}{27}$

    $(0,0)$ and $(\frac{2}{3},\frac{16}{27})$ are pts. of infl.

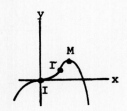

25. $y = x - 0.00025x^2$

    $y' = 1 - 0.0005x$

    $y'' = -0.0005$

    $y' = 1 - 0.0005x = 0$ for $x=2000$

    $x<2000$, $y'>0$, y inc.

    $x>2000$, $y'<0$, y dec.

    For $x=2000$, $y=1000$

    $(2000,1000)$ is rel. max.

    $y''< 0$ for all x

    conc. down for all x

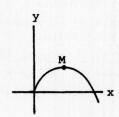

27. $F = x^4 - 12x^3 + 46x^2 - 60x + 25$ ($1 \leq x \leq 5$)

$F' = 4x^3 - 36x^2 + 92x - 60$

$F'' = 12x^2 - 72x + 92$

$F' = 4x^3 - 36x^2 + 92x - 60 = 4(x-1)(x-3)(x-5)$

$\quad = 0$ for $x = 1, 3, 5$

$x < 1$, $F' < 0$, F dec.

$1 < x < 3$, $F' > 0$, F inc.   (1,0) rel. min.

$3 < x < 5$, $F' < 0$, F dec.   (3,16) rel. max.

$x > 5$, $F' > 0$, F inc.   (5,0) rel. min.

$F'' = 12x^2 - 72x + 92 = 4(3x^2 - 18x + 23)$

$\quad = 0$ for $x = 3 \pm \frac{2}{3}\sqrt{3} = 1.85, 4.15$

$\quad$ (use quadratic formula)

$x < 1.85$, $F'' > 0$, conc. up

$1.85 < x < 4.15$, $F'' < 0$, conc. down

$x > 4.15$, $F'' > 0$, conc. up

(1.85, 7.17), (4.15, 7.17) pts. of infl.

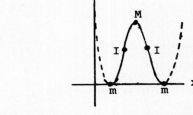

29. $V = (12 - 2x)(8 - 2x)(x)$

$\quad = 96x - 40x^2 + 4x^3$

$V' = 96 - 80x + 12x^2$

$V'' = -80 + 24x$

$V' = 12x^2 - 80x + 96 = 4(3x^2 - 20x + 24)$

$\quad = 0$ for $x = 1.57, 5.10$

$x < 1.57$, $V' > 0$, V inc.

$1.57 < x < 5.10$, $V' < 0$, V dec.

$x > 5.10$, $V' > 0$, V inc.

(1.57, 67.6) rel. max.

(5.10, -20.2) rel. min.

$V'' = -80 + 24x = 0$ for $x = 3.33$

$x < 3.33$, $V'' < 0$, conc. down

$x > 3.33$, $V'' > 0$, conc. up

(3.33, 23.8) point of infl.

Values of x must be $0 < x < 4$ since $x = 0$ means no square cut, and $x = 4$ means cut edges meet along 8 in. dimension.

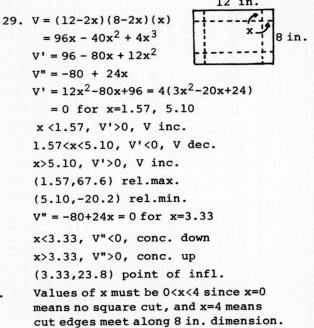

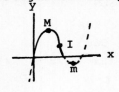

31. $f(1) = 0$ means curve passes through (1,0)

$f'(x) > 0$ for all x means $f(x)$ is increasing for all x

$f''(x) < 0$ for all x means $f(x)$ is concave down for all x

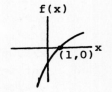

Exercises 3-6, p. 150

1. $y = \frac{4}{x^2}$

Intercepts: None

For $x = 0$, y undefined

$y \neq 0$ since $\frac{4}{x^2} > 0$ for all x ($x \neq 0$)

Symmetry: To y-axis

Replacing x by -x produces no change in equation.

Behavior as x becomes large:

$y \to 0$ as $x \to +\infty$ and as $x \to -\infty$

$y = 0$ (x-axis) is asymptote

Vertical asymptotes:

$x = 0$, y is undefined

Derivatives:

$y' = \frac{-8}{x^3}$

$x < 0$, $y' > 0$, y inc.

$x > 0$, $y' < 0$, y dec.

$y'' = \frac{24}{x^4}$

$y'' > 0$ all x ($x \neq 0$)

Conc. up all x

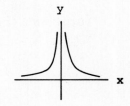

3. $y = \dfrac{2}{x + 1}$

<u>Intercepts</u>: (0,2)

For x=0, y=2

$y \neq 0$ for any x

<u>Symmetry</u>: None

To axes or origin

<u>Behavior as x becomes large</u>:

$y \to 0$ as $x \to +\infty$ and as $x \to -\infty$

(y>0, x>-1 and y<0, x<-1)

y=0 (x-axis) is asymptote

<u>Vertical asymptotes</u>:

x+1=0, x=-1 (y undefined)

<u>Derivatives</u>:

$y' = \dfrac{-2}{(x + 1)^2}$

y'<0 all x (x≠-1)

y dec. all x

$y'' = \dfrac{4}{(x + 1)^3}$

x<-1, y''<0, conc. down

x>-1, y''>0, conc. up

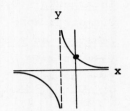

5. $y = x^2 + \dfrac{2}{x}$

<u>Intercepts</u>: $(-\sqrt[3]{2},0)$

For x=0, y undefined

$y = x^2 + \dfrac{2}{x} = \dfrac{x^3 + 2}{x}$

$x^3 + 2 = 0$, $x = -\sqrt[3]{2}$

<u>Symmetry</u>: None

<u>Behavior as x becomes large</u>:

$y \to +\infty$ as $x \to +\infty$ and as $x \to -\infty$

since $x^2 \to +\infty$ and $2/x \to 0$

<u>Vertical asymptotes</u>:

x=0, y is undefined

<u>Derivatives</u>:

$y' = 2x - \dfrac{2}{x^2} = \dfrac{2x^3 - 2}{x^2} = \dfrac{2(x^3 - 1)}{x^2}$

$x^3 - 1 = 0$, x = 1

x<1 (x≠0), y'<0, y dec.

x>1, y'>0, y inc.

(1,3) rel. min.

$y'' = 2 + \dfrac{4}{x^3} = \dfrac{2(x^3 + 2)}{x^3}$

$x^3 + 2 = 0$, $x = -\sqrt[3]{2}$

$x < -\sqrt[3]{2}$, y''>0, conc. up

$-\sqrt[3]{2} < x < 0$, y''<0, conc. down

x>0, y''>0, conc. up

$(-\sqrt[3]{2},0)$ point of inflection

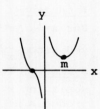

7. $y = x - \dfrac{1}{x}$

<u>Intercepts</u>: (1,0), (-1,0)

For x=0, y undefined

For $y = \dfrac{x^2 - 1}{x} = 0$, x=1,-1

<u>Symmetry</u>: To origin

(replacing x by -x and y

by -y produces same

equation)

<u>Behavior as x becomes large</u>:

$y \to x$ as $x \to +\infty$ and as $x \to -\infty$

since $1/x \to 0$

y=x is an asymptote

<u>Vertical asymptotes</u>:

x=0, y undefined

<u>Derivatives</u>:

$y' = 1 + \dfrac{1}{x^2} = \dfrac{x^2 + 1}{x^2}$

y'>0 all x (x≠0)

y inc. all x

$y'' = -\dfrac{2}{x^3}$

x<0, y''>0, conc. up

x>0, y''<0, conc. down

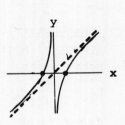

9. $y = \dfrac{x^2}{x+1}$

Intercepts: $(0,0)$

Symmetry: None

Behavior as x becomes large:

$y \to +\infty$ as $x \to +\infty$

$y \to -\infty$ as $x \to -\infty$

($y = \dfrac{x^2}{x+1} = x - \dfrac{x}{x+1}$ shows that

$y = x$ is an asymptote)

Vertical asymptotes:

$x+1=0$, $x=-1$

Derivatives:

$y' = \dfrac{(x+1)(2x) - x^2(1)}{(x+1)^2}$

$\quad = \dfrac{x^2 + 2x}{(x+1)^2} = \dfrac{x(x+2)}{(x+1)^2}$

$x(x+2) = 0$ for $x=0,-2$

$x<-2$, $y'>0$, y inc.

$-2<x<0$ $(x \neq -1)$, $y'<0$, y dec.

$x>0$, $y'>0$, y inc.

$(-2,-4)$ rel. max.

$(0,0)$ rel. min.

$y'' = \dfrac{(x+1)^2(2x+2) - (x^2+2x)(2)(x+1)}{(x+1)^4} = \dfrac{2}{(x+1)^3}$

$x<-1$, $y''<0$, conc. down; $x>-1$, $y''>0$, conc. up

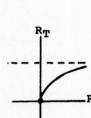

11. $y = \dfrac{1}{x^2 - 1}$

Intercepts: $(0,-1)$

For $x=0$, $y=-1$

$y \neq 0$ since $\dfrac{1}{x^2-1} \neq 0$

Symmetry: To y-axis

Replacing x by $-x$
produces no change in
equation

Behavior as x becomes large:

$y \to 0$ as $x \to +\infty$ and as $x \to -\infty$

$y=0$ is an asymptote

Vertical asymptotes:

$x^2-1=0$, $x=-1$, $x=1$

Derivatives:

$y' = -(x^2-1)^{-2}(2x) = \dfrac{-2x}{(x^2-1)^2}$

$-2x = 0$, $x = 0$

$x<0$ $(x \neq -1)$, $y'>0$, y inc.

$x>0$ $(x \neq 1)$, $y'<0$, y dec.

$(0,-1)$ rel. max.

$y'' = \dfrac{(x^2-1)^2(-2) + 2x(2)(x^2-1)(2x)}{(x^2-1)^4}$

$\quad = \dfrac{6x^2 + 2}{(x^2-1)^3}$

$x<-1$, $y''>0$, conc. up

$-1<x<1$, $y''<0$, conc. down

$x>1$, $y''>0$, conc. up

13. $R_T = \dfrac{5R}{5 + R}$ ($R \geq 0$ for physical meaning. Consider only values $R \geq 0$)

Intercepts: $(0,0)$

Symmetry: None

Behavior as R becomes large:

$R_T = \dfrac{5}{\dfrac{5}{R} + 1}$, $R_T \to 5$ as $R \to +\infty$

$R_T = 5$ is an asymptote

Vertical asymptotes:

$5+R=0$, $(R=-5)$

Derivatives:

$R_T' = \dfrac{(5+R)(5) - 5R(1)}{(5+R)^2}$

$\quad = \dfrac{25}{(5+R)^2}$

R_T inc. all $R>0$

$R_T'' = -50(5+R)^{-3} = \dfrac{-50}{(5+R)^3}$

$R>0$, $R_T'' < 0$, conc. down

15. $V = \pi r^2 h$, $V = 20$ kL

$20 = \pi r^2 h$, $h = \dfrac{20}{\pi r^2}$

$A = 2\pi r^2 + 2\pi rh = 2\pi r^2 + 2\pi r(\dfrac{20}{\pi r^2})$

$\quad = 2\pi r^2 + \dfrac{40}{r}$ (Consider values r>0)

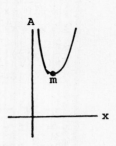

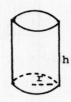

Intercepts: None

(There would be an intercept
for r<0.)

Symmetry: None

Behavior as r becomes large:

$A \to +\infty$ as $r \to +\infty$

Vertical asymptotes:

r=0

Derivatives:

$A' = 4\pi r - \dfrac{40}{r^2} = \dfrac{4(\pi r^3 - 10)}{r^2}$

$\pi r^3 - 10 = 0$, r=1.47

0<r<1.47, A'<0, A dec.
r>1.47, A'>0, A inc.
(1.47, 40.8) rel. min.

$A'' = 4\pi + \dfrac{80}{r^3}$, A''>0 if r>0

Conc. up all r>0

Exercises 3-7, p. 155

1. $s = 112t - 16t^2$

$\dfrac{ds}{dt} = s' = 112 - 32t$

$s'' = -32$

$112 - 32t = 0$, t = 3.5 s
s''<0 for all t
s is max. for t=3.5 s
$s = 112(3.5) - 16(3.5)^2$
$\quad = 196$ ft

3. $v = k(100x - x^2)$

$\dfrac{dv}{dt} = v' = k(100 - 2x)$

$v'' = -2k$

$k(100 - 2x) = 0$, x = 50
v''<0 for all x
v is max. for x = 50 mg

5. $s = \sqrt{t - 4t^2}$

$\dfrac{ds}{dt} = s' = \dfrac{1}{2}(t - 4t^2)^{-1/2}(1-8t)$

$\quad = \dfrac{1 - 8t}{2(t - 4t^2)^{1/2}}$

$1 - 8t = 0$, t = 0.125 min

$s = \sqrt{0.125 - 4(0.125)^2}$
$\quad = 0.250$ ft/min

This is rel. max. since
s'>0 for t<0.125 and
s'<0 for t>0.125.

7. Let x and y be the numbers.

$xy = 64$, $y = \dfrac{64}{x}$

Let $S = x + y = x + \dfrac{64}{x}$

$\dfrac{dS}{dx} = S' = 1 - \dfrac{64}{x^2}$, $S'' = \dfrac{128}{x^3}$

$1 - \dfrac{64}{x^2} = 0$, $\dfrac{x^2 - 64}{x^2} = 0$

$x^2 - 64 = 0$, x = 8 (-8<0)
S''>0 for x>0
Min. at x=8; $y = \dfrac{64}{8} = 8$

9.

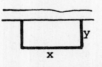

$x + 2y = 2000$, $y = \dfrac{2000 - x}{2}$

$A = xy = x(\dfrac{2000 - x}{2})$

$\quad = 1000x - \dfrac{1}{2}x^2$

$\dfrac{dA}{dx} = 1000 - x$

1000 - x = 0, x=1000 ft

$A = 1000(1000) - \dfrac{1}{2}(1000)^2$
$\quad = 5.00 \times 10^5$ ft^2

This is max. since
A''=-1 for all x.

11.

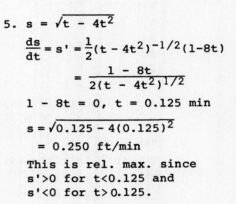

$h = \sqrt{144 - x^2}$

$A = \dfrac{1}{2}(2x)\sqrt{144 - x^2} = x\sqrt{144 - x^2}$

$\dfrac{dA}{dx} = x(\dfrac{1}{2})(144 - x^2)^{-1/2}(-2x)$

$\quad + (144 - x^2)^{1/2}$

$\quad = \dfrac{144 - 2x^2}{(144 - x^2)^{1/2}}$

$144 - 2x^2 = 0$, x=8.49 cm

2x = 17.0 cm

This is rel. max. since
A'>0 for x<8.49 and
A'<0 for x>8.49.

13. $x^2 + y^2 = 16$

$y = \sqrt{16 - x^2}$

$A = xy = x\sqrt{16-x^2}$

$\frac{dA}{dx} = x(\frac{1}{2})(16-x^2)^{-1/2}(-2x) + (16-x^2)^{1/2}$

$\qquad = \frac{16 - 2x^2}{(16 - x^2)^{1/2}}$

$16 - 2x^2 = 0, \quad x = \sqrt{8} = 2\sqrt{2}$

$y = \sqrt{16 - 8} = \sqrt{8} = 2\sqrt{2}$

These are for max. since A'>0 for $0<x<\sqrt{8}$ and A'<0 for $x>\sqrt{8}$.

15.

$y = \sqrt{100 - x^2}$

$A = \frac{1}{2} x \sqrt{100 - x^2}$

$\frac{dA}{dx} = \frac{1}{2} x(\frac{1}{2})(100-x^2)^{-1/2}(-2x)$

$\qquad + \frac{1}{2}(100-x^2)^{1/2} = \frac{100 - 2x^2}{2(100-x^2)^{1/2}}$

$100 - 2x^2 = 0, \quad x = \sqrt{50} = 5\sqrt{2}$

$y = \sqrt{100 - 50} = \sqrt{50} = 5\sqrt{2}$

These are for max. since A'>0 for $0<x<\sqrt{50}$ and A'<0 for $x>\sqrt{50}$.

17. $y = 6x^2 - x^3$

$y' = 12x - 3x^2$

This is the equation for slope, and it is the slope which is to be a maximum. Therefore, let

$m = 12x - 3x^2$

$\frac{dm}{dx} = 12 - 6x$

$12 - 6x = 0, \quad x = 2$

$m = 12(2) - 3(2)^2 = 12$

Max. slope is 12. This is max. since m'>0 for x<2 and m'<0 for x>2.

19. See Fig. 3-41.

Let L be landing point between P and A.

Distance = speed × time

time = distance ÷ speed

Total time (t) = time(boat to L)

$\qquad\qquad\qquad$ + time(L to A)

$t = \frac{\sqrt{16 + x^2}}{3} + \frac{5 - x}{5}$

$\frac{dt}{dx} = \frac{1}{3}(\frac{1}{2})(16 + x^2)^{-1/2}(2x) - \frac{1}{5}$

$\frac{x}{3(16 + x^2)^{1/2}} - \frac{1}{5} = 0$

$5x = 3\sqrt{16+x^2} \; ; \; 25x^2 = 9(16 + x^2)$

$16x^2 = 144, \quad x^2 = 9, \quad x = 3 \text{ km}$

This is for rel. min. since t=2.07 h for x=3 km which is less than t=2.13 h for x = 5 km.

21. See Fig. 3-42

$V = (8 - 2x)^2(x) = 64x - 32x^2 + 4x^3$

$V' = 64 - 64x + 12x^2$

$12x^2 - 64x + 64 = 0$

$3x^2 - 16x + 16 = 0$

$(3x - 4)(x - 4) = 0, \quad x = \frac{4}{3}, \; 4$

$x = \frac{4}{3}$ in. for max. V

(x=4 for min. V)

23. See Fig. 3-43

Let r = radius of circular ends

$p = 2\pi r + 2x$

$8 = 2\pi r + 2x, \quad r = \frac{4 - x}{\pi}$

$A_{rect} = x(2r) = 2x(\frac{4 - x}{\pi})$

$\qquad = \frac{2}{\pi}(4x - x^2)$

$A' = \frac{2}{\pi}(4 - 2x) \; ; \; 4 - 2x = 0, \quad x = 2$

This is for max. since $A'' = -\frac{4}{\pi}$ for all x.

25. $y = k(2x^4 - 5Lx^3 + 3L^2x^2)$

$y' = k(8x^3 - 15Lx^2 + 6L^2x)$

$8x^3 - 15Lx^2 + 6L^2x = 0$

$x(8x^2 - 15Lx + 6L^2) = 0$

(x=0 gives y=0)

$8x^2 - 15Lx + 6L^2 = 0$

$x = \frac{15L \pm \sqrt{225L^2 - 4(8)6L^2}}{16} = \frac{15 \pm \sqrt{33}}{16}L = 0.58L, \; 1.30L$

(Discard 1.30L since x cannot exceed L)

x = 0.58L (At max. since y''<0 for x = 0.58L)

27. See Fig. 3-44

$d^2 + w^2 = 256$

$S = kwd^2 = kw(256 - w^2)$

$= k(256w - w^3)$

$S' = k(256 - 3w^2)$

$256 - 3w^2 = 0$

$w = \sqrt{\dfrac{256}{3}} = \dfrac{16}{3}\sqrt{3}$

$d^2 = 256 - \dfrac{256}{3} = \dfrac{2}{3}(256)$, $d = \dfrac{16}{3}\sqrt{6}$

$w = 9.24$ in., $d = 13.1$ in.

This is for max. since for min. $w=0$ or $d=0$.)

Review Exercises for Chapter 3, p. 157

1. $y = 3x - x^2$ at $(-1,-4)$

$y' = 3 - 2x$

$y'\big|_{x=-1} = 3-2(-1) = 5$

$m=5$, $(x_1,y_1) = (-1,-4)$

$y - y_1 = m(x - x_1)$

$y + 4 = 5(x + 1)$

$5x - y + 1 = 0$

3. $y = \dfrac{x}{4x - 1}$, at $(1,\tfrac{1}{3})$

$y'=\dfrac{(4x-1)(1)-x(4)}{(4x-1)^2} = \dfrac{-1}{(4x-1)^2}$

$y'\big|_{x=1} = \dfrac{-1}{(4-1)^2} = -\dfrac{1}{9}$

$m=9$, $(x_1,y_1) = (1,\tfrac{1}{3})$

$y - y_1 = m(x - x_1)$

$y - \dfrac{1}{3} = 9(x - 1)$, $27x - 3y - 26 = 0$

5. $y = \sqrt{x^2 + 3}$, $m_{tan} = \dfrac{1}{2}$

$y' = \dfrac{1}{2}(x^2+3)^{-1/2}(2x)$

$= \dfrac{x}{(x^2 + 3)^{1/2}}$

$\dfrac{1}{2} = \dfrac{x}{(x^2 + 3)^{1/2}}$

$\sqrt{x^2+3} = 2x$, $x^2+3=4x^2$

$3x^2 = 3$, $x = -1, 1$

(Discard $x=-1$ since $y'<0$ for $x=-1$)

For $x=1$, $y=2$

$y - 2 = \dfrac{1}{2}(x - 1)$

$x - 2y + 3 = 0$

7. $x = t^4 - t$, $y = \dfrac{1}{t}$, $t = 2$

$v_x = \dfrac{dx}{dt} = 4t^3 - 1$

$v_y = \dfrac{dy}{dt} = -\dfrac{1}{t^2}$

$v_x\big|_{t=2} = 31$, $v_y\big|_{t=2} = -\dfrac{1}{4}$

$v\big|_{t=2} = \sqrt{31^2+(-\tfrac{1}{4})^2} = 31.0$

$\tan\theta = \dfrac{-0.25}{31.0} = -0.0081$

$\theta = -0.5°$ or $359.5°$

($v_y<0$, $v_x>0$)

9. $y=0.5x^2 + x$ at $2,4)$, $v_x=0.5\sqrt{x}$

$\dfrac{dy}{dt} = 0.5(2x\dfrac{dx}{dt}) + \dfrac{dx}{dt} = (x+1)\dfrac{dx}{dt}$

$v_y = (x+1)v_x = (x+1)(0.5\sqrt{x})$

$= 0.5(x^{3/2} + x^{1/2})$

$v_y\big|_{x=2} = 0.5(2\sqrt{2} + \sqrt{2})$

$= 1.5\sqrt{2} = 2.12$

11. $x = t^4 - t$, $y = \dfrac{1}{t}$, $t=2$

$v_x = 4t^3 - 1$, $v_y = -\dfrac{1}{t^2}$

$a_x = 12t^2$, $a_y = \dfrac{2}{t^3}$

$a_x\big|_{t=2} = 48$, $a_y\big|_{t=2} = \dfrac{1}{4}$

$a\big|_{t=2} = \sqrt{48^2 + 0.25^2} = 48.0$

$\tan\theta = \dfrac{0.25}{48.0} = 0.0052$

$\theta = 0.3°$ ($v_x>0$, $v_y>0$)

13. $x^3-3x^2-x+2 = 0$, $f(x) = x^3-3x^2-x+2$, $f'(x) = 3x^2-6x-1$

$f(0) = 2$, $f(1) = -1$, Let $x_1 = 0.7$

n	x_n	$f(x_n)$	$f'(x_n)$	$x_n - \dfrac{f(x_n)}{f'(x_n)}$
1	0.7	0.173	-3.73	0.7463807
2	0.7463807	-0.0018363	-3.8070318	0.7458984
3	0.7458984	-0.0000002	-3.8062971	0.7458983
4	0.7458983			

$x_4 = x_3 = 0.745898$ (to 6 decimal places)

15. $3x^3 - x^2 - 8x - 2 = 0$

$f(x) = 3x^3 - x^2 - 8x - 2$

$f'(x) = 9x^2 - 2x - 8$

$f(1) = -8$, $f(2) = 2$

Let $x_1 = 1.8$

n	x_n	$f(x_n)$	$f'(x_n)$	$x_n - \dfrac{f(x_n)}{f'(x_n)}$
1	1.8	-2.144	17.56	1.9220957
2	1.9220957	0.2320521	21.405875	1.9112551
3	1.9112551	0.0019116	21.053554	1.9111643
4	1.9111643	0.0000001	21.050612	1.9111643

$x_5 = x_4 = 1.9111643$ (to 7 decimal places)

17. $y = 4x^2 + 16x$

Intercepts: $(0,0)$, $(-4,0)$
$y=0$, $4x^2+16x = 4x(x+4) = 0$
Derivatives:
$y' = 8x + 16$, $y'' = 8$
$8x + 16 = 0$, $x = -2$
$x<-2$, $y'<0$, y dec.
$x>-2$, $y'>0$, y inc.
$(-2,-16)$ rel. min.
Conc. up all x

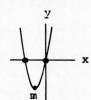

19. $y = 27x - x^3$

Intercepts:
$(0,0)$, $(3\sqrt{3},0)$, $(-3\sqrt{3},0)$
$y=0$, $27x-x^3 = x(27-x^2) = 0$
Symmetry: To origin
Replace x by $-x$ and y by $-y$
Derivatives:
$y' = 27 - 3x^2$, $y'' = -6x$
$27 - 3x^2 = 3(9 - x^2) = 0$, $x = -3$, 3
$x<-3$, $y'<0$, y dec.
$-3<x<3$, $y'>0$, y inc. $(-3,-54)$ rel min.
$x>3$, $y'<0$, y dec. $(3,54)$ rel. max.

$x<0$, $y''>0$, conc. up $(0,0)$ pt. of infl.
$x>0$, $y''<0$, conc. down

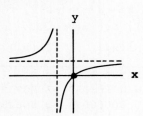

21. $y = x^4 - 32x$

Intercepts: $(0,0)$, $(2\sqrt[3]{4},0)$
$y=0$, $x^4 - 32x = x(x^3-32) = 0$
Derivatives:
$y' = 4x^3 - 32$, $y'' = 12x^2$
$4x^3-32 = 4(x^3-8) = 0$, $x=2$
$x<2$, $y'<0$, y dec.
$x>2$, $y'>0$, y inc. $(2,-48)$ rel. min.
$x<0$, $y''>0$, conc. up
$x>0$, $y''>0$, conc. up   (No infl.)

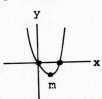

23. $y = \dfrac{x}{x + 1}$

Intercepts: $(0,0)$
Symmetry: None
Behavior as x becomes large:
$y \to 1$ as $x \to +\infty$ and as $x \to -\infty$
$y = 1$ is an asymptote
Vertical asymptotes:
$x+1 = 0$, $x = -1$

Derivatives:
$y' = \dfrac{(x+1)(1)-x(1)}{(x+1)^2}$
$= \dfrac{1}{(x + 1)^2}$
$y'>0$ all x $(x \neq -1)$
y inc. all x
$y'' = \dfrac{-2}{(x + 1)^3}$
$x<-1$, $y''>0$, conc. up
$x>-1$, $y''<0$, conc. dn

25. $y_1 = x^2 + 2$, $y_2 = 4x - x^2$

Since curves are tangent, they have
a point in common, and they have
equal slopes at that point.
$x^2 + 2 = 4x - x^2$, $2x^2 - 4x + 2 = 0$
$x^2 - 2x + 1 = (x - 1)^2 = 0$, $x = 1$
$(1,3)$ is the common point

$y_1' = 2x$, $y_1'\big|_{x=1} = 2$
$y_2' = 4 - 2x$, $y_2'\big|_{x=1} = 2$
The slope is 2.
$y - 3 = 2(x - 1)$
$2x - y + 1 = 0$

27. $y = k(x^4 - 30x^3 + 1000x)$

   $k(x^4 - 30x^3 + 1000x) = 0$

   $x(x^3 - 30x^2 + 1000) = 0$

   $x = 0, \quad x^3 - 30x^2 - 1000 = 0$

   $f(x) = x^3 - 30x^2 + 1000$

   $f'(x) = 3x^2 - 60x$

x	0	6	7	10
f(x)	1000	136	-127	-1000

   Let $x_1 = 6.5$

n	x_n	$f(x_n)$	$f'(x_n)$	$x_n - \dfrac{f(x_n)}{f'(x_n)}$
1	6.5	7.125	-263.25	6.5270655
2	6.5270655	-0.0076718	-263.81618	6.5270365
3	6.5270365			

   $x = 6.527$ m

29. $x = 1200t, \quad y = -490t^2$

   $\dfrac{dx}{dt} = 1200, \quad \dfrac{dy}{dt} = -980t$

   At $t = 3.00$ s

   $v_x = 1200$ cm/s, $v_y = -980(3.00) = -2940$ cm/s

   $v = \sqrt{1200^2 + (-2940)^2} = 3180$ cm/s

   $\tan\theta = \dfrac{-2940}{1200} = -2.45, \quad \theta = -67.8° \text{ or } 292.2°$

   $(v_x > 0, \; v_y < 0)$

31. $d = \dfrac{1000}{\sqrt{T^2 - 400}} = 1000(T^2 - 400)^{-1/2}$

   $\dfrac{dd}{dt} = 1000(-\frac{1}{2})(T^2 - 400)^{-3/2}(2T\frac{dT}{dt})$

   $= \dfrac{-1000T \, dT/dt}{(T^2 - 400)^{3/2}}$

   $\dfrac{dT}{dt} = 2.00$ N/s, $T = 28.0$ N

   $\dfrac{dd}{dt} = \dfrac{-1000(28.0)(2.00)}{(28.0^2 - 400)^{3/2}} = -7.44$ cm/s

33. $f(0) = 2$ means curve passes through $(0,2)$

   $f'(x) < 0$ means $f(x)$ decreasing for $x < 0$

   $f'(x) > 0$ means $f(x)$ increasing for $x > 0$

   $f''(x) > 0$ means curve concave up for all x

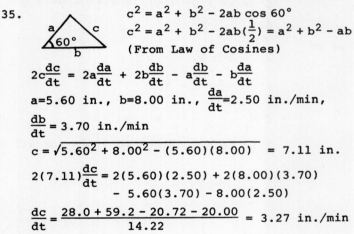

35. $c^2 = a^2 + b^2 - 2ab\cos 60°$

   $c^2 = a^2 + b^2 - 2ab(\frac{1}{2}) = a^2 + b^2 - ab$

   (From Law of Cosines)

   $2c\dfrac{dc}{dt} = 2a\dfrac{da}{dt} + 2b\dfrac{db}{dt} - a\dfrac{db}{dt} - b\dfrac{da}{dt}$

   $a = 5.60$ in., $b = 8.00$ in., $\dfrac{da}{dt} = 2.50$ in./min,

   $\dfrac{db}{dt} = 3.70$ in./min

   $c = \sqrt{5.60^2 + 8.00^2 - (5.60)(8.00)} = 7.11$ in.

   $2(7.11)\dfrac{dc}{dt} = 2(5.60)(2.50) + 2(8.00)(3.70)$

   $\qquad\qquad - 5.60(3.70) - 8.00(2.50)$

   $\dfrac{dc}{dt} = \dfrac{28.0 + 59.2 - 20.72 - 20.00}{14.22} = 3.27$ in./min

37. See Fig. 3-46

   $A = \frac{1}{2} r^2\theta$ (See Sec. 6-1)

   $1 = \frac{1}{2} r^2\theta, \quad \theta = \dfrac{2}{r^2}$

   $p = 2r + r\theta = 2r + r(\dfrac{2}{r^2})$

   $\qquad = 2r + \dfrac{2}{r}$

   $\dfrac{dp}{dr} = 2 - \dfrac{2}{r^2}; \quad 2 - \dfrac{2}{r^2} = \dfrac{2r - 2}{r^2} = 0$

   $2r^2 - 2 = 0, \quad r = 1$

   $\theta = \dfrac{2}{1} = 2$

39. $(P + \frac{a}{V^2})(V - b) = RT$

$R = T = a = 1, \ b = 0$

$(P + \frac{1}{V^2})V = 1$

$P = \frac{1}{V} - \frac{1}{V^2} = \frac{V - 1}{V^2}$

<u>Intercepts</u>: (1,0)

$\frac{V - 1}{V^2} = 0, \ V = 1$

(If $V<1$, $P<0$. Therefore, meaningful values of V are $V \geq 1$. Also, meaningful values of P are $P \geq 0$.)

<u>Symmetry</u>: None

<u>Behavior as V becomes large</u>:

$P \to 0$ as $V \to +\infty$

$P = 0$ is an asymptote

<u>Vertical asymptotes</u>:

$(V = 0)$

<u>Derivatives</u>: $\frac{dP}{dV} = P' = -\frac{1}{V^2} + \frac{2}{V^3}$

$-\frac{1}{V^2} + \frac{2}{V^3} = \frac{2 - V}{V^3} = 0, \ V = 2$

$1<V<2$, $P'>0$, P inc.

$V>2$, $P'<0$, P dec. $(2, \frac{1}{4})$ rel. max.

$P'' = \frac{2}{V^3} - \frac{6}{V^4}$; $\frac{2}{V^3} - \frac{6}{V^4} = \frac{2(V - 3)}{V^4} = 0, \ V = 3$

$1<V<3$, $P''<0$, conc. down

$V>3$, $P''>0$, conc. up $(3, \frac{2}{9})$ point of infl.

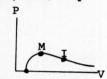

41.

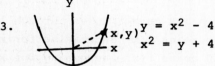

$\frac{dx}{dt} = 5$ km/h

$\frac{dy}{dt} = 12$ km/h

$z^2 = x^2 + y^2$

$2z\frac{dz}{dt} = 2x\frac{dx}{dt} + 2y\frac{dy}{dt}$

$z\frac{dz}{dt} = x\frac{dx}{dt} + y\frac{dy}{dt}$

For $t = 2$ h, $x = 10$ km, $y = 24$ km

$z = \sqrt{10^2 + 24^2} = 26$ km

$26\frac{dz}{dt} = 10(5) + 24(12)$

$\frac{dz}{dt} = 13$ km/h

43.

$y = x^2 - 4$

$x^2 = y + 4$

$d^2 = x^2 + y^2 = y + 4 + y^2 = y^2 + y + 4$

$\frac{dd^2}{dy} = 2y + 1$

$2y + 1 = 0, \ y = -\frac{1}{2}$

$x^2 = -\frac{1}{2} + 4 = \frac{7}{2}$

$d = \sqrt{\frac{7}{2} + \frac{1}{4}} = \sqrt{\frac{15}{4}} = 1.94$

This is min. since $(d^2)' = 2$ for all y.

Exercises 4-1, p. 163

1. $y = x^5 + x$

$\dfrac{dy}{dx} = 5x^4 + 1$

$dy = (5x^4+1)dx$

3. $y = \dfrac{2}{x^5} + 3 = 2x^{-5} + 3$

$\dfrac{dy}{dx} = -10x^{-6} = -\dfrac{10}{x^6}$

$dy = -\dfrac{10dx}{x^6}$

5. $y = (x^2 - 1)^4$

$\dfrac{dy}{dx} = 4(x^2-1)^3(2x)$

$dy = 8x(x^2-1)^3dx$

7. $y = \dfrac{2}{3x^2+1} = 2(3x^2+1)^{-1}$

$\dfrac{dy}{dx} = -2(3x^2+1)^{-2}(6x)$

$dy = \dfrac{-12xdx}{(3x^2+1)^2}$

9. $y = x^2(1 - x)^3$

$\dfrac{dy}{dx} = x^2(3)(1-x)^2(-1) + (1-x)^3(2x)$

$= x(1-x)^2[-3x + 2(1-x)]$

$dy = x(1-x)^2(-5x + 2)dx$

11. $y = \dfrac{x}{x + 1}$

$\dfrac{dy}{dx} = \dfrac{(x+1)(1) - x(1)}{(x+1)^2} = \dfrac{1}{(x+1)^2}$

$dy = \dfrac{dx}{(x+1)^2}$

13. $y = 7x^2 + 4x \; ; \; x = 4, \; \Delta x = 0.2$

$y + \Delta y = 7(x+\Delta x)^2 + 4(x + \Delta x)$

$\Delta y = 14x\Delta x + 7(\Delta x)^2 + 4\Delta x$

$dy = (14x + 4)dx$

For $x = 4$, $\Delta x = dx = 0.2$

$\Delta y = 14(4)(0.2)+7(0.2)^2+4(0.2) = 12.28$

$dy = [14(4) + 4](0.2) = 12$

15. $y = 2x^3 - 4x \; ; \; x = 2.5, \Delta x = 0.05$

$y + \Delta y = 2(x+\Delta x)^3 - 4(x+\Delta x)$

$\Delta y = 6x^2\Delta x + 6x(\Delta x)^2 + 2(\Delta x)^3 - 4\Delta x$

$dy = (6x^2 - 4)dx$

For $x = 2.5$, $\Delta x = dx = 0.05$

$\Delta y = 6(2.5)^2(0.05) + 6(2.5)(0.05)^2$
$+ 2(0.05)^3 - 4(0.05) = 1.71275$

$dy = [6(2.5)^2 - 4](0.05) = 1.675$

17. $y = (1 - 3x)^5 \; ; \; x = 1, \; \Delta x = 0.01$

$dy = 5(1-3x)^4(-3)dx = -15(1-3x)^4dx$

For $x = 1$, $\Delta x = dx = 0.01$

$dy = -15[1-3(1)]^4(0.01) = -2.4$

$f(1.01)-f(1) = [1-3(1.01)]^5$

$- [1-3(1)]^5 = -2.4730881$

19. $y = x\sqrt{1 + 4x} = x(1 + 4x)^{1/2} \; ; x=12, \Delta x=0.06$

$dy = [x(\tfrac{1}{2})(1+4x)^{-1/2}(4)+(1+4x)^{1/2}(1)]dx$

$= \dfrac{(1 + 6x) \, dx}{(1 + 4x)^{1/2}}$

For $x = 12$, $\Delta x = dx = 0.06$

$dy = \dfrac{[1 + 6(12)](0.06)}{[1 + 4(12)]^{1/2}} = 0.6257$

$f(12.06) - f(12) = 0.6264903$

21. See Fig. 4-2.

$A = s^2$

$dA = 2sds$

For $s = 16.0$ in., $\Delta s = 0.50$ in.

$dA = 2(16.0)(0.50) = 16$ in.^{2}

23. $E = 6.2T + 0.0002T^3$

$dE = (6.2 + 0.0006T^2)dT$

For $T = 100°C$, $\Delta T = 1°C$

$dE = [6.2 + 0.0006(100)^2](1)$

$= 12.2$ V

25. $V = \pi r^2 h$

$dV = 2\pi rhdr + \pi r^2 dh$

For $r = 3.00$ in., $h = 4.00$ in.,

$dr = 0.02$ in., $dh = 0.04$ in.

$dV = 2\pi(3.00)(4.00)(0.02)$
$+ \pi(3.00)^2(0.04)$

$= 2.64$ in.^{3}

27. $A = s^2$

$dA = 2sds$

$\dfrac{dA}{A} = \dfrac{2sds}{s^2} = 2\dfrac{ds}{s}$

$\dfrac{ds}{s} = 2\% = 0.02$

$\dfrac{dA}{A} = 2(0.02) = 0.04 = 4\%$

Exercises 4-2, p. 167

1. For the function $3x^2$, the required power of x in the antiderivative is 3. This also means that we must divide out a factor of 3, which we can do by multiplying by $\frac{1}{3}$. Therefore, the antiderivative of $3x^2$ is $\frac{1}{3}(3x^3) = x^3$. Checking, the derivative of x^3 is $3x^2$.

3. For the function $6x^5$, the required power of x in the antiderivative is 6. This also means that we must multiply by $\frac{1}{6}$. Therefore, the antiderivative of $6x^5$ is $\frac{1}{6}(6x^6) = x^6$. Checking, the derivative of x^6 is $6x^5$.

5. For the function $6x^3$, the required power of x in the antiderivative is 4. Thus, we must also multiply by $\frac{1}{4}$. For $1=1x^0$, the required power of x in the antiderivative is 1, and we multiply it by $\frac{1}{1}=1$. Therefore, the antiderivative of $6x^3 + 1$ is $\frac{1}{4}(6x^4) + 1(x^1) = \frac{3}{2}x^4 + x$. Checking, the derivative of $\frac{3}{2}x^4 + x$ is $6x^3 + 1$.

7. For $2x^2$, the required power of x in the antiderivative is 3, and we must also therefore multiply by $\frac{1}{3}$. For $-x$, the required power of x in the antiderivative is 2, and we must multiply by $\frac{1}{2}$. Therefore, the antiderivative of $2x^2 - x$ is $\frac{1}{3}(2x^3) - \frac{1}{2}x^2 = \frac{2}{3}x^3 - \frac{1}{2}x^2$. Checking, the derivative of $\frac{2}{3}x^3 - \frac{1}{2}x^2$ is $2x^2 - x$.

9. For $\frac{5}{2}x^{3/2}$, the required power of x in the antiderivative is $\frac{5}{2}$, and we must also therefore divide by $\frac{5}{2}$, or multiply by $\frac{2}{5}$. Therefore, the antiderivative of $\frac{5}{2}x^{3/2}$ is $\frac{2}{5}(\frac{5}{2}x^{5/2}) = x^{5/2}$. Checking, the derivative of $x^{5/2}$ is $\frac{5}{2}x^{3/2}$.

11. For $2\sqrt{x} = 2x^{1/2}$, the required power of in the antiderivative is $\frac{3}{2}$, and we must therefore multiply by $\frac{2}{3}$. For $3 = 3x^0$, the required power of x in the antiderivative is 1, and we must multiply by 1. Therefore, the antiderivative of $2\sqrt{x} + 3$ is $\frac{2}{3}(2x^{3/2}) + 3x = \frac{4}{3}x^{3/2} + 3x$. Checking, the derivative of $\frac{4}{3}x^{3/2} + 3x$ is $2x^{1/2} + 3$.

13. For $-1/x^2 = -x^{-2}$, the required power of x in the antiderivative is $-2+1=-1$. We must then also multiply by $\frac{1}{-1} = -1$. Therefore, the antiderivative of $-x^{-2}$ is $-(-x^{-1}) = \frac{1}{x}$. Checking, the derivative of $\frac{1}{x} = x^{-1}$ is $-x^{-2} = -\frac{1}{x^2}$.

15. For $-6/x^4 = -6x^{-4}$, the required power of x in the antiderivative is $-4+1=-3$. We must then also multiply by $\frac{1}{-3} = -\frac{1}{3}$. Therefore, the antiderivative of $-6x^{-4}$ is $-\frac{1}{3}(-6x^{-3}) = \frac{2}{x^3}$. Checking, the derivative of $\frac{2}{x^3} = 2x^{-3}$ is $-6x^{-4} = -\frac{6}{x^4}$.

17. For $2x^4$, the required power of x in the antiderivative is 5. We must then multiply by $\frac{1}{5}$. For $1 = 1x^0$, the required power of x is 1, and we multiply by $\frac{1}{1} = 1$. Therfore, the antiderivative of $2x^4 + 1$ is $\frac{1}{5}(2x^5) + 1(x^1) = \frac{2}{5}x^5 + x$. Checking, the derivative of $\frac{2}{5}x^5 + x$ is $2x^4 + 1$.

19. For $6x$, the required power of x in the antiderivative is 2, and we must also multiply by $\frac{1}{2}$. For $1/x^4 = x^{-4}$, the required power of x is $-4+1=-3$, and we must also multiply by $\frac{1}{-3} = -\frac{1}{3}$. Therefore, the antiderivative of $6x + \frac{1}{x^4}$ is $\frac{1}{2}(6x^2) - \frac{1}{3}(\frac{1}{x^3}) = 3x^2 - \frac{1}{3x^3}$. Checking, the derivative of $3x^2 - \frac{1}{3x^3}$ is $6x + \frac{1}{x^4}$.

21. For x^2, the required exponent of x in the antiderivative is 3, and we must therefore also multiply by $\frac{1}{3}$. For $2 = 2x^0$, the power is 1 and we multiply by 1. For x^{-2}, the power is -2+1=-1 and we multiply by -1. Therefore, the anti-derivative of $x^2 + 2 + x^{-2}$ is $\frac{1}{3}x^3 + 2x - x^{-1} = \frac{1}{3}x^3 + 2x - \frac{1}{x}$. Checking, the derivative of $\frac{1}{3}x^3 + 2x - \frac{1}{x}$ is $x^2 + 2 + x^{-2}$.

23. For $6(2x+1)^5(2)$, the required power of 2x+1 in the antiderivative is 6, and we must therefore also multiply by $\frac{1}{6}$. Also, since the derivative of 2x+1 is 2, a factor of 2 must appear in the function, but not in the anti-derivative. Therefore, the antiderivative of $6(2x+1)^5(2)$ is $\frac{1}{6}(6)(2x+1)^6 = (2x+1)^6$. Checking, the derivative of $(2x+1)^6$ is $6(2x+1)^5(2)$.

25. For $4(x^2-1)^3(2x)$, the required power of x^2-1 in the antiderivative is 4, and we must therefore also multiply by $\frac{1}{4}$. Also, since the derivative of x^2-1 is 2x, a factor of 2x must appear in the function, but not in the antiderivative. Therefore, the antiderivative of $4(x^2-1)^3(2x)$ is $\frac{1}{4}(4)(x^2-1)^4 = (x^2-1)^4$. Checking, the derivative of $(x^2-1)^4$ is $4(x^2-1)^3(2x)$.

27. For $x^3(2x^4+1)^4$, the required power of $2x^4+1$ in the antiderivative is 5, and we must therefore also multiply by $\frac{1}{5}$. Also, since the derivative of $2x^4+1$ is $8x^3$, a factor of $8x^3$ must appear in the function, but not in the antiderivative. Therefore, the antiderivative of $x^3(2x^4+1)^4 = \frac{1}{8}(2x^4+1)^4(8x^3)$ is $\frac{1}{5}(\frac{1}{8})(2x^4+1)^5 = \frac{1}{40}(2x^4+1)^5$. Checking, the derivative of $\frac{1}{40}(2x^4+1)^5$ is $\frac{1}{40}(5)(2x^4+1)^4(8x^3) = x^3(2x^4+1)^4$.

29. For $\frac{3}{2}(6x+1)^{1/2}(6)$, the required power of 6x+1 in the antiderivative is $\frac{3}{2}$, and we must divide by $\frac{3}{2}$, or multiply by $\frac{1}{3/2}=\frac{2}{3}$. Also, since the derivative of 6x+1 is 6, a factor of 6 must appear in the function, but not in the antiderivative. Therefore, the antiderivative of $\frac{3}{2}(6x+1)^{1/2}(6)$ is $\frac{2}{3}(\frac{3}{2})(6x+1)^{3/2} = (6x+1)^{3/2}$. Checking, the derivative of $(6x+1)^{3/2}$ is $\frac{3}{2}(6x+1)^{1/2}(6)$.

31. For $(3x+1)^{1/3}$, the required power of 3x+1 in the antiderivative is $\frac{4}{3}$, and we must multiply by $\frac{1}{4/3} = \frac{3}{4}$. Also, since the derivative of 3x+1 is 3, a factor of 3 must appear in the function, but not in the antiderivative. Therefore, the antiderivative of $(3x+1)^{1/3} = \frac{1}{3}(3x+1)^{1/3}(3)$ is $\frac{3}{4}(\frac{1}{3})(3x+1)^{4/3} = \frac{1}{4}(3x+1)^{4/3}$. Checking, the derivative of $\frac{1}{4}(3x+1)^{4/3}$ is $\frac{1}{4}(\frac{4}{3})(3x+1)^{1/3}(3) = (3x+1)^{1/3}$.

Exercises 4-3, p. 172

1. $\int 2x\,dx = 2\int x\,dx$; u = x, du = dx ; n = 1, n+1 = 2

$2\int x\,dx = 2(\frac{x^2}{2}) + C = x^2 + C$

3. $\int x^7\,dx$; u = x, du = dx ; n = 7, n+1 = 8

$\int x^7\,dx = \frac{x^8}{8} + C = \frac{1}{8}x^8 + C$

5. $\int x^{3/2}dx$; $u = x$, $du = dx$; $n = \frac{3}{2}$, $n+1 = \frac{5}{2}$ 7. $\int x^{-4}dx$; $u = x$, $du = dx$; $n = -4$, $n+1 = -3$

$\int x^{3/2}dx = \frac{x^{5/2}}{5/2} + C = \frac{2}{5}x^{5/2} + C$ $\int x^{-4}dx = \frac{x^{-3}}{-3} = -\frac{1}{3x^3} + C$

9. $\int(x^2 - x^5)dx = \int x^2dx - \int x^5dx = \frac{x^3}{3} - \frac{x^6}{6} + C = \frac{1}{3}x^3 - \frac{1}{6}x^6 + C$

11. $\int(9x^2 + x + 3)dx = \int 9x^2dx + \int xdx + \int 3dx = 9\int x^2dx + \int xdx + 3\int dx$

$= 9(\frac{x^3}{3}) + \frac{x^2}{2} + 3x + C = 3x^3 + \frac{1}{2}x^2 + 3x + C$

13. $\int(\frac{1}{x^3} + \frac{1}{2})dx = \int\frac{1}{x^3}dx + \int\frac{1}{2}dx = \int x^{-3}dx + \frac{1}{2}\int dx = \frac{x^{-2}}{-2} + \frac{1}{2}x + C = -\frac{1}{2x^2} + \frac{1}{2}x + C$

15. $\int\sqrt{x}(x^2 - x)dx = \int x^{1/2}(x^2 - x)dx = \int(x^{5/2} - x^{3/2})dx = \int x^{5/2}dx - \int x^{3/2}dx$

$= \frac{x^{7/2}}{7/2} - \frac{x^{5/2}}{5/2} + C = \frac{2}{7}x^{7/2} - \frac{2}{5}x^{5/2} + C$

17. $\int(2x^{-2/3} + 3^{-2})dx = \int 2x^{-2/3}dx + \int 3^{-2}dx = 2\int x^{-2/3}dx + \frac{1}{9}\int dx = 2\frac{x^{1/3}}{1/3} + \frac{1}{9}x + C$

$= 6x^{1/3} + \frac{1}{9}x + C$

19. $\int(1 + 2x)^2dx$; $u = 1 + 2x$, $du = 2dx$; $n = 2$, $n+1 = 3$

$\int(1 + 2x)^2dx = \frac{1}{2}\int(1 + 2x)^2(2dx) = \frac{1}{2}\frac{(1 + 2x)^3}{3} + C = \frac{1}{6}(1 + 2x)^3 + C$

21. $\int(x^2 - 1)^5(2xdx)$; $u = x^2 - 1$, $du = 2xdx$; $n = 5$, $n+1 = 6$

$\int(x^2 - 1)^5(2xdx) = \frac{(x^2 - 1)^6}{6} + C = \frac{1}{6}(x^2 - 1)^6 + C$

23. $\int(x^4 + 3)^4(4x^3dx)$; $u = x^4 + 3$, $du = 4x^3dx$; $n = 4$, $n+1 = 5$

$\int(x^4 + 3)^4(4x^3dx) = \frac{(x^4 + 3)^5}{5} + C = \frac{1}{5}(x^4 + 3)^5 + C$

25. $\int(x^5 + 4)^7x^4dx$; $u = x^5 + 4$, $du = 5x^4dx$; $n = 7$, $n+1 = 8$

$\int(x^5 + 4)^7x^4dx = \frac{1}{5}\int(x^5 + 4)^7(5x^4dx) = \frac{1}{5}\frac{(x^5 + 4)^8}{8} + C = \frac{1}{40}(x^5 + 4)^8 + C$

27. $\int\sqrt{8x + 1}\,dx = \int(8x + 1)^{1/2}dx$; $u = 8x + 1$, $du = 8dx$; $n = \frac{1}{2}$, $n+1 = \frac{3}{2}$

$\int(8x+1)^{1/2}dx = \frac{1}{8}\int(8x+1)^{1/2}(8xdx) = \frac{1}{8}\frac{(8x + 1)^{3/2}}{3/2} = \frac{1}{8}(\frac{2}{3})(8x+1)^{3/2}+C = \frac{1}{12}(8x + 1)^{3/2}+C$

29. $\int\frac{xdx}{\sqrt{6x^2 + 1}} = \int(6x^2 + 1)^{-1/2}xdx$; $u = 6x^2 + 1$, $du = 12xdx$; $n = -\frac{1}{2}$, $n+1 = \frac{1}{2}$

$\int(6x^2 + 1)^{-1/2}xdx = \frac{1}{12}\int(6x^2 + 1)^{-1/2}(12xdx) = \frac{1}{12}\frac{(6x^2 + 1)^{1/2}}{1/2} + C = \frac{1}{6}(6x^2+1)^{1/2} + C$

31. $\int\frac{x - 1}{\sqrt{x^2 - 2x}}dx = \int(x^2 - 2x)^{-1/2}(x-1)dx$; $u = x^2 - 2x$, $du = (2x-2)dx = 2(x-1)dx$; $n=-\frac{1}{2}$, $n+1=\frac{1}{2}$

$\int(x^2 - 2x)^{-1/2}(x-1)dx = \frac{1}{2}\int(x^2 - 2x)^{-1/2}[2(x-1)dx] = \frac{1}{2}\frac{(x^2 - 2x)^{1/2}}{1/2} + C = (x^2 - 2x)^{1/2}+C$

33. $\frac{dy}{dx} = 6x^2$; $dy = 6x^2dx$; $\int dy = \int 6x^2dx = 6\int x^2dx$; $y = 6\frac{x^3}{3} + C = 2x^3 + C$

Curve passes through $(0,2)$; $2 = 2(0)^3 + C$, $C = 2$; $y = 2x^3 + 2$

35. $\dfrac{dy}{dx} = x^2(1 - x^3)^5$

$dy = x^2(1 - x^3)^5 dx$

$\int dy = \int x^2(1 - x^3)^5 dx$

$\qquad = -\dfrac{1}{3}\int (1-x^3)^5(-3x^2 dx)$

$y = -\dfrac{1}{3}\dfrac{(1 - x^3)^6}{6} + C$

$\qquad = -\dfrac{1}{18}(1 - x^3)^6 + C$

Curve passes thru (1,5).

$5 = -\dfrac{1}{18}[1-1^3]^6 + C, \quad C = 5$

$y = -\dfrac{1}{18}(1 - x^3)^6 + 5$

37. $\dfrac{dy}{dx} = -x\sqrt{1 - 4x^2}$

$dy = -x(1 - 4x^2)^{1/2} dx$

$\int dy = \int -x(1-4x^2)^{1/2} dx$

$\qquad = \dfrac{1}{8}\int (1-4x^2)^{1/2}(-8x dx)$

$y = \dfrac{1}{8}\dfrac{(1 - 4x^2)^{3/2}}{3/2} + C$

$\qquad = \dfrac{1}{12}(1 - 4x^2)^{3/2} + C$

Curve passes thru (0,7).

$7 = \dfrac{1}{12}[1-4(0)^2]^{3/2} + C, \quad C = \dfrac{83}{12}$

$y = \dfrac{1}{12}(1 - 4x^2)^{3/2} + \dfrac{83}{12}$

$12y = 83 + (1 - 4x^2)^{3/2}$

39. $\dfrac{d^2y}{dx^2} = \dfrac{d(y')}{dx} = 6$

$dy' = 6dx$

$y' = 6x + C$

$y' = 8$ for $x = 1$

$\qquad 8 = 6(1) + C; \quad C = 2$

$y' = 6x + 2 = \dfrac{dy}{dx}$

$dy = (6x + 2)dx$

$y = 6\dfrac{x^2}{2} + 2x + C_1$

$\qquad = 3x^2 + 2x + C_1$

Curve thru (1,2).

$2 = 3(1)^2 + 2(1) + C_1$

$C_1 = -3; \quad y = 3x^2 + 2x - 3$

Exercises 4-4 , p. 179

1. $y = 3x$, between $x=0$ and $x=3$

(a) $n = 3, \; \Delta x = 1$

x	0	1	2
y	0	3	6

$A = 1(0 + 3 + 6)$
$\quad = 9$

(b) $n = 10, \; \Delta x = 0.3$

x	0.0	0.3	0.6	0.9	1.2	1.5	1.8	2.1	2.4	2.7
y	0.0	0.9	1.8	2.7	3.6	4.5	5.4	6.3	7.2	8.1

$A = 0.3(0.0+0.9+1.8+2.7+3.6+4.5+5.4+6.3+7.2+8.1) = 12.15$

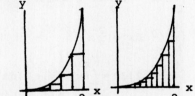

3. $y = x^2$, between $x=0$ and $x=2$

(a) $n = 5, \; \Delta x = 0.4$

x	0.0	0.4	0.8	1.2	1.6
y	0.0	0.16	0.64	1.44	2.56

$A = 0.4(0.0+0.16+0.64+1.44+2.56) = 1.92$

(b) $n = 10, \; \Delta x = 0.2$

x	0.0	0.2	0.4	0.6	0.8	1.0	1.2	1.4	1.6	1.8
y	0.0	0.04	0.16	0.36	0.64	1.00	1.44	1.96	2.56	3.24

$A = 0.2(0.0+0.04+0.16+0.36+0.64+1.00+1.44+1.96+2.56+3.24) = 2.28$

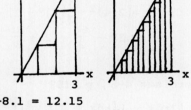

5. $y = 4x - x^2$, between $x=1$ and $x=4$

(a) $n = 6, \; \Delta x = 0.5$

x	1.0	1.5	2.5	3.0	3.5	4.0
y	3.00	3.75	3.75	3.00	1.75	0.00

$A = 0.5(3.00+3.75+3.75+3.00+1.75+0.00)$
$\quad = 7.625$

(b) $n = 10, \; \Delta x = 0.3$

x	1.0	1.3	1.6	2.2	2.5	2.8	3.1	3.4	3.7	4.0
y	3.00	3.51	3.84	3.96	3.75	3.36	2.79	2.04	1.11	0.00

$A = 0.3(3.00+3.51+3.84+3.96+3.75+3.36+2.79+2.04+1.11+0.00) = 8.208$

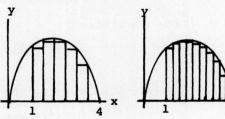

7. $y = \frac{1}{x^2}$, between x=1 and x=5

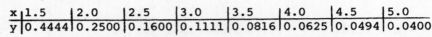

(a) n = 4, $\Delta x = 1$

x	2	3	4	5
y	0.2500	0.1111	0.0625	0.0400

A = 1(0.2500+0.1111+0.0625+0.0400)
= 0.4636

(b) n = 8, $\Delta x = 0.5$

x	1.5	2.0	2.5	3.0	3.5	4.0	4.5	5.0
y	0.4444	0.2500	0.1600	0.1111	0.0816	0.0625	0.0494	0.0400

A = 0.5(0.4444+0.2500+0.1600+0.1111+0.0816+0.0625+0.0494+0.0400) = 0.5995

9. $F(x) = \int 3x\,dx = \frac{3}{2}x^2 + C$

$A_{0,3} = F(3) - F(0)$

$= [\frac{3}{2}(3)^2 + C] - [\frac{3}{2}(0)^2 + C] = \frac{27}{2} = 13.5$

11. $F(x) = \int x^2 dx = \frac{1}{3}x^3 + C$

$A_{0,2} = F(2) - F(0)$

$= [\frac{1}{3}(2)^3 + C] - [\frac{1}{3}(0)^3 + C] = \frac{8}{3} = 2.667$

13. $F(x) = \int(4x - x^2)dx = 2x^2 - \frac{1}{3}x^3 + C$

$A_{1,4} = F(4) - F(1)$

$= [2(4)^2 - \frac{1}{3}(4)^3 + C] - [2(1)^2 - \frac{1}{3}(1)^3 + C]$

$= 32 - \frac{64}{3} - 2 + \frac{1}{3} = 9$

15. $F(x) = \int \frac{1}{x^2}dx = \int x^{-2}dx$

$= -x^{-1} + C = -\frac{1}{x} + C$

$A_{1,5} = F(5) - F(1)$

$= (-\frac{1}{5} + C) - (-\frac{1}{1} + C) = \frac{4}{5} = 0.8$

Exercises 4-5 , p. 182

1. $\int_0^1 2x\,dx = 2\int_0^1 x\,dx = 2(\frac{1}{2}x^2)\Big|_0^1 = x^2\Big|_0^1 = 1 - 0 = 1$

3. $\int_1^4 x^{5/2}\,dx = \frac{2}{7}x^{7/2}\Big|_1^4 = \frac{2}{7}(4)^{7/2} - \frac{2}{7}(1)^{7/2} = \frac{2}{7}(128) - \frac{2}{7} = \frac{254}{7}$

5. $\int_3^6 (\frac{1}{\sqrt{x}} + 2)dx = \int_3^6 (x^{-1/2} + 2)dx = 2x^{1/2} + 2x\Big|_3^6 = [2\sqrt{6} + 2(6)] - [2\sqrt{3} + 2(3)]$

$= 6 + 2\sqrt{6} - 2\sqrt{3} = 7.435$

7. $\int_{-1}^1 (1-x)^{1/3}dx = -\int_{-1}^1 (1-x)^{1/3}(-dx) = -\frac{3}{4}(1-x)^{4/3}\Big|_{-1}^1 = [-\frac{3}{4}(1-1)^{4/3}] - [-\frac{3}{4}(1+1)^{4/3}]$

$= \frac{3}{4}2^{4/3} = \frac{3}{4}(2)(2^{1/3}) = \frac{3}{2}\sqrt[3]{2} = 1.890$

9. $\int_0^3 (x^4 - x^3 + x^2)dx = \frac{1}{5}x^5 - \frac{1}{4}x^4 + \frac{1}{3}x^3\Big|_0^3 = [\frac{1}{5}(3)^5 - \frac{1}{4}(3)^4 + \frac{1}{3}(3)^3] - [\frac{1}{5}(0)^5 - \frac{1}{4}(0)^4 + \frac{1}{3}(0)^3]$

$= \frac{243}{5} - \frac{81}{4} + 9 = \frac{747}{20} = 37.35$

11. $\int_{0.5}^{2.2} (\sqrt[3]{x} - 2)dx = \int_{0.5}^{2.2} (x^{1/3} - 2)dx = \frac{3}{4}x^{4/3} - 2x\Big|_{0.5}^{2.2} = [\frac{3}{4}(2.2)^{4/3} - 2(2.2)]$

$- [\frac{3}{4}(0.5)^{4/3} - 2(0.5)] = \frac{3}{4}(\frac{22}{10})(\frac{22}{10})^{1/3} - 4.4 - \frac{3}{4}(\frac{1}{2})(\frac{1}{2})^{1/3} + 1$

$= \frac{33}{20}(\frac{11}{5})^{1/3} - \frac{3}{8}(\frac{1}{2})^{1/3} - \frac{17}{5} = -1.552$

13. $\int_0^4 (1 - \sqrt{x})^2 dx = \int_0^4 (1 - 2\sqrt{x} + x)dx = x - 2(\frac{2}{3}x^{3/2}) + \frac{1}{2}x^2\Big|_0^4 = x - \frac{4}{3}x^{3/2} + \frac{1}{2}x^2\Big|_0^4$

$= [4 - \frac{4}{3}(4)^{3/2} + \frac{1}{2}(4)^2] - [0 - \frac{4}{3}(0)^{3/2} + \frac{1}{2}(0)^2] = 4 - \frac{32}{3} + 8 = \frac{4}{3}$

15. $\int_1^2 2x(4-x^2)^3 dx = -\int_1^2 (4-x^2)^3(-2xdx) = -\frac{1}{4}(4-x^2)^4\Big|_1^2 = [-\frac{1}{4}(4-2^2)^4]-[-\frac{1}{4}(4-1^2)^4]$

$= 0 + \frac{1}{4}(81) = \frac{81}{4}$

17. $\int_0^4 \frac{xdx}{\sqrt{x^2+9}} = \frac{1}{2}\int_0^4 (x^2+9)^{-1/2}(2xdx) = \frac{1}{2}[2(x^2+9)^{1/2}]\Big|_0^4 = (x^2+9)^{1/2}\Big|_0^4$

$= (4^2+9)^{1/2} - (0^2+9)^{1/2} = 5 - 3 = 2$

19. $\int_{2.75}^{3.25} \frac{dx}{\sqrt[3]{6x+1}} = \frac{1}{6}\int_{2.75}^{3.25}(6x+1)^{-1/3}(6dx) = \frac{1}{6}[\frac{3}{2}(6x+1)^{2/3}]\Big|_{2.75}^{3.25} = \frac{1}{4}(6x+1)^{2/3}\Big|_{2.75}^{3.25}$

$= \frac{1}{4}[6(3.25)+1]^{2/3} - \frac{1}{4}[6(2.75)+1]^{2/3} = \frac{1}{4}(20.5^{2/3} - 17.5^{2/3}) = 0.1875$

21. $\int_1^3 \frac{2xdx}{(2x^2+1)^3} = 2(\frac{1}{4})\int_1^3 (2x^2+1)^{-3}(4xdx) = \frac{1}{2}(-\frac{1}{2})(2x^2+1)^{-2}\Big|_1^3 = -\frac{1}{4(2x^2+1)^2}\Big|_1^3$

$= -\frac{1}{4[2(3)^2+1]^2} - [-\frac{1}{4[2(1)^2+1]^2}] = -\frac{1}{1444} + \frac{1}{36} = \frac{88}{3249} = 0.0271$

23. $\int_3^7 3\sqrt{4x-3}\,dx = \frac{3}{4}\int_3^7 (4x-3)^{1/2}(4dx) = \frac{3}{4}[\frac{2}{3}(4x-3)^{3/2}]\Big|_3^7 = \frac{1}{2}(4x-3)^{3/2}\Big|_3^7$

$= \frac{1}{2}[4(7)-3]^{3/2} - \frac{1}{2}[4(3)-3]^{3/2} = \frac{125}{2} - \frac{27}{2} = 49$

25. $\int_0^2 2x(9-2x^2)^2 dx = 2(-\frac{1}{4})\int_0^2 (9-2x^2)^2(-4xdx) = -\frac{1}{2}[\frac{1}{3}(9-2x^2)^3]\Big|_0^2$

$= -\frac{1}{6}[9-2(2)^2]^3 - [-\frac{1}{6}(9-2(0)^2)^3] = -\frac{1}{6} + \frac{729}{6} = \frac{728}{6} = \frac{364}{3}$

27. $\int_0^1 (x^2+3)(x^3+9x+6)dx = \frac{1}{3}\int_0^1 (x^3+9x+6)[3(x^2+3)dx] = \frac{1}{3}[\frac{1}{2}(x^3+9x+6)^2]\Big|_0^1$

$= \frac{1}{6}[1^3+9(1)+6]^2 - \frac{1}{6}[0^3+9(0)+6]^2 = \frac{256}{6} - \frac{36}{6} = \frac{110}{3}$

Exercises 4-6, p. 187

1. (a) $\int_0^2 2x^2 dx$; $n=4$; $f(x) = 2x^2$; $\Delta x = \frac{2-0}{4} = \frac{1}{2}$

$y_0 = 2(0)^2 = 0$; $y_1 = 2(\frac{1}{2})^2 = \frac{1}{2}$; $y_2 = 2(1)^2 = 2$; $y_3 = 2(\frac{3}{2})^2 = \frac{9}{2}$; $y_4 = 2(2)^2 = 8$

$\int_0^2 2x^2 dx = A_T = [\frac{1}{2}(0) + \frac{1}{2} + 2 + \frac{9}{2} + \frac{1}{2}(8)](\frac{1}{2}) = \frac{11}{2} = 5.50$

(b) $\int_0^2 2x^2 dx = \frac{2}{3}x^3\Big|_0^2 = \frac{2}{3}(8-0) = \frac{16}{3} = 5.33$

3. (a) $\int_1^4 (1+\sqrt{x})dx$; $n=6$; $f(x) = 1+x^{1/2}$; $\Delta x = \frac{4-1}{6} = \frac{1}{2}$

$y_0 = f(1) = 2.0000$; $y_1 = f(1.5) = 2.2247$; $y_2 = f(2) = 2.4142$; $y_3 = f(2.5) = 2.5811$;
$y_4 = f(3) = 2.7321$; $y_5 = f(3.5) = 2.8708$; $y_6 = f(4) = 3.0000$

$\int_1^4 (1+\sqrt{x})dx = A_T = [\frac{1}{2}(2.0000) + 2.2247 + 2.4142 + 2.5811 + 2.7321 + 2.8708$

$+ \frac{1}{2}(3.0000)](\frac{1}{2}) = 7.661$

(b) $\int_1^4 (1+\sqrt{x})dx = x + \frac{2}{3}x^{3/2}\Big|_1^4 = 4 + \frac{2}{3}(4)^{3/2} - [1 + \frac{2}{3}(1)^{3/2}]$

$= 4 + \frac{16}{3} - 1 - \frac{2}{3} = \frac{23}{3} = 7.667$

5. $\int_2^3 \frac{1}{2x} dx$; $n = 2$; $f(x) = \frac{1}{2x}$; $\Delta x = \frac{3-2}{2} = \frac{1}{2}$

$y_0 = f(2) = 0.2500$; $y_1 = f(2.5) = 0.2000$; $y_2 = = f(3) = 0.1667$

$\int_2^3 \frac{1}{2x} dx = A_T = [\frac{1}{2}(0.2500) + 0.2000 + \frac{1}{2}(0.1667)](\frac{1}{2}) = 0.2042$

7. $\int_0^2 \sqrt{4-x^2} dx$; $n = 4$; $f(x) = (4-x^2)^{1/2}$; $\Delta x = \frac{2-0}{4} = \frac{1}{2}$

$y_0 = f(0) = 2.0000$; $y_1 = f(0.5) = 1.9365$; $y_2 = f(1) = 1.7321$; $y_3 = f(1.5) = 1.3229$; $y_4 = f(2) = 0.0000$

$\int_0^2 \sqrt{4-x^2} dx = A_T = [\frac{1}{2}(2.0000) + 1.9365 + 1.7321 + 1.3229 + \frac{1}{2}(0.0000)](\frac{1}{2}) = 2.996$

9. $\int_1^5 \frac{1}{x^2+x} dx$; $n = 10$; $f(x) = \frac{1}{x^2+x}$; $\Delta x = \frac{5-1}{10} = 0.4$

x	1.0	1.4	1.8	2.2	2.6	3.0	3.4	3.8	4.2	4.6	5.0
y	0.5000	0.2976	0.1984	0.1420	0.1068	0.0833	0.0668	0.0548	0.0458	0.0388	0.0333

$\int_1^5 \frac{1}{x^2+x} dx = A_T = [\frac{1}{2}(0.5000) + 0.2976 + 0.1984 + 0.1420 + 0.1068 + 0.0833 + 0.0668$

$+ 0.0548 + 0.0458 + 0.0388 + \frac{1}{2}(0.0333)](0.4) = 0.5204$

11. $\int_0^4 2^x dx$; $n = 12$; $f(x) = 2^x$; $\Delta x = \frac{4-0}{12} = \frac{1}{3}$

x	0.0000	0.3333	0.6667	1.0000	1.3333	1.6667	2.0000	2.3333	2.6667
y	1.0000	1.2599	1.5874	2.0000	2.5198	3.1749	4.0000	5.0396	6.3498

x	3.0000	3.3333	3.6667	4.0000
y	8.0000	10.0791	12.6995	16.0000

$\int_0^4 2^x dx = A_T = [\frac{1}{2}(1.0000) + 1.2599 + 1.5874 + 2.0000 + 2.5198 + 3.1749 + 4.0000 + 5.0396$

$+ 6.3498 + 8.0000 + 10.0791 + 12.6995 + \frac{1}{2}(16.0000)](0.3333) = 21.74$

13.

x	3	4	5	6	7
y	1.08	1.65	3.23	3.67	2.97

; $f(x)$ defined by table ; $\Delta x = 1$

$\int_3^7 f(x) dx = A_T = [\frac{1}{2}(1.08) + 1.65 + 3.23 + 3.67 + \frac{1}{2}(2.97)](1) = 10.58$

15.

x	1.4	1.7	2.0	2.3	2.6	2.9	3.2
y	0.18	7.87	18.23	23.53	24.62	20.93	20.76

; $f(x)$ defined by table ; $\Delta x = 0.3$

$\int_{1.4}^{3.2} f(x) dx = A_T = [\frac{1}{2}(0.18) + 7.87 + 18.23 + 23.53 + 24.62 + 20.93 + \frac{1}{2}(20.76)](0.3) = 31.70$

Exercises 4-7, p. 190

1. (a) $\int_0^2 (1+x^3) dx$; $n = 2$; $f(x) = 1 + x^3$; $\Delta x = \frac{2-0}{2} = 1$

$y_0 = f(0) = 1$; $y_1 = f(1) = 2$; $y_2 = f(2) = 9$

$\int_0^2 (1+x^3) dx = \frac{1}{3}[1 + 4(2) + 9] = \frac{18}{3} = 6$

(b) $\int_0^2 (1+x^3) dx = x + \frac{1}{4} x^4 \Big|_0^2 = 2 + \frac{1}{4}(2)^4 - 0 = 2 + 4 = 6$

3. (a) $\int_1^4 (2x + \sqrt{x})dx$; $n = 6$; $f(x) = 2x + x^{1/2}$; $\Delta x = \dfrac{4-1}{6} = \dfrac{1}{2}$

$y_0 = f(1) = 3.0000$; $y_1 = f(1.5) = 4.2247$; $y_2 = f(2) = 5.4142$; $y_3 = f(2.5) = 6.5811$
$y_4 = f(3) = 7.7321$; $y_5 = f(3.5) = 8.8708$; $y_6 = f(4) = 10.0000$

$\int_1^4 (2x + \sqrt{x})dx = \dfrac{0.5}{3}[3.0000 + 4(4.2247) + 2(5.4142) + 4(6.5811) + 2(7.7321)$
$\qquad\qquad + 4(8.8708 + 10.0000] = 19.67$

(b) $\int_1^4 (2x + \sqrt{x})dx = x^2 + \dfrac{2}{3}x^{3/2}\Big|_1^4 = 4^2 + \dfrac{2}{3}(4)^{3/2} - [1^2 + \dfrac{2}{3}(1)^{3/2}] = \dfrac{59}{3} = 19.67$

5. $\int_2^3 \dfrac{1}{2x}dx$; $n = 2$; $f(x) = \dfrac{1}{2x}$; $\Delta x = \dfrac{3-2}{2} = \dfrac{1}{2}$

From #5 of Exer. 4-6: $y_0 = 0.2500$; $y_1 = 0.2000$; $y_2 = 0.1667$

$\int_2^3 \dfrac{1}{2x}dx = \dfrac{0.5}{3}[0.2500 + 4(0.2000) + 0.1667] = 0.2028$

7. $\int_0^2 \sqrt{4-x^2}dx$; $n = 4$; $f(x) = (4-x^2)^{1/2}$; $\Delta x = \dfrac{2-0}{4} = \dfrac{1}{2}$

From #7 of Exer. 4-6: $y_0 = 2.0000$; $y_1 = 1.9365$; $y_2 = 1.7321$; $y_3 = 1.3229$; $y_4 = 0.0000$

$\int_0^2 \sqrt{4-x^2}dx = \dfrac{0.5}{3}[2.0000 + 4(1.9365) + 2(1.7321) + 4(1.3229) + 0.0000] = 3.084$

9. $\int_1^5 \dfrac{1}{x^2+x}dx$; $n = 10$; $f(x) = \dfrac{1}{x^2+x}$; $\Delta x = \dfrac{5-1}{10} = 0.4$

See table for #9 of Exer. 4-6 for values of y_n.

$\int_0^5 \dfrac{1}{x^2+x}dx = \dfrac{0.4}{3}[0.5000 + 4(0.2976) + 2(0.1984) + 4(0.1420) + 2(0.1068) + 4(0.0833)$
$\qquad\qquad + 2(0.0668) + 4(0.0548) + 2(0.0458) + 4(0.0388) + 0.0333] = 0.5113$

11.

x	3	4	5	6	7
y	1.08	1.65	3.23	3.67	2.97

; $f(x)$ defined by table ; $\Delta x = 1$

$\int_3^7 f(x)dx = \dfrac{1}{3}[1.08 + 4(1.65) + 2(3.23) + 4(3.67) + 2.97] = 10.60$

Review Exercises for Chapter 4, p. 191

1. $\int (4x^3 - x)dx = 4\int x^3 dx - \int x\,dx = 4\left(\dfrac{x^4}{4}\right) - \dfrac{x^2}{2} + C = x^4 - \dfrac{1}{2}x^2 + C$

3. $\int 2(x - x^{3/2})dx = 2\int x\,dx - 2\int x^{3/2}dx = 2\left(\dfrac{x^2}{2}\right) - 2\left(\dfrac{x^{5/2}}{5/2}\right) + C = x^2 - \dfrac{4}{5}x^{5/2} + C$

5. $\int_1^4 \left(\sqrt{x} + \dfrac{1}{\sqrt{x}}\right)dx = \int_1^4 x^{1/2}dx + \int_1^4 x^{-1/2}dx = \dfrac{2}{3}x^{3/2} + 2x^{1/2}\Big|_1^4$
$\qquad = \dfrac{2}{3}(4)^{3/2} + 2(4)^{1/2} - [\dfrac{2}{3}(1)^{3/2} + 2(1)^{1/2}] = \dfrac{16}{3} + 4 - \dfrac{2}{3} - 2 = \dfrac{20}{3}$

7. $\int_0^2 x(4-x)dx = 4\int_0^2 x\,dx - \int_0^2 x^2 dx = 4\left(\dfrac{x^2}{2}\right) - \dfrac{x^3}{3}\Big|_0^2 = 2x^2 - \dfrac{1}{3}x^3\Big|_0^2$
$\qquad = 2(2)^2 - \dfrac{1}{3}(2)^3 - [2(0)^2 - \dfrac{1}{3}(0)^3] = 8 - \dfrac{8}{3} = \dfrac{16}{3}$

9. $\int \left(3 + \dfrac{2}{x^3}\right)dx = 3\int dx + 2\int x^{-3}dx = 3x + \dfrac{2x^{-2}}{-2} + C = 3x - \dfrac{1}{x^2} + C$

11. $\int_{-1.5}^{0.5} \frac{dx}{\sqrt{1-x}} = -\int_{-1.5}^{0.5}(1-x)^{-1/2}(-dx) = -2(1-x)^{1/2}\Big|_{-1.5}^{0.5} = -2[1-0.5]^{1/2} -[-2(1+1.5)^{1/2}]$

$= -2\sqrt{0.5} + 2\sqrt{2.5} = -2\sqrt{\frac{1}{2}} + 2\sqrt{\frac{5}{2}} = -\sqrt{2} + \sqrt{10} = 1.748$

13. $\int \frac{dx}{(2-5x)^2} = -\frac{1}{5}\int(2-5x)^{-2}(-5dx) = -\frac{1}{5}\frac{(2-5x)^{-1}}{-1} = \frac{1}{5(2-5x)} + C$

15. $\int(7-2x)^{3/4}dx = -\frac{1}{2}\int(7-2x)^{3/4}(-2dx) = -\frac{1}{2}\frac{(7-2x)^{7/4}}{7/4} + C = -\frac{2}{7}(7-2x)^{7/4} + C$

17. $\int_0^2 \frac{3xdx}{\sqrt[3]{1+2x^2}} = 3\int_0^2(1+2x^2)^{-1/3}xdx = \frac{3}{4}\int_0^2(1+2x^2)^{-1/3}(4xdx) = \frac{3}{4}\frac{(1+2x^2)^{2/3}}{2/3}\Big|_0^2$

$= \frac{9}{8}(1+2x^2)^{2/3}\Big|_0^2 = \frac{9}{8}[1+2(2)^2]^{2/3} - \frac{9}{8}[1+2(0)^2]^{2/3} = \frac{9}{8}(9)^{2/3} - \frac{9}{8}$

$= \frac{9}{8}[3\sqrt[3]{3} - 1] = 3.743$

19. $\int x^2(1-2x^3)^4dx = -\frac{1}{6}\int(1-2x^3)^4dx(-6x^2dx) = -\frac{1}{6}\frac{(1-2x^3)^5}{5} + C = -\frac{1}{30}(1-2x^3)^5 + C$

21. $\int \frac{(2-3x^2)dx}{(2x-x^3)^2} = \int(2x-x^3)^{-2}[(2-3x^2)dx] = \frac{(2x-x^3)^{-1}}{-1} + C = -\frac{1}{2x-x^3} + C$

23. $\int_1^3(x^2+x+2)(2x^3+3x^2+12x)dx = \frac{1}{6}\int_1^3(2x^3+3x^2+12x)[6(x^2+x+2)dx]$

$= \frac{1}{6}\frac{(2x^3+3x^2+12x)^2}{2}\Big|_1^3 = \frac{1}{12}(2x^3+3x^2+12x)^2\Big|_1^3$

$= \frac{1}{12}[2(3)^3+3(3)^2+12(3)]^2 - \frac{1}{12}[2(1)^3+3(1)^2+12(1)]^2 = \frac{117^2}{12} - \frac{17^2}{12} = \frac{3350}{3}$

25. $y = \frac{1}{(x^2-1)^3} = (x^2-1)^{-3}$

$dy = -3(x^2-1)^{-4}(2x)dx$

$= \frac{-6xdx}{(x^2-1)^4}$

27. $y = x\sqrt[3]{1-3x} = x(1-3x)^{1/3}$

$dy = [x(\frac{1}{3})(1-3x)^{-2/3}(-3) + (1-3x)^{1/3}(1)]dx$

$= \frac{-x+1-3x}{(1-3x)^{2/3}}dx = \frac{(1-4x)dx}{(1-3x)^{2/3}}$

29. $y = x^3$; $x = 2$, $\Delta x = 0.1$

$y + \Delta y = (x+\Delta x)^3$

$\Delta y = 3x^2\Delta x + 3x(\Delta x)^2 + (\Delta x)^3$; $dy = 3x^2dx$

For $x = 2$, $\Delta x = dx = 0.1$

$\Delta y = 3(2)^2(0.1) + 3(2)(0.1)^2 + (0.1)^3 = 1.261$

$dy = 3(2)^2(0.1) = 1.2$

$\Delta y - dy = 0.061$

31. $\frac{dy}{dx} = 3 - x^2$

$dy = (3-x^2)dx$

$y = 3x - \frac{1}{3}x^3 + C$

Curve passes thru $(-1,3)$.

$3 = 3(-1) - \frac{1}{3}(-1)^3 + C$, $C = \frac{17}{3}$

$y = 3x - \frac{1}{3}x^3 + \frac{17}{3}$

33. (a) $\int(1-2x)dx = \int dx - 2\int xdx = x - 2(\frac{x^2}{2}) + C_1 = x - x^2 + C_1$

(b) $\int(1-2x)dx = -\frac{1}{2}\int(1-2x)(-2dx) = -\frac{1}{2}\frac{(1-2x)^2}{2} + C_2 = -\frac{(1-2x)^2}{4} + C_2$

$= C_2 - \frac{1}{4} + x - x^2$; $C_1 = C_2 - \frac{1}{4}$

35. $F(x) = \int(6x-1)dx = 3x^2 - x + C$

$A_{1,3} = F(3) - F(1) = [3(3)^2 - 3 + C] - [3(1)^2 - 1 + C] = 22$

37. $\int_1^3 \dfrac{dx}{2x-1}$; $n=4$; $f(x) = \dfrac{1}{2x-1}$; $\Delta x = \dfrac{3-1}{4} = \dfrac{1}{2}$

x	1.0	1.5	2.0	2.5	3.0
y	1.0000	0.5000	0.3333	0.2500	0.2000

$\int_1^3 \dfrac{dx}{2x-1} = A_T = [\dfrac{1}{2}(1.0000) + 0.5000 + 0.3333 + 0.2500 + \dfrac{1}{2}(0.2000)](\dfrac{1}{2}) = 0.842$

39. $\int_1^3 \dfrac{dx}{2x-1}$; $n=4$; $f(x) = \dfrac{1}{2x-1}$; $\Delta x = \dfrac{3-1}{4} = \dfrac{1}{2}$; See #37 for values of y_n.

$\int_1^3 \dfrac{dx}{2x-1} = \dfrac{0.5}{3}[1.0000 + 4(0.5000) + 2(0.3333) + 4(0.2500) + 0.2000] = 0.811$

41. $y = \dfrac{x}{x^2+2}$, between $x=0$ [$y=0$ when $x=0$] and $x=5$; $n=5$, $\Delta x = 1$

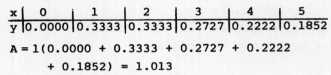

x	0	1	2	3	4	5
y	0.0000	0.3333	0.3333	0.2727	0.2222	0.1852

$A = 1(0.0000 + 0.3333 + 0.2727 + 0.2222 + 0.1852) = 1.013$

43. $A = \int_1^3 x\sqrt{x^3+1}\,dx$; $n=10$; $f(x) = x\sqrt{x^3+1}$; $\Delta x = \dfrac{3-1}{10} = 0.2$

x	1.0	1.2	1.4	1.6	1.8	2.0	2.2	2.4	2.6	2.8	3.0
y	1.4142	1.9820	2.7089	3.6119	4.7049	6.0000	7.5084	9.2405	11.2060	13.4143	15.8745

$A = \int_1^3 x\sqrt{x^3+1}\,dx = \dfrac{0.2}{3}[1.4142 + 4(1.9820) + 2(2.7089) + 4(3.6119) + 2(4.7049) + 4(6.000)$
$+ 2(7.5084) + 4(9.2405) + 2(11.2060) + 4(13.4143) + 15.8745)]$
$= 13.77$

45. $V = \dfrac{4}{3}\pi r^3$

$dV = \dfrac{4}{3}\pi(3r^2 dr) = 4\pi r^2 dr$

For $r = 4.00$ mm, $dr = 0.02$ mm

$dV = 4\pi(4.00)^2(0.02) = 4.02$ mm^3

47. $z = \sqrt{R^2 + X^2} = (R^2 + X^2)^{1/2}$

$dz = \dfrac{1}{2}(R^2 + X^2)^{-1/2}(2RdR)$

$= \dfrac{RdR}{(R^2 + X^2)^{1/2}}$

$\dfrac{dz}{z} = \dfrac{RdR}{\sqrt{R^2 + X^2}\sqrt{R^2 + X^2}} = \dfrac{RdR}{R^2 + X^2}$

Exercises 5-1, p. 199

1. $a = -32$ ft/s^2, $t = 3$ s
   $v = 0$ when $t = 0$ (dropped)
   $v = \int -32\,dt = -32t + C$
   $0 = -32(0) + C$, $C = 0$
   $v = -32t$
   $v|_{t=3} = -32(3) = -96$ ft/s

3. $a = 6t$, $v = 8$ when $t = 1$
   $v = \int a\,dt = \int 6t\,dt = 3t^2 + C$
   $8 = 3(1)^2 + C$, $C = 5$
   $v = 3t^2 + 5$

5. $v = 80$ cm/s, $t = 7$ s
   $s = 0$ when $t = 0$
   (s measured from starting point)
   $s = \int v\,dt = \int 80\,dt$
   $= 80t + C$
   $0 = 80(0) + C$, $C = 0$
   $s = 80t$, $s|_{t=7} = 560$ cm

7. $a = -9.80$ m/s^2, $s = 16500$ m and $v = 450$ m/s when $t = 0$
   $v = \int a\,dt = \int -9.80\,dt = -9.80t + C_1$, $450 = -9.80(0) + C_1$, $C_1 = 450$
   $v = -9.80t + 450$; $s = \int v\,dt = \int (-9.80t + 450)\,dt = -4.90t^2 + 450t + C_2$
   $16500 = -4.90(0)^2 + 450(0) + C_2$, $C_2 = 16500$; $s = -4.90t^2 + 450t + 16500$
   $s|_{t=3.00} = -4.90(3.00)^2 + 450(3.00) + 16500 = 17,800$ m

9. $v = 64$ ft/s when $t = 0$ (s measured
   $s = 0$ when $t = 7$ s   vertically from
                                   base of cliff)
   $v = \int a\,dt = \int -32\,dt$
   $= -32t + C_1$; $64 = -32(0) + C_1$, $C_1 = 64$
   $v = -32t + 64$
   $s = \int v\,dt = \int (-32t + 64)\,dt = -16t^2 + 64t + C_2$
   $0 = -16(7)^2 + 64(7) + C_2$, $C_2 = 336$
   $s = -16t^2 + 64t + 336$
   $s|_{t=0} = 336$ ft (height of cliff)

11. $a = 24t$, $s = 0$ and $v = 0$ when $t = 0$
    $v = \int a\,dt = \int 24t\,dt = 12t^2 + C_1$
    $0 = 12(0) + C_1$, $C_1 = 0$
    $v = 12t^2$
    $s = \int v\,dt = \int 12t^2\,dt = 4t^3 + C$
    $0 = 4(0) + C_2$, $C_2 = 0$
    $s = 4t^3$
    $s|_{t=1.5} = 4(1.5)^3 = 13.5$ m

13. $i = 0.500$ mA $= 5.00 \times 10^{-4}$ A
    $q = 0$ when $t = 0$, $t = 2.00$ s
    $q = \int i\,dt = \int 5.00 \times 10^{-4}\,dt = 5.00 \times 10^{-4}t + C$
    $0 = 5.00 \times 10^{-4}(0) + C$, $C = 0$
    $q = 5.00 \times 10^{-4}t$
    $q|_{t=2.00} = 5.00 \times 10^{-4}(2.00)$
    $= 1.00 \times 10^{-3}$ C $= 1.00$ mC

15. $i = \sqrt[3]{1 + 3t}$, $q = 0$ when $t = 0$
    $q = \int i\,dt = \int (1 + 3t)^{1/3}\,dt$
    $= \frac{1}{3}\int (1 + 3t)^{1/3}(3\,dt) = \frac{1}{3}\frac{3}{4}(1 + 3t)^{4/3} + C$
    $= \frac{1}{4}(1 + 3t)^{4/3} + C$
    $0 = \frac{1}{4}(1 + 0)^{4/3} + C$, $C = -\frac{1}{4}$
    $q = \frac{1}{4}[(1 + 3t)^{4/3} - 1]$
    $q|_{t=3} = \frac{1}{4}[(1 + 3(3))^{4/3} - 1] = 5.14$ C

17. $C = 3.0$ μF $= 3.0 \times 10^{-6}$ F, $i = 0.20$ A
    $V = 0$ when $t = 0$, $t = 10$ ms $= 0.010$ s
    $V_C = \frac{1}{C}\int i\,dt = \frac{1}{3.0 \times 10^{-6}}\int 0.20\,dt$
    $= 6.67 \times 10^4 \int dt = 6.67\ 10^4 t + C_1$
    $0 = 6.67 \times 10^4(0) + C_1$, $C_1 = 0$
    $V_C = 6.67 \times 10^4 t$
    $V_C|_{t=0.010} = 6.67 \times 10^4(0.010) = 667$ V

19. $C = 150$ μF $= 1.50 \times 10^{-4}$ F
    $V = 50$ V when $t = 0$, $t = 0.012t^{1/5}$
    $V_C = \frac{1}{C}\int i\,dt = \frac{1}{1.50 \times 10^{-4}}\int 0.012t^{1/5}\,dt$
    $= 80\int t^{1/5}\,dt = 80(\frac{5}{6})t^{6/5} + C_1$
    $= 66.7t^{6/5} + C_1$
    $50 = 66.7t^{6/5}(0) + C_1$, $C_1 = 50$
    $V_C = 66.7t^{6/5} + 50$
    $V_C|_{t=1} = 66.7(1)^{6/5} + 50 = 117$ V

21. $\omega = \frac{d\theta}{dt}$, $\omega = 4t$, $\theta = 0$ when $t = 0$
    $\theta = \int \omega\,dt = \int 4t\,dt = 2t^2 + C$
    $0 = 2(0) + C$, $C = 0$, $\theta = 2t^2$

23. $V_L = L\dfrac{di}{dt}$, $V_L = 12.0 - 0.2t$, $L = 3\,H$, $i = 0$ when $t = 0$

$di = \dfrac{1}{L}\int V_L dt = \dfrac{1}{3}\int(12.0 - 0.2t)dt$

$i = 4.0t - 0.0333t^2 + C$

$0 = 4.0(0) - 0.0333(0)^2 + C$, $C = 0$

$i = 4.0t - 0.0333t^2$

$i\,\big|\,t=20 = 4.0(20) - 0.0333(20)^2 = 66.7\,A$

25. $\dfrac{dV}{dx} = -E$, $E = \dfrac{k}{x^2}$

$\dfrac{dV}{dx} = -\dfrac{k}{x^2}$, $V \to 0$ as $x \to \infty$

$V = -k\int\dfrac{dx}{x^2} = \dfrac{k}{x} + C$

$0 = 0 + C$, $C = 0$

$V = \dfrac{k}{x}$, $V\big|_{x=x_1} = \dfrac{k}{x_1}$

27. $\dfrac{dm}{dt} = -\dfrac{1}{\sqrt{t+1}}$, $m = 1000\,g$ when $t = 0$

$m = -\int\dfrac{dt}{\sqrt{t+1}} = -2(t+1)^{1/2} + C$, $1000 = -2(0+1)^{1/2} + C$, $C = 1002$

$m = 1002 - 2\sqrt{t+1}$; For $m = 0$, $0 = 1002 - 2\sqrt{t+1}$, $\sqrt{t+1} = 501$

$t + 1 = 501^2$, $t = 2.51\times10^5$ min

Exercises 5-2, p. 205

1. $y = 4x$, $y = 0$, $x = 1$

$A = \int_0^1 y\,dx = \int_0^1 4x\,dx$

$= 2x^2\big|_0^1$

$= 2(1)^2 - 2(0)^2 = 2$

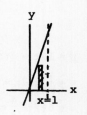

3. $y = x^2$, $y = 0$, $x = 2$

$A = \int_0^2 y\,dx = \int_0^2 x^2 dx$

$= \dfrac{1}{3}x^3\big|_0^2$

$= \dfrac{1}{3}(2)^3 - \dfrac{1}{3}(0)^3 = \dfrac{8}{3}$

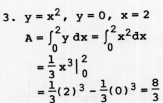

5. $y = 6x$, $x = 0$, $y = 3$

$A = \int_0^3 x\,dy = \int_0^3 \dfrac{y}{6}\,dy$

$= \dfrac{1}{6}(\dfrac{1}{2}y^2)\big|_0^3 = \dfrac{1}{12}y^2\big|_0^3$

$= \dfrac{1}{12}(3)^2 - \dfrac{1}{12}(0)^2 = \dfrac{3}{4}$

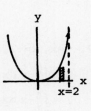

7. $y = x^2$, $x = 0$, $y = 2$

$A = \int_0^2 x\,dy = \int_0^2 y^{1/2}dy$

$= \dfrac{2}{3}y^{3/2}\big|_0^2$

$= \dfrac{2}{3}(2)^{3/2} - \dfrac{2}{3}(0)^{3/2} = \dfrac{4}{3}\sqrt{2}$

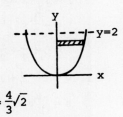

9. $y = 2 - x$, $y = 0$, $x = 1$

$A = \int_1^2 y\,dx = \int_1^2(2-x)dx$

$= 2x - \dfrac{x^2}{2}\big|_1^2$

$= 4 - 2 - (2 - \dfrac{1}{2}) = \dfrac{1}{2}$

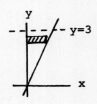

11. $y = x^{-2}$, $y = 0$, $x = 2$, $x = 3$

$A = \int_2^3 y\,dx = \int_2^3 x^{-2}dx$

$= -\dfrac{1}{x}\big|_2^3 = -\dfrac{1}{3} + \dfrac{1}{2} = \dfrac{1}{6}$

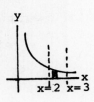

13. $y = \dfrac{1}{2}x$, $x = 0$, $y = 2$, $y = 4$

$A = \int_2^4 x\,dy = \int_2^4 2y\,dy$

$= y^2\big|_2^4 = 4^2 - 2^2 = 12$

15. $y = \sqrt{x}$, $x = 0$, $y = 2$, $y = 3$

$A = \int_2^3 x\,dy = \int_2^3 y^2 dy$

$= \dfrac{1}{3}y^3\big|_2^3 = 9 - \dfrac{8}{3} = \dfrac{19}{3}$

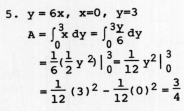
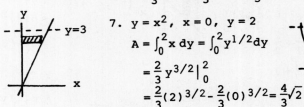
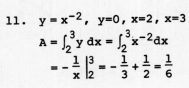
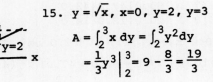
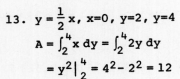

17. $y = 4-2x$, $x=0$, $y=0$, $y=3$

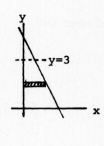

$$A = \int_0^3 x\, dy = \int_0^3 \frac{4-y}{2} dy$$

$$= \int_0^3 \left(2 - \frac{1}{2} y\right) dy$$

$$= 2y - \frac{1}{4} y^2 \Big|_0^3$$

$$= 6 - \frac{9}{4} - (0 - 0) = \frac{15}{4}$$

19. $y = x^2$, $y = 2-x$, $x=0$
(smaller)

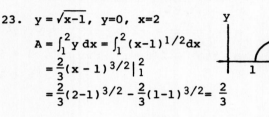

To find pts. of intersection, $y_1 = y_2$

$x^2 = 2-x$, $x^2 + x - 2 = 0$

$(x+2)(x-1) = 0$, $x = -2, 1$

$$A = \int_0^1 (y_1 - y_2) dx = \int_0^1 (2 - x - x^2) dx$$

$$= 2x - \frac{x^2}{2} - \frac{x^3}{3} \Big|_0^1 = 2 - \frac{1}{2} - \frac{1}{3} - 0 = \frac{7}{6}$$

21. $y = x^4$, $y = 16$

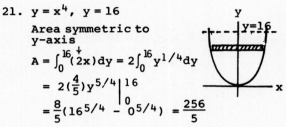

Area symmetric to
y-axis

$$A = \int_0^{16} (2x) dy = 2\int_0^{16} y^{1/4} dy$$

$$= 2\left(\frac{4}{5}\right) y^{5/4} \Big|_0^{16}$$

$$= \frac{8}{5}(16^{5/4} - 0^{5/4}) = \frac{256}{5}$$

23. $y = \sqrt{x-1}$, $y=0$, $x=2$

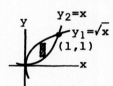

$$A = \int_1^2 y\, dx = \int_1^2 (x-1)^{1/2} dx$$

$$= \frac{2}{3}(x-1)^{3/2} \Big|_1^2$$

$$= \frac{2}{3}(2-1)^{3/2} - \frac{2}{3}(1-1)^{3/2} = \frac{2}{3}$$

25. $y = x^2$, $y = \sqrt{x}$

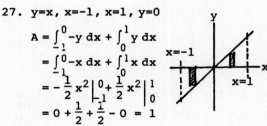

$$A = \int_0^1 (y_1 - y_2) dx = \int_0^1 (\sqrt{x} - x^2) dx$$

$$= \frac{2}{3} x^{3/2} - \frac{1}{3} x^3 \Big|_0^1 = \frac{2}{3}(1)^{3/2} - \frac{1}{3}(1)^3 - 0 = \frac{1}{3}$$

27. $y=x$, $x=-1$, $x=1$, $y=0$

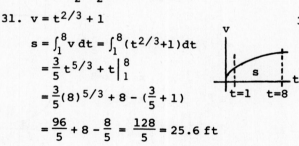

$$A = \int_{-1}^0 -y\, dx + \int_0^1 y\, dx$$

$$= \int_{-1}^0 -x\, dx + \int_0^1 x\, dx$$

$$= -\frac{1}{2} x^2 \Big|_{-1}^0 + \frac{1}{2} x^2 \Big|_0^1$$

$$= 0 + \frac{1}{2} + \frac{1}{2} - 0 = 1$$

29. $p = 12t - 4t^2$, See Fig. 5-10

$$w = \int_0^3 p\, dt = \int_0^3 (12t - 4t^2) dt$$

$$= 6t^2 - \frac{4}{3} t^3 \Big|_0^3$$

$$= 6(3)^2 - \frac{4}{3}(3)^3 - 0$$

$$= 54 - 36 = 18 \; ; \; w = 18.0\, J$$

31. $v = t^{2/3} + 1$

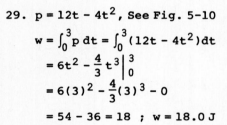

$$s = \int_1^8 v\, dt = \int_1^8 (t^{2/3}+1) dt$$

$$= \frac{3}{5} t^{5/3} + t \Big|_1^8$$

$$= \frac{3}{5}(8)^{5/3} + 8 - \left(\frac{3}{5} + 1\right)$$

$$= \frac{96}{5} + 8 - \frac{8}{5} = \frac{128}{5} = 25.6 \text{ ft}$$

33. $y_1 = x^3 - 2x^2 - x + 2$, $y_2 = x^2 - 1$

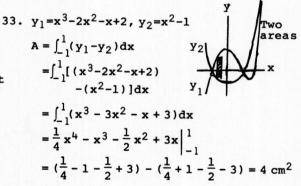

$$A = \int_{-1}^1 (y_1 - y_2) dx$$

$$= \int_{-1}^1 [(x^3 - 2x^2 - x + 2) - (x^2 - 1)] dx$$

$$= \int_{-1}^1 (x^3 - 3x^2 - x + 3) dx$$

$$= \frac{1}{4} x^4 - x^3 - \frac{1}{2} x^2 + 3x \Big|_{-1}^1$$

$$= \left(\frac{1}{4} - 1 - \frac{1}{2} + 3\right) - \left(\frac{1}{4} + 1 - \frac{1}{2} - 3\right) = 4 \text{ cm}^2$$

35.

p(kPa)	102	77.0	60.0	47.0	43.0
V(cm^3)	6.0	8.0	10.0	12.0	14.0

$w = \int p\, dV$ (See Exer. 32), $\Delta V = 2.0 \text{ cm}^3$

$A_T = 2.0[\frac{1}{2}(102) + 77.0 + 60.0 + 47.0) + \frac{1}{2}(43.0)] = 513$; $w = 513 \times 10^{-3}\, J = 0.513\, J$

37. See Fig. 5-11

$(x - h)^2 = 4p(y - k)$

$(h,k) = (0, 0.640)$

$(x - 0)^2 = 4p(y - 0.640)$

$x^2 = 4p(y - 0.640)$

$(0.800)^2 = 4p(0 - 0.640)$

$0.640 = 4p(-0.640), \quad 4p = -1$

$x^2 = -(y - 0.640)$

$y = 0.640 - x^2$

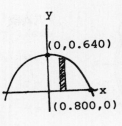

(0, 0.640)

(0.800, 0)

Area symmetrical to y-axis

$A = 2\int_0^{0.800}(0.640 - x^2)dx$

$= 1.28x - \dfrac{2}{3}x^3 \Big|_0^{0.800}$

$= 1.28(0.800) - \dfrac{2}{3}(0.800)^3 - 0$

$= 0.683 \text{ m}^2$

Exercises 5-3, p. 212

1. $y = 1 - x, \quad x = 0, \quad y = 0$

disk: $r = y, \quad t = dx$

$V = \pi\int_0^1 y^2 dx = \pi\int_0^1 (1-x)^2 dx$

$= -\dfrac{\pi}{3}(1-x)^3 \Big|_0^1$

$= -\dfrac{\pi}{3}(0)^3 + \dfrac{\pi}{3}(1)^3 = \dfrac{\pi}{3}$

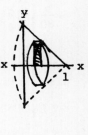

3. $y = 1 - x, \quad x = 0, \quad y = 0$

shell: $r = y, \quad h = x, \quad t = dy$

$V = 2\pi\int_0^1 xy\, dy = 2\pi\int_0^1 (1-y)y\, dy$

$= 2\pi\int_0^1 (y - y^2)dy$

$= 2\pi(\dfrac{1}{2}y^2 - \dfrac{1}{3}y^3)\Big|_0^1 = 2\pi(\dfrac{1}{2} - \dfrac{1}{3}) - 0 = \dfrac{\pi}{3}$

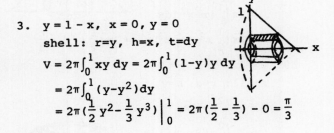

5. $y = x^2 + 1, \quad x = 0, \quad x = 3, \quad y = 0$

disk: $r = y, \quad t = dx$

$V = \pi\int_0^3 y^2 dx = \pi\int_0^3 (x^2 + 1)^2 dx = \pi\int_0^3 (x^4 + 2x^2 + 1)dx = \pi(\dfrac{1}{5}x^5 + \dfrac{2}{3}x^3 + x)\Big|_0^3$

$= \pi[\dfrac{1}{5}(3)^5 + \dfrac{2}{3}(3)^3 + 3] - \pi(0) = \dfrac{348\pi}{5}$

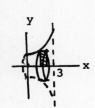

7. $y = x^3, \quad y = 8, \quad x = 0$

shell: $r = y, \quad h = x, \quad t = dy$

$V = 2\pi\int_0^8 xy\, dy = 2\pi\int_0^8 y^{1/3}y\, dy$

$= 2\pi\int_0^8 y^{4/3}dy = 2\pi(\dfrac{3}{7}y^{7/3})\Big|_0^8$

$= \dfrac{6\pi}{7}(8^{7/3} - 0) = \dfrac{768\pi}{7}$

9. $y = 3\sqrt{x}, \quad y = 0, \quad x = 4$

disk: $r = y, \quad t = dx$

$V = \pi\int_0^4 y^2 dx = \pi\int_0^4 (3\sqrt{x})^2 dx$

$= 9\pi\int_0^4 x\, dx = \dfrac{9\pi}{2}x^2\Big|_0^4$

$= \dfrac{9\pi}{2}(16) - 0 = 72\pi$

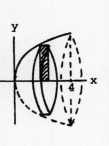

11. $x = 4y - y^2 - 3$, $x = 0$
shell: $r=y$, $h=x$, $t=dy$

$V = 2\pi \int_1^3 xy \, dy$

$= 2\pi \int_1^3 (4y - y^2 - 3)y \, dy$

$= 2\pi \int_1^3 (4y^2 - y^3 - 3y) \, dy$

$= 2\pi (\frac{4}{3}y^3 - \frac{1}{4}y^4 - \frac{3}{2}y^2) \Big|_1^3$

$= 2\pi(36 - \frac{81}{4} - \frac{27}{2}) - 2\pi(\frac{4}{3} - \frac{1}{4} - \frac{3}{2}) = \frac{16\pi}{3}$

13. $y = 1 - x$, $x=0$, $y=0$
disk: $r=x$, $t=dy$

$V = \pi \int_0^1 x^2 \, dy$

$= \pi \int_0^1 (1-y)^2 \, dy$

$= -\frac{\pi}{3}(1-y)^3 \Big|_0^1$

$= -\frac{\pi}{3}(0)^3 + \frac{\pi}{3}(1)^3 = \frac{\pi}{3}$

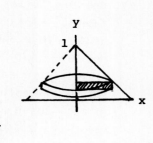

15. $y = 1 - x$, $x=0$, $y=0$
shell: $r=x$, $h=y$, $t=dx$

$V = 2\pi \int_0^1 xy \, dx$

$= 2\pi \int_0^1 x(1-x) \, dx$

$= 2\pi \int_0^1 (x - x^2) \, dx$

$= 2\pi(\frac{1}{2}x^2 - \frac{1}{3}x^3) \Big|_0^1 = 2\pi(\frac{1}{2} - \frac{1}{3}) = \frac{\pi}{3}$

17. $y = 2\sqrt{x}$, $x=0$, $y=2$
disk: $r=x$, $t=dy$

$V = \pi \int_0^2 x^2 \, dy$

$= \pi \int_0^2 \frac{y^4}{16} \, dy$

$= \frac{\pi}{80} y^5 \Big|_0^2 = \frac{\pi}{80}(32-0) = \frac{2\pi}{5}$

19. $y = 2x - x^2$, $y = 0$
shell: $r=x$, $h=y$, $t=dx$

$V = 2\pi \int_0^2 xy \, dx$

$= 2\pi \int_0^2 x(2x - x^2) \, dx$

$= 2\pi \int_0^2 (2x^2 - x^3) \, dx$

$= 2\pi(\frac{2}{3}x^3 - \frac{1}{4}x^4) \Big|_0^2$

$= 2\pi(\frac{16}{3} - 4) - 0 = \frac{8\pi}{3}$

21. $x = 4y - y^2 - 3$, $x=0$
disk: $r=x$, $t=dy$

$V = \pi \int_1^3 x^2 \, dy$

$= \pi \int_1^3 (4y - y^2 - 3)^2 \, dy$

$= \pi \int_1^3 (y^4 - 8y^3 + 22y^2 - 24y + 9) \, dy$

$= \pi(\frac{1}{5}y^5 - 2y^4 + \frac{22}{3}y^3 - 12y^2 + 9y) \Big|_1^3$

$= \pi(\frac{243}{5} - 162 + 198 - 108 + 27)$

$\quad - \pi(\frac{1}{5} - 2 + \frac{22}{3} - 12 + 9) = \frac{16\pi}{15}$

23. $y = \sqrt{4 - x^2}$ (quadrant I)
shell: $r=x$, $h=y$, $t=dx$

$V = 2\pi \int_0^2 xy \, dx$

$= 2\pi \int_0^2 x(4 - x^2)^{1/2} \, dx$

$= -\frac{2\pi}{3}(4 - x^2)^{3/2} \Big|_0^2$

$= -\frac{2\pi}{3}(0) + \frac{2\pi}{3}(4)^{3/2} = \frac{16\pi}{3}$

25. $y = 2x - x^2$, $y=0$
shell: $r=2-x$, $h=y$, $t=dx$

$V = 2\pi \int_0^2 (2-x)y \, dx$

$= 2\pi \int_0^2 (2-x)(2x - x^2) \, dx$

$= 2\pi \int_0^2 (4x - 4x^2 + x^3) \, dx = 2\pi(2x^2 - \frac{4}{3}x^3 + \frac{1}{4}x^4) \Big|_0^2$

$= 2\pi(8 - \frac{32}{3} + 4) - 0 = \frac{8\pi}{3}$

27. $y = \frac{r}{h}x$, $y=0$, $x=h$

disk: $r=y$, $t=dx$

$V = \pi \int_0^h y^2 dx = \pi \int_0^h (\frac{r}{h}x)^2 dx$

$= \frac{\pi r^2}{h^2} \int_0^h x^2 dx$

$= \frac{\pi r^2}{3h^2} x^3 \Big|_0^h = \frac{\pi r^2 h}{3}$

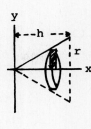

29. See Fig. 5-24, $y = x^4 + 1.5$

shell: $r=x$, $h=y$, $t=dx$

$V = 2\pi \int_0^{1.1} xy \, dx = 2\pi \int_0^{1.1} x(x^4 + 1.5) dx$

$= 2\pi \int_0^{1.1} (x^5 + 1.5x) dx$

$= 2\pi (\frac{x^6}{6} + 0.75x^2) \Big|_0^{1.1}$

$= 2\pi [\frac{1.1^6}{6} + 0.75(1.1)^2] - 0 = 7.56 \text{ mm}^3$

31. $x^2 + y^2 = 400$

disk: $r=x$, $t=dy$ $V = \pi \int_{-20}^{-5} x^2 dy = \pi \int_{-20}^{-5} (400 - y^2) dy = \pi (400y - \frac{1}{3}y^3) \Big|_{-20}^{-5}$

$= \pi (-2000 + \frac{125}{3} + 8000 - \frac{8000}{3}) = 3375\pi = 10,600 \text{ m}^3$

Exercises 5-4, p. 219

1. Origin is reference point.

$m_1 = 5$, $d_1 = 1$, $m_2 = 10$, $d_2 = 4$, $m_3 = 3$, $d_3 = 5$

$m_1 d_1 + m_2 d_2 + m_3 d_3 = (m_1 + m_2 + m_3)\bar{d}$

$(5)(1) + (10)(4) + (3)(5) = (5+10+3)\bar{d}$

$\bar{d} = \frac{60}{18} = \frac{10}{3}$; Center of mass $(\frac{10}{3}, 0)$

3. Origin is reference point. $m_1 = 4$, $d_1 = -3$,

$m_2 = 2$, $d_2 = 0$, $m_3 = 1$, $d_3 = 2$, $m_4 = 8$, $d_4 = 3$

$m_1 d_1 + m_2 d_2 + m_3 d_3 + m_4 d_4 = (m_1 + m_2 + m_3 + m_4)\bar{d}$

$4(-3) + 2(0) + 1(2) + 8(3) = (4+2+1+8)\bar{d}$

$\bar{d} = \frac{14}{15}$; Center of mass $(\frac{14}{15}, 0)$

5. See Fig. 5-37(a)

Area left rect. $= 4(2) = 8$

Center left rect. at $(-2, 0)$

Area rt. rect. $= 2(4) = 8$

Center rt. rect. at $(1, 1)$

$8(-2) + 8(1) = (8 + 8)\bar{x}$, $\bar{x} = -\frac{1}{2}$

$8(0) + 8(1) = (8 + 8)\bar{y}$, $\bar{y} = \frac{1}{2}$

Centroid at $(-\frac{1}{2}, \frac{1}{2})$

7. See Fig. 5-37(c)

Area left rect. $= 2(3) = 6$

Center left rect. $(-1, -\frac{1}{2})$

Area mid. rect. $= 2(1) = 2$

Center mid. rect. $(1, \frac{1}{2})$

Area rt. rect. $= 1(3) = 3$

Center rt. rect. $(\frac{5}{2}, \frac{3}{2})$

$6(-1) + 2(1) + 3(\frac{5}{2}) = (6+2+3)\bar{x}$, $\bar{x} = \frac{7/2}{11} = \frac{7}{22}$

$6(-\frac{1}{2}) + 2(\frac{1}{2}) + 3(\frac{3}{2}) = (6+2+3)\bar{y}$, $\bar{y} = \frac{5/2}{11} = \frac{5}{22}$

Centroid $(\frac{7}{22}, \frac{5}{22})$

9. $y = x^2$, $y = 2$

Area symmetric to y-axis, $\bar{x} = 0$

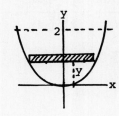

$\bar{y} = \frac{\int_0^2 y(2x \, dy)}{\int_0^2 2x \, dy} = \frac{\int_0^2 2y(y^{1/2}) dy}{\int_0^2 2y^{1/2} dy} = \frac{\int_0^2 2y^{3/2} dy}{\int_0^2 2y^{1/2} dy}$

$= \frac{2(\frac{2}{5})y^{5/2} \Big|_0^2}{2(\frac{2}{3})y^{3/2} \Big|_0^2} = \frac{\frac{4}{5}(2^{5/2})}{\frac{4}{3}(2^{3/2})} = \frac{4}{5}(\frac{3}{4})(2) = \frac{6}{5}$

Centroid $(0, \frac{6}{5})$

11. $y = 4 - x$, $x=0$, $y=0$

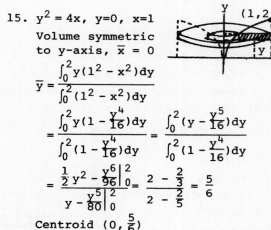

$$\bar{x} = \frac{\int_0^4 x(y\,dx)}{\int_0^4 y\,dx}$$

$$= \frac{\int_0^4 x(4-x)dx}{\int_0^4 (4-x)dx} = \frac{\int_0^4 (4x-x^2)dx}{\int_0^4 (4-x)dx}$$

$$= \frac{2x^2 - \frac{1}{3}x^3 \Big|_0^4}{4x - \frac{1}{2}x^2 \Big|_0^4} = \frac{32 - \frac{64}{3}}{16 - 8} = \frac{\frac{32}{3}}{8} = \frac{4}{3}$$

$$\bar{y} = \frac{\int_0^4 y(x\,dy)}{8} = \frac{\int_0^4 y(4-y)dy}{8} = \frac{4}{3}$$

(Integration as for $\bar{x}$)

Centroid $\left(\frac{4}{3}, \frac{4}{3}\right)$

13. $y = x^2$, $y = x^3$

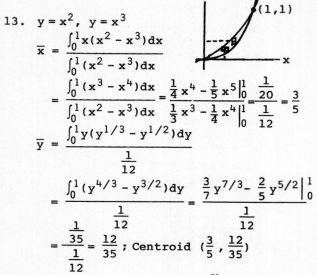

$$\bar{x} = \frac{\int_0^1 x(x^2 - x^3)dx}{\int_0^1 (x^2 - x^3)dx}$$

$$= \frac{\int_0^1 (x^3 - x^4)dx}{\int_0^1 (x^2 - x^3)dx} = \frac{\frac{1}{4}x^4 - \frac{1}{5}x^5\Big|_0^1}{\frac{1}{3}x^3 - \frac{1}{4}x^4\Big|_0^1} = \frac{\frac{1}{20}}{\frac{1}{12}} = \frac{3}{5}$$

$$\bar{y} = \frac{\int_0^1 y(y^{1/3} - y^{1/2})dy}{\frac{1}{12}}$$

$$= \frac{\int_0^1 (y^{4/3} - y^{3/2})dy}{\frac{1}{12}} = \frac{\frac{3}{7}y^{7/3} - \frac{2}{5}y^{5/2}\Big|_0^1}{\frac{1}{12}}$$

$$= \frac{\frac{1}{35}}{\frac{1}{12}} = \frac{12}{35} ; \text{ Centroid } \left(\frac{3}{5}, \frac{12}{35}\right)$$

15. $y^2 = 4x$, $y=0$, $x=1$

Volume symmetric to y-axis, $\bar{x} = 0$

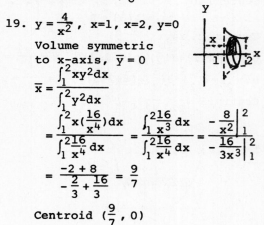

$$\bar{y} = \frac{\int_0^2 y(1^2 - x^2)dy}{\int_0^2 (1^2 - x^2)dy}$$

$$= \frac{\int_0^2 y(1 - \frac{y^4}{16})dy}{\int_0^2 (1 - \frac{y^4}{16})dy} = \frac{\int_0^2 (y - \frac{y^5}{16})dy}{\int_0^2 (1 - \frac{y^4}{16})dy}$$

$$= \frac{\frac{1}{2}y^2 - \frac{y^6}{96}\Big|_0^2}{y - \frac{y^5}{80}\Big|_0^2} = \frac{2 - \frac{2}{3}}{2 - \frac{2}{5}} = \frac{5}{6}$$

Centroid $\left(0, \frac{5}{6}\right)$

17. $y^2 = 4x$, $x=1$

Volume symmetric to x-axis, $\bar{y}=0$

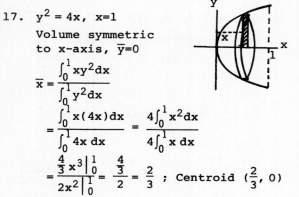

$$\bar{x} = \frac{\int_0^1 xy^2 dx}{\int_0^1 y^2 dx}$$

$$= \frac{\int_0^1 x(4x)dx}{\int_0^1 4x\,dx} = \frac{4\int_0^1 x^2 dx}{4\int_0^1 x\,dx}$$

$$= \frac{\frac{4}{3}x^3\Big|_0^1}{2x^2\Big|_0^1} = \frac{\frac{4}{3}}{2} = \frac{2}{3} ; \text{ Centroid } \left(\frac{2}{3}, 0\right)$$

19. $y = \frac{4}{x^2}$, $x=1$, $x=2$, $y=0$

Volume symmetric to x-axis, $\bar{y} = 0$

$$\bar{x} = \frac{\int_1^2 xy^2 dx}{\int_1^2 y^2 dx}$$

$$= \frac{\int_1^2 x(\frac{16}{x^4})dx}{\int_1^2 \frac{16}{x^4}dx} = \frac{\int_1^2 \frac{16}{x^3}dx}{\int_1^2 \frac{16}{x^4}dx} = \frac{-\frac{8}{x^2}\Big|_1^2}{-\frac{16}{3x^3}\Big|_1^2}$$

$$= \frac{-2 + 8}{-\frac{2}{3} + \frac{16}{3}} = \frac{9}{7}$$

Centroid $\left(\frac{9}{7}, 0\right)$

21. $y = \frac{b}{a}x$

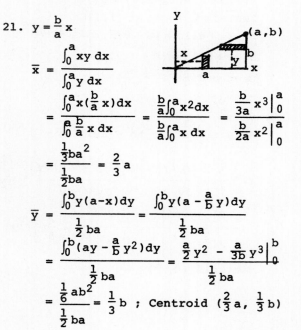

$$\bar{x} = \frac{\int_0^a xy\,dx}{\int_0^a y\,dx}$$

$$= \frac{\int_0^a x(\frac{b}{a}x)dx}{\int_0^a \frac{b}{a}x\,dx} = \frac{\frac{b}{a}\int_0^a x^2 dx}{\frac{b}{a}\int_0^a x\,dx} = \frac{\frac{b}{3a}x^3\Big|_0^a}{\frac{b}{2a}x^2\Big|_0^a}$$

$$= \frac{\frac{1}{3}ba^2}{\frac{1}{2}ba} = \frac{2}{3}a$$

$$\bar{y} = \frac{\int_0^b y(a-x)dy}{\frac{1}{2}ba} = \frac{\int_0^b y(a - \frac{a}{b}y)dy}{\frac{1}{2}ba}$$

$$= \frac{\int_0^b (ay - \frac{a}{b}y^2)dy}{\frac{1}{2}ba} = \frac{\frac{a}{2}y^2 - \frac{a}{3b}y^3\Big|_0^b}{\frac{1}{2}ba}$$

$$= \frac{\frac{1}{6}ab^2}{\frac{1}{2}ba} = \frac{1}{3}b ; \text{ Centroid } \left(\frac{2}{3}a, \frac{1}{3}b\right)$$

23.

$$\bar{x} = \frac{\int_{2.5}^{7.5} xy^2 dx}{\int_{2.5}^{7.5} y^2 dx} = \frac{\int_{2.5}^{7.5} x(\frac{4}{5}x)^2 dx}{\int_{2.5}^{7.5}(\frac{4}{5}x)^2 dx} = \frac{\frac{16}{25}\int_{2.5}^{7.5} x^3 dx}{\frac{16}{25}\int_{2.5}^{7.5} x^2 dx}$$

$$= \frac{\frac{1}{4}x^4\big|_{2.5}^{7.5}}{\frac{1}{3}x^3\big|_{2.5}^{7.5}} = \frac{\frac{1}{4}(7.5^4 - 2.5^4)}{\frac{1}{3}(7.5^3 - 2.5^3)} = \frac{781.25}{135.42} = 5.769$$

$\bar{y} = 0$

$m = \dfrac{6-2}{5} = \dfrac{4}{5}$; $y = \dfrac{4}{5}x$

For y=2, x=2.5
For y=6, x=7.5

$= 5.769$ in. $(7.500 - 5.769 = 1.731$ in.
from larger base)

Exercises 5-5, p. 224

1. Origin is reference point.
$m_1=5, d_1=2, m_2=3, d_2=6$
$I = m_1 d_1^2 + m_2 d_2^2$
$= 5(2)^2 + 3(6)^2 = 128$
$(m_1 + m_2)R^2 = I$
$(5 + 3)R^2 = 128$
$R^2 = 16,\ R = 4$

3. Origin is reference point.
$m_1=4, d_1=-4, m_2=10, d_2=0, m_3=6, d_3=5$
$I = m_1 d_1^2 + m_2 d_2^2 + m_3 d_3^2$
$= 4(-4)^2 + 10(0)^2 + 6(5)^2 = 214$
$(m_1 + m_2 + m_3)R^2 = I$
$(4 + 10 + 6)R^2 = 214,\ R^2 = 10.7,\ R = 3.27$

5. $y^2 = x$, x=4, y=0
(with k=1)
$I_x = \int_0^2 y^2(4-x)dy$
$= \int_0^2 y^2(4-y^2)dy$
$= \int_0^2 (4y^2 - y^4)dy$
$= \frac{4}{3}y^3 - \frac{1}{5}y^5\big|_0^2 = \frac{32}{3} - \frac{32}{5} = \frac{64}{15}$

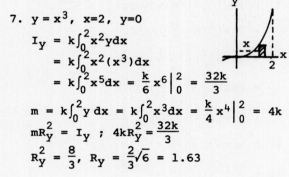

7. $y = x^3$, x=2, y=0
$I_y = k\int_0^2 x^2 y\,dx$
$= k\int_0^2 x^2(x^3)dx$
$= k\int_0^2 x^5 dx = \frac{k}{6}x^6\big|_0^2 = \frac{32k}{3}$
$m = k\int_0^2 y\,dx = k\int_0^2 x^3 dx = \frac{k}{4}x^4\big|_0^2 = 4k$
$mR_y^2 = I_y$; $4kR_y^2 = \frac{32k}{3}$
$R_y^2 = \frac{8}{3}$, $R_y = \frac{2}{3}\sqrt{6} = 1.63$

9. See Example C.
$I_x = \frac{ba^3}{12}$

From geometry
$m = \frac{1}{2}ab\ (k = 1)$
$I_x = \frac{1}{2}ab(\frac{1}{6}a^2) = m(\frac{1}{6}a^2)$
$= \frac{1}{6}ma^2$

11. $y = x^2$, x=2, y=0
$I_x = k\int_0^4 y^2(2-x)dy$
$= k\int_0^4 y^2(2-\sqrt{y})dy$
$= k\int_0^4 (2y^2 - y^{5/2})dy$
$= k(\frac{2}{3}y^3 - \frac{2}{7}y^{7/2})\big|_0^4 = k(\frac{128}{3} - \frac{256}{7}) = \frac{128k}{21}$
$m = k\int_0^4 (2 - x)dy = k\int_0^4 (2 - y^{1/2})dy$
$= 2y - \frac{2}{3}y^{3/2}\big|_0^4 = k(8 - \frac{16}{3}) = \frac{8k}{3}$
$mR_x^2 = I_x$; $\frac{8k}{3}R_x^2 = \frac{128k}{21}$; $R_x^2 = \frac{16}{7}$; $R_x = \frac{4\sqrt{7}}{7} = 1.51$

13. $y^2 = x^3$, $y=8$, $x=0$

$I_y = k\int_0^8 y^2(x\,dy)$

$= k\int_0^8 y^2(y^{2/3})dy$

$= k\int_0^8 y^{8/3}dy$

$= \frac{3k}{11}y^{11/3}\Big|_0^8 = \frac{6144k}{11}$

$m = k\int_0^8 x\,dy = k\int_0^8 y^{2/3}dy$

$= \frac{3k}{5}y^{5/3}\Big|_0^8 = \frac{96k}{5}$

$mR_y^2 = I_y$; $\frac{96k}{5}R_y^2 = \frac{6144k}{11}$

$R_y^2 = \frac{320}{11}$, $R_y = \frac{8}{11}\sqrt{55} = 5.39$

19. See Fig. 5-47

$y = \frac{r}{h}x$

$I_x = 2\pi k\int_0^r y^2[(h-x)y\,dy]$

$= 2\pi k\int_0^r (h - \frac{h}{r}y)y^3 dy$

$= 2\pi kh\int_0^r (y^3 - \frac{1}{r}y^4)dy$

$= 2\pi kh(\frac{1}{4}y^4 - \frac{1}{5r}y^5)\Big|_0^r$

$= 2\pi kh(\frac{r^4}{4} - \frac{r^4}{5}) = \frac{\pi khr^4}{10}$

From geometry: $m = k(\frac{1}{3}\pi r^2 h) = \frac{\pi kr^2 h}{3}$

$I_x = \frac{\pi kr^2 h}{3}(\frac{3r^2}{10}) = \frac{3mr^2}{10}$

15. $y^2 = x$, $y=2$, $x=0$

(with $k=1$)

$I_x = 2\pi\int_0^2 xy^3 dy$

$= 2\pi\int_0^2 y^2(y^3)dy$

$= 2\pi\int_0^2 y^5 dy = \frac{\pi}{3}y^6\Big|_0^2 = \frac{64\pi}{3}$

17. $y = 2x - x^2$, $y=0$

$I_y = 2\pi k\int_0^2 yx^3 dx$

$= 2\pi k\int_0^2 (2x-x^2)x^3 dx$

$= 2\pi k\int_0^2 (2x^4-x^5)dx$

$= 2\pi k(\frac{2}{5}x^5 - \frac{1}{6}x^6)\Big|_0^2$

$= 2\pi k(\frac{64}{5} - \frac{64}{6}) = \frac{64\pi k}{15}$

$m = 2\pi k\int_0^2 yx\,dx = 2\pi k\int_0^2 (2x - x^2)x\,dx$

$= 2\pi k\int_0^2 (2x^2 - x^3)dx = 2\pi k(\frac{2}{3}x^3 - \frac{1}{4}x^4)\Big|_0^2$

$= 2\pi k(\frac{16}{3} - 4) = \frac{8\pi k}{3}$

$mR_y^2 = I_y$, $\frac{8\pi k}{3}R_y^2 = \frac{64\pi k}{15}$, $R_y^2 = \frac{8}{5}$

$R_y = \frac{2}{5}\sqrt{10} = 1.26$

21. See Exercise 19.

$I = \frac{3}{10}mr^2$; $r = 0.6$ cm, $h = 0.8$ cm, $m = 3$ g

$I = \frac{3}{10}(3)(0.6)^2 = 0.324$ g·cm^2

(Value of h not needed)

Exercises 5-6, p. 228

1. 12 lb stretch spring 4 in.

$F = kx$, $F = 12$ lb, $x = 4$ in.

$12 = 4k$, $k = 3$ lb/in.; $F = 3x$

To stretch spring 5 in.

$W = \int_0^5 3x\,dx = \frac{3}{2}x^2\Big|_0^5$

$= 37.5$ lb·in.

5. $q_1 = 2$ μC $= 2\times10^{-6}$ C

$q_2 = -4$ μC $= -4\times10^{-6}$ C

$k = 9.0\times10^9$ N·m^2/C^2

2 cm $= 0.02$ m, 6 cm $= 0.06$ m

3. 500 lb stretch spring 1 ft

$F = kx$, $F = 500$ lb, $x = 1$ ft

$500 = k(1)$, $k = 500$ lb/ft

$F = 500x$, $F = 300$ lb

$300 = 500x$, $x = 0.6$ ft

$W = \int_0^{0.6} 500x\,dx = 250x^2\Big|_0^{0.6} = 90.0$ ft·lb

$W = \int_{0.02}^{0.06} \frac{9.0\times10^9(2\times10^{-6})(-4\times10^{-6})}{x}\,dx$

$= -0.072\int_{0.02}^{0.06}\frac{dx}{x} = \frac{0.072}{x}\Big|_{0.02}^{0.06}$

$= \frac{0.072}{0.06} - \frac{0.072}{0.02} = 1.20 - 3.60 = -2.40$ J

(Negative shows work done by system)

7. $F = \dfrac{k}{x^2}$, 1 cm = 0.01 m

$$W = \int_{0.01}^{1} \frac{k}{x^2} dx = -\frac{k}{x}\bigg|_{0.01}^{1}$$

$$= -\frac{k}{1} + \frac{k}{0.01} = 99k \text{ J}$$

11. See Example C

Let x = length of rope wound up

Weight/length = 6.00 N/m

$F = 6.00(25 - x)$

$$W = \int_{0}^{20} 6.00(25 - x)dx$$

$$= 150x - 3.00x^2\bigg|_{0}^{20}$$

$$= 3000 - 1200 = 1800 \text{ N·m}$$

15. Force = Weight of element
   $= (62.4)(\pi)(3^2)(dh)$

Distance = $10 - h$

$$W = \int_{0}^{10} (10-h)(62.4)(9\pi dh)$$

$$= 9\pi(62.4)\left(10h - \frac{1}{2}h^2\right)\bigg|_{0}^{10}$$

$$= 9\pi(62.4)(100 - 50)$$

$$= 8.82\times10^4 \text{ ft·lb}$$

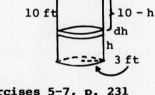

9. See Example C.

Let x = length of cable wound up

Total weight = 100 lb, total length = 200 ft

weight/length = 0.500 lb/ft, $F = 0.5(200 - x)$

$$W = \int_{0}^{200} 0.500(200 - x)dx = 100x - 0.25x^2\bigg|_{0}^{200}$$

$$= 20,000 - 10,000 = 10,000 \text{ ft·lb}$$

13. Let x = distance ascended

Weight loss $= \dfrac{1.25 \text{ ton}}{1000 \text{ ft}} = 1.25\times10^{-3}$ ton/ft

Weight $= 32.5 - 1.25\times10^{-3}x$

$$W = \int_{0}^{12000} (32.5 - 1.25\times10^{-3}x)dx$$

$$= 32.5x - 0.625\times10^{-3}x^2\bigg|_{0}^{12000}$$

$$= 3.90\times10^5 - 0.90\times10^5 = 3.00\times10^5 \text{ ft·ton}$$

17. See Example D

Force = Weight of element = $9800\pi r^2 dh$

Distance = $6 - h$; Limits: 0 m to 4 m

$$W = \int_{0}^{4} (9800\pi r^2 dh)(6 - h)$$

$$= 9800\pi\int_{0}^{4} (0.8h)^2(6-h)dh = 6270\pi\int_{0}^{4} (6h^2 - h^3)dh$$

$$= 6270\pi\left(2h^3 - \frac{1}{4}h^4\right)\bigg|_{0}^{4} = 6270\pi(128 - 64)$$

$$= 1.26\times10^6 \text{ N·m} = 1.26 \text{ MJ}$$

Exercises 5-7, p. 231

All solutions use Eq.(5-27), $F = w\int_{a}^{b} \ell h\, dh$.

1. $w = 62.4$ lb/ft^3

$\ell = 12$ ft

$$F = 62.4\int_{0}^{2.5} 12h\, dh$$

$$= 62.4(6h^2)\bigg|_{0}^{2.5}$$

$$= 2340 \text{ lb}$$

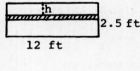

3. $w = 9800$ N/m^3

$\ell = 2$ m

$$F = 9800\int_{4}^{6} 2h\, dh$$

$$= 9800h^2\bigg|_{4}^{6}$$

$$= 9800(36-16) = 196,000 \text{ N} = 196 \text{ kN}$$

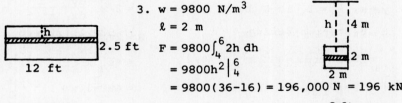

5. $w = 64.0$ lb/ft^3

$\ell = 10$ ft

$$F = 64.0\int_{4}^{9} 10h\, dh$$

$$= 64.0(5h^2)\bigg|_{4}^{9}$$

$$= 320(81-16) = 20,800 \text{ lb}$$

7. $w = 62.4$ lb/ft^3

$\ell = x$, $h = y$

$$F = 62.4\int_{0}^{8} xy\, dy$$

$$= 62.4\int_{0}^{8} (8-y)y\, dy$$

$$= 62.4\int_{0}^{8} (8y - y^2)dy = 62.4\left(4y^2 - \frac{1}{3}y^3\right)\bigg|_{0}^{8}$$

$$= 62.4\left(256 - \frac{512}{3}\right) = 5320 \text{ lb}$$

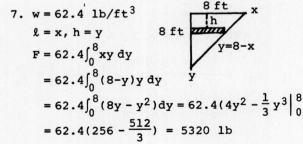

9. See Fig. 5-56

$w = 9800$ N/m^3

$\ell = x$, $h = y$

$F = 9800\int_3^4 xy\,dy$

$= 9800\int_3^4 (2y - 6)y\,dy$

$= 19600\int_3^4 (y^2 - 3y)dy = 19600\left(\frac{1}{3}y^3 - \frac{3}{2}y^2\right)\Big|_3^4$

$= 19600[\frac{64}{3} - 24 - (9 - \frac{27}{2})]$

$= 35{,}900$ N $= 35.9$ kN

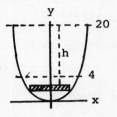

11. $x^2 + y^2 = 16$

$w = 60.0$ lb/ft^3

$\ell = 2x$, $h = y$

$F = 60.0\int_0^4 (2x)y\,dy$

$= 120\int_0^4 (16 - y^2)^{1/2}y\,dy$

$= -\frac{120}{3}(16 - y^2)^{3/2}\Big|_0^4$

$= \frac{120}{3}(64) = 2560$ lb

13. $y = x^2$, $y = 20$

$w = 62.4$ lb/ft^3

$\ell = 2x$, $h = 20-y$

$dh = -dy$

Top element:
  $h=16$, $y=4$
Bottom element:
  $h=20$, $y=0$

$F = 62.4\int_4^0 (2x)(20-y)(-dy)$

$= -62.4\int_4^0 (2y^{1/2})(20-y)dy$

$= -124.8\int_4^0 (20y^{1/2}-y^{3/2})dy$

$= -124.8\left(\frac{40}{3}y^{3/2} - \frac{2}{5}y^{5/2}\right)\Big|_4^0$

$= 124.8(93.87) = 11{,}700$ lb

Exercises 5-8, p. 234

1. $s = \int_a^b \sqrt{1 + \left(\frac{dy}{dx}\right)^2}\,dx$, $a=3$, $b=8$

$y = \frac{2}{3}x^{3/2}, \frac{dy}{dx} = x^{1/2}$

$s = \int_3^8 [1 + (x^{1/2})^2]^{1/2}dx$

$= \int_3^8 (1 + x)^{1/2}dx = \frac{2}{3}(1 + x)^{3/2}\Big|_3^8$

$= \frac{2}{3}(27 - 8) = \frac{38}{3}$

3. $s = \int_a^b \sqrt{1 + \left(\frac{dy}{dx}\right)^2}\,dx$, $a=0$, $b=3$

$y = \frac{2}{3}(x^2 + 1)^{3/2}$, $\frac{dy}{dx} = 2x(x^2 + 1)^{1/2}$

$s = \int_0^3 [1 + (2x(x^2+1)^{1/2})^2]^{1/2}dx$

$= \int_0^3 (1 + 4x^4 + 4x^2)^{1/2}dx = \int_0^3 [(1+2x^2)^2]^{1/2}dx$

$= \int_0^3 (1 + 2x^2)dx = x + \frac{2}{3}x^3\Big|_0^3 = 3 + 18 = 21$

5. $S = 2\pi\int_a^b y\sqrt{1 + \left(\frac{dy}{dx}\right)^2}\,dx$

$y = x$, $\frac{dy}{dx} = 1$, $a = 0$, $b = 2$

$S = 2\pi\int_0^2 x(1 + 1^2)^{1/2}dx$

$= 2\pi\sqrt{2}\int_0^2 x\,dx = \pi\sqrt{2}\,x^2\Big|_0^2 = 4\pi\sqrt{2}$

7. $S = 2\pi\int_a^b y\sqrt{1 + \left(\frac{dy}{dx}\right)^2}\,dx$

$y = 2\sqrt{x}$, $\frac{dy}{dx} = \frac{1}{x^{1/2}}$, $a = 0$, $b = 3$

$S = 2\pi\int_0^3 (2x^{1/2})[1 + (\frac{1}{x^{1/2}})^2]^{1/2}dx$

$= 4\pi\int_0^3 x^{1/2}(1 + \frac{1}{x})^{1/2}dx = 4\pi\int_0^3 (x + 1)^{1/2}dx$

$= 4\pi(\frac{2}{3})(x + 1)^{3/2}\Big|_0^3 = \frac{8\pi}{3}(8 - 1) = \frac{56\pi}{3}$

9. $i_{av} = \frac{\int_a^b i\,dt}{b - a}$

$i = 4t - t^2$, $a = 0$, $b = 4$

$i_{av} = \frac{\int_0^4 (4t - t^2)dt}{4 - 0} = \frac{1}{4}(2t^2 - \frac{1}{3}t^3)\Big|_0^4$

$= \frac{1}{4}(32 - \frac{64}{3}) = \frac{8}{3} = 2.67$ A

11. $v_{av} = \frac{\int_a^b v\,dt}{b - a}$, $x = t^3 - 3t$, $v = 3t^2 - 3$

$a = 1$, $b = 3$

$v_{av} = \frac{\int_1^3 (3t^2 - 3)dt}{3 - 1} = \frac{1}{2}(t^3 - 3t)\Big|_1^3$

$= \frac{1}{2}(27 - 9 - 1 + 3) = 10$

Review Exercises for Chapter 5, p. 235

1. $a = -32$ ft/s^2, $t = 5$ s
   $v = 0$ when $t = 0$ (dropped)
   $v = \int -32\,dt = -32t + C$
   $0 = -32(0) + C$, $C = 0$
   $v = -32t$
   $\dot{v}\big|_{t=5} = -32(5) = -160$ ft/s

3. $a = -32$ ft/s^2; $v = 20$ ft/s, $s = 200$ ft when $t = 0$
   $v = \int -32\,dt = -32t + C_1$; $20 = -32(0) + C_1$, $C_1 = 20$
   $v = -32t + 20$
   $s = \int v\,dt = \int (-32t + 20)\,dt = -16t^2 + 20t + C_2$
   $200 = -16(0) + 20(0) + C_2$, $C_2 = 200$
   $s = -16t^2 + 20t + 200$
   Hits ground for $t = 0$: $16t^2 - 20t - 200 = 0$
   $4t^2 - 5t - 50 = 0$, $t = \dfrac{5 \pm \sqrt{25 + 4(4)(50)}}{8} = -2.97, 4.22$
   $t = 4.22$ s $(t > 0)$

5. $i = 5t^{2/3}$, $q = 0$ when $t = 0$
   $q = \int i\,dt = \int 5t^{2/3}\,dt$
   $= 5(\tfrac{3}{5})t^{5/3} + C_1 = 3t^{5/3} + C_1$
   $0 = 0 + C_1$, $C_1 = 0$; $q = 3t^{5/3}$
   For $t = 2$: $q_2 = 3(2^{5/3})$ C
   For $t = 3$: $q_3 = 3(3^{5/3})$ C
   $q_3 - q_2 = 3(3^{5/3} - 2^{5/3})$
   $\quad\quad = 9.20$ C

7. $C = 55.0$ nF $= 55.0 \times 10^{-9}$ F ; $i = 10$ mA $= 0.010$ A
   $V = 0$ when $t = 0$
   $V = \dfrac{1}{C}\int i\,dt = \dfrac{1}{55.0 \times 10^{-9}}\int 0.010\,dt = \dfrac{10^7}{55.0}\int dt = \dfrac{10^7 t}{55.0} + C_1$
   $0 = 0 + C_1$, $C_1 = 0$; $V = \dfrac{10^7 t}{55.0}$
   $V\big|_{t=0.020} = \dfrac{10^7(0.020)}{55.0} = 3640$ V

9. $\dfrac{dy}{dx} = 20 + \dfrac{1}{40}x^2$, $y = 0$ for $x = 0$
   $y = \int (20 + \dfrac{1}{40}x^2)\,dx = 20x + \dfrac{1}{120}x^3 + C$
   $0 = 0 + 0 + C$, $C = 0$
   $y = 20x + \dfrac{1}{120}x^3$

11. $y = \sqrt{1 - x}$, $x = 0$, $y = 0$
    $A = \int_0^1 y\,dx = \int_0^1 (1-x)^{1/2}\,dx$
    $= -\dfrac{2}{3}(1 - x)^{3/2}\Big|_0^1 = \dfrac{2}{3}$

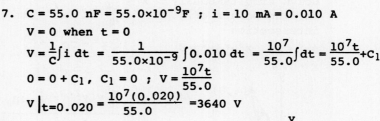

13. $y^2 = 2x$, $y = x - 4$
    Solve simultaneously to find pts. of intersection
    $y^2 = 2(y+4)$, $y^2 - 2y - 8 = 0$
    $(y-4)(y+2) = 0$, $y = 4, -2$
    $A = \int_{-2}^4 (x_2 - x_1)\,dy = \int_{-2}^4 (y + 4 - \tfrac{1}{2}y^2)\,dy$
    $= \dfrac{1}{2}y^2 + 4y - \dfrac{1}{6}y^3\Big|_{-2}^4$
    $= 8 + 16 - \dfrac{32}{3} - (2 - 8 + \dfrac{4}{3}) = 18$

15. $y = x^2$, $y = x^3 - 2x^2$
    Solve simultaneously to find points of intersection
    $x^3 - 2x^2 = x^2$
    $x^3 - 3x^2 = 0$, $x^2(x-3) = 0$, $x = 0, 3$
    $A = \int_0^3 (y_2 - y_1)\,dx = \int_0^3 [x^2 - (x^3 - 2x^2)]\,dx$
    $= \int_0^3 (3x^2 - x^3)\,dx = x^3 - \dfrac{1}{4}x^4\Big|_0^3$
    $= 27 - \dfrac{81}{4} = \dfrac{27}{4}$

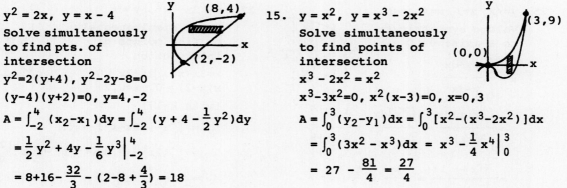

17. $y = 3 + x^2$, $y = 4$
    disks: $r_1 = y$, $r_2 = 4$, $t = dx$

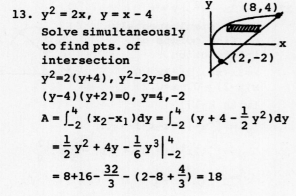

    Volume symmetric to y-axis
    $V = 2\pi\int_0^1 (4^2 - y^2)\,dx = 2\pi\int_0^1 [16 - (3 + x^2)^2]\,dx$
    $= 2\pi\int_0^1 (7 - 6x^2 - x^4)\,dx$
    $= 2\pi(7x - 2x^3 - \dfrac{1}{5}x^5)\Big|_0^1 = 2\pi(7 - 2 - \dfrac{1}{5}) = \dfrac{48\pi}{5}$

19. $y = 3x^2 - x^3$, $y=0$
shell: $r=x$, $h=y$, $t=dx$
$V = 2\pi \int_0^3 xy\,dx$
$\quad = 2\pi \int_0^3 x(3x^2 - x^3)dx$
$\quad = 2\pi \int_0^3 (3x^3 - x^4)dx = 2\pi(\frac{3}{4}x^4 - \frac{1}{5}x^5)\Big|_0^3$
$\quad = 2\pi(\frac{243}{4} - \frac{243}{5}) = \frac{243\pi}{10}$

21. $\frac{x^2}{a^2} + \frac{y^2}{b^2} = 1$
disk: $r=y$, $t=dx$
$\qquad\downarrow$ Symmetry
$V = 2\pi \int_0^a y^2 dx$
$\quad = 2\pi \int_0^a \frac{b^2}{a^2}(a^2 - x^2)dx$
$\quad = \frac{2\pi b^2}{a^2}(a^2 x - \frac{1}{3}x^3)\Big|_0^a = \frac{2\pi b^2}{a^2}(\frac{2a^3}{3}) = \frac{4\pi ab^2}{3}$

23. $y^2 = x^3$, $y = 2x$
Solve simultaneously to find points of intersection
$4x^2 = x^3$, $x^3 - 4x^2 = 0$
$x = 0, 4$

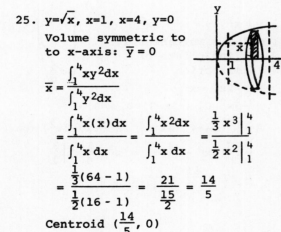

Centroid $(\frac{40}{21}, \frac{10}{3})$

$\overline{x} = \dfrac{\int_0^4 x(2x - x^{3/2})dx}{\int_0^4 (2x - x^{3/2})dx} = \dfrac{\int_0^4 (2x^2 - x^{5/2})dx}{\int_0^4 (2x - x^{3/2})dx}$

$= \dfrac{\frac{2}{3}x^3 - \frac{2}{7}x^{7/2}\Big|_0^4}{x^2 - \frac{2}{5}x^{5/2}\Big|_0^4} = \dfrac{\frac{128}{3} - \frac{256}{7}}{16 - \frac{64}{5}} = \frac{128}{21}\cdot\frac{5}{16} = \frac{40}{21}$

$\overline{y} = \dfrac{\int_0^8 y(y^{2/3} - \frac{1}{2}y)dy}{\frac{16}{5}} = \dfrac{\int_0^8 (y^{5/3} - \frac{1}{2}y^2)dy}{\frac{16}{5}}$

$= \dfrac{\frac{3}{8}y^{8/3} - \frac{1}{6}y^3\Big|_0^8}{\frac{16}{5}} = \dfrac{96 - \frac{256}{3}}{\frac{16}{5}} = \frac{32}{3}\cdot\frac{5}{16} = \frac{10}{3}$

25. $y=\sqrt{x}$, $x=1$, $x=4$, $y=0$
Volume symmetric to to x-axis: $\overline{y} = 0$

$\overline{x} = \dfrac{\int_1^4 xy^2 dx}{\int_1^4 y^2 dx}$

$= \dfrac{\int_1^4 x(x)dx}{\int_1^4 x\,dx} = \dfrac{\int_1^4 x^2 dx}{\int_1^4 x\,dx} = \dfrac{\frac{1}{3}x^3\Big|_1^4}{\frac{1}{2}x^2\Big|_1^4}$

$= \dfrac{\frac{1}{3}(64 - 1)}{\frac{1}{2}(16 - 1)} = \dfrac{21}{\frac{15}{2}} = \frac{14}{5}$

Centroid $(\frac{14}{5}, 0)$

27. $y = 3x - x^2$, $y = x$
Solve simultaneously to find points of intersection
$x = 3x - x^2$, $x^2 - 2x = 0$
$x(x-2) = 0$, $x = 0, 2$
(with k=1)
$I_y = \int_0^2 x^2[(3x - x^2) - x]dx$
$\quad = \int_0^2 (2x^3 - x^4)dx = \frac{1}{2}x^4 - \frac{1}{5}x^5\Big|_0^2$
$\quad = 8 - \frac{32}{5} = \frac{8}{5}$

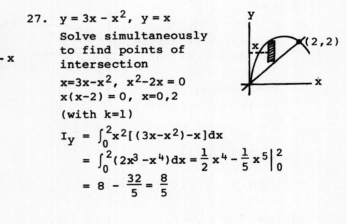

29. $y = x^{1/2}$, $y=0$, $x=8$
(with k=1)

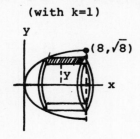

$I_x = 2\pi \int_0^{\sqrt{8}} (8 - x)y^3 dy = 2\pi \int_0^{\sqrt{8}} (8 - y^2)y^3 dy$
$\quad = 2\pi \int_0^{\sqrt{8}} (8y^3 - y^5)dy = 2\pi(2y^4 - \frac{1}{6}y^6)\Big|_0^{\sqrt{8}}$
$\quad = 2\pi(128 - \frac{256}{3}) = \frac{256\pi}{3}$

31. $\dfrac{dC}{dx} = 2 + 0.004x$, C=2500 when x=0

$C = \int (2 + 0.004x)dx$

$= 2x + 0.002x^2 + C$

$2500 = 0 + 0 + C$, $C = 2500$

$C = 2x + 0.002x^2 + 2500$

$C\big|_{x=300} = 600 + 180 + 2500 = \$3280$

33. See Fig. 5-60

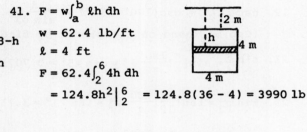

(8,-10)

$x^2 = 4py$

$8^2 = 4p(-10)$

$4p = -6.4$

$x^2 = -6.4y$

$A = \int_{-10}^{0} 2x\,dy = 2\int_{-10}^{0}\sqrt{-6.4y}\,dy$

$= 2\sqrt{6.4}\int_{-10}^{0}(-y)^{1/2}dy = -2\sqrt{6.4}(\tfrac{2}{3})(-y)^{3/2}\big|_{-10}^{0}$

$= 0 + \tfrac{4}{3}\sqrt{6.4}(10)^{3/2} = 107 \text{ ft}^2$

35. 8 lb stretch spring 4 in.
(use differences)

$F = kx$, $F = 8$ lb, $x = 4$ in.

$8 = 4k$, $k = 2$ lb/in., $F = 2x$

For F=6 lb, x = 3 in.

$W = \int_{0}^{3} 2x\,dx = x^2\big|_{0}^{3} = 9.0$ ft·lb

37. $F = \dfrac{10^{11}}{x^2}$; $W = \int_{3960}^{5960}\dfrac{10^{11}}{x^2}dx = -\dfrac{10^{11}}{x}\bigg|_{3960}^{5960}$

$= 10^{11}(-\dfrac{1}{5960} + \dfrac{1}{3960}) = 8.47\times10^6$ mi·lb

39. $w = 9800$ N/m^3, r = 2 m

Force = weight of element
$= 9800\pi r^2 dh$

$W = 9800\int_{0}^{3} 4\pi(3 - h)dh$

$= 39200\pi\int_{0}^{3}(3 - h)dh$

$= 39200\pi(3h - \tfrac{1}{2}h^2)\big|_{0}^{3}$

$= 39200\pi(9 - \tfrac{9}{2}) = 5.54\times10^5$ J $= 554$ kJ

41. $F = w\int_{a}^{b}\ell h\,dh$

$w = 62.4$ lb/ft

$\ell = 4$ ft

$F = 62.4\int_{2}^{6} 4h\,dh$

$= 124.8h^2\big|_{2}^{6} = 124.8(36 - 4) = 3990$ lb

43. $s = \int_{a}^{b}\sqrt{1 + (y')^2}\,dx$, a=1, b=3

$y = \tfrac{2}{3}(x^2 - 1)^{3/2}$, $y' = 2x(x^2 - 1)^{1/2}$

$s = \int_{1}^{3}[1 + (2x(x^2-1)^{1/2})^2]^{1/2}dx$

$= \int_{1}^{3}(1 + 4x^4 - 4x^2)^{1/2}dx$

$= \int_{1}^{3}[(2x^2-1)^2]^{1/2}dx = \int_{1}^{3}(2x^2-1)dx$

$= \tfrac{2}{3}x^3 - x\big|_{1}^{3} = 18 - 3 - \tfrac{2}{3} + 1 = \dfrac{46}{3}$

45. $R = \dfrac{k}{r^2}$, R = 0.3 Ω for r = 2.0 mm

$0.3 = \dfrac{k}{2.0^2}$, $k = 1.2$ Ω·mm^2, $R = \dfrac{1.2}{r^2}$

$R_{av} = \dfrac{\int_{2.0}^{2.1}\frac{1.2}{r^2}dr}{2.1 - 2.0} = \dfrac{1.2}{0.1}(-\dfrac{1}{r})\bigg|_{2.0}^{2.1}$

$= 12(-\dfrac{1}{2.1} + \dfrac{1}{2.0}) = 0.286$ Ω

47.

Time	12 M	4 AM	8 AM	12 N	4 PM	8 PM	12 M
Production rate	2000	2000	7000	12000	12000	8000	2000

$P = 4[\tfrac{1}{2}(2000) + 2000 + 7000 + 12000 + 12000 + 8000 + \tfrac{1}{2}(2000)]$

$= 172,000$ lb

Exercises 6-1, p. 245

1. $15°12' = 15\frac{12°}{60} = 15.2°$ 3. $0.8° = 0.8(60') = 48'$ 5. $15° = 15(\frac{\pi}{180}) = \frac{\pi}{12}$

  $315.8° = 315°48'$

  $150° = 150(\frac{\pi}{180}) = \frac{5\pi}{6}$

7. $75° = 75(\frac{\pi}{180}) = \frac{5\pi}{12}$ 9. $\frac{2\pi}{5} = \frac{2\pi}{5}(\frac{180°}{\pi}) = 72°$ 11. $\frac{\pi}{18} = \frac{\pi}{18}(\frac{180°}{\pi}) = 10°$

  $330° = 330(\frac{\pi}{180}) = \frac{11\pi}{6}$ $\frac{3\pi}{2} = \frac{3\pi}{2}(\frac{180°}{\pi}) = 270°$ $\frac{7\pi}{4} = \frac{7\pi}{4}(\frac{180°}{\pi}) = 315°$

13. $23.0° = 23.0(\frac{\pi}{180}) = 0.401$ 15. $0.750 = 0.750(\frac{180°}{\pi}) = 43.0°$

17. $\cos 214° = -\cos(214°-180°) = -\cos 34° = -0.8290$ (Can be found directly on calculator.)

19. $\csc 137° = \csc(180°-137°) = \frac{1}{\sin 43°} = 1.4663$

  (Can be found on calculator by finding reciprocal of $\sin 137°$.)

21. $\sin\frac{\pi}{4} = \sin[\frac{\pi}{4}(\frac{180°}{\pi})] = \sin 45° = 0.7071$ (Can be found directly on calculator

  in radian mode.)

23. $\tan\frac{5\pi}{12} = \tan[\frac{5\pi}{12}(\frac{180°}{\pi})] = \tan 75° = 3.732$ (Can be found directly on calculator

  in radian mode.)

25. $\cot\frac{5\pi}{6} = \cot[\frac{5\pi}{6}(\frac{180°}{\pi})] = \cot 150° = -\cot(180°-150°) = -\cot 30° = -\frac{1}{\tan 30°} = -1.732$

  (Can be found on calculator using radian mode and reciprocal.)

27. $\cos 4.596 = -\cos(4.596 - \pi) = -\cos 1.4544 = -0.1161$ (Can be found directly on calcu-

  lator in radian mode.)

29. $\sin\frac{3}{4}\pi = \sin(\pi - \frac{3}{4}\pi) = \sin\frac{\pi}{4} = \frac{1}{2}\sqrt{2}$ 31. $\cos\frac{5}{3}\pi = \cos(2\pi - \frac{5}{3}\pi) = \cos\frac{\pi}{3} = \frac{1}{2}$

33. $\sin\theta = 0.3420$ 35. $\tan\theta = -1.192$

  $\theta = 0.3490$ (radians) (Quad. 1)   Using $\tan\theta = 1.192$, $\theta_{ref} = 0.8728$ (rad)

  $\theta = \pi - 0.3490 = 2.7925$ (Quad. 2) $\theta = \pi - \theta_{ref} = 2.2689$ (Quad. 2)

  $\theta = 2\pi - \theta_{ref} = 5.4104$ (Quad. 4)

37. $y = 3\sin x$ 39. $y = -4\sin 3x$ 41. $y = 2\sin(3x - \frac{\pi}{2})$ 43. $y = \cos(\pi x - \frac{\pi}{2})$

  $a=3, b=1, c=0$ $a=-4, b=3, c=0$ $a=2, b=3, c=-\frac{\pi}{2}$ $a=1, b=\pi, c=-\frac{\pi}{2}$

  ampl.: $|3| = 3$ ampl.: $|-4| = 4$ ampl.: $|2| = 2$ ampl.: $|1| = 1$

  per.: $\frac{2\pi}{1} = 2\pi$ per.: $\frac{2\pi}{3}$ per.: $\frac{2\pi}{3}$ per.: $\frac{2\pi}{\pi} = 2$

  disp.: $-\frac{0}{1} = 0$ disp.: $-\frac{0}{3} = 0$ disp.: $-\frac{-\pi/2}{3} = \frac{\pi}{6}$ disp.: $-\frac{-\pi/2}{\pi} = \frac{1}{2}$

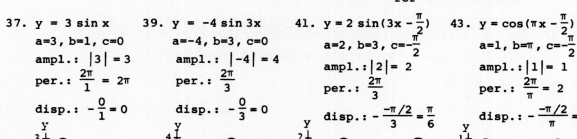

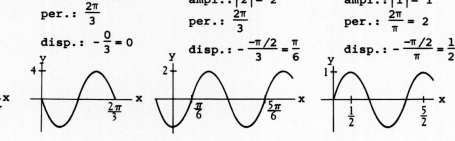

45. $i = 10 \sin 120\pi t$

a=10, b=120π, c=0

ampl.: 10

per.: $\dfrac{2\pi}{120\pi} = \dfrac{1}{60}$

disp.: 0

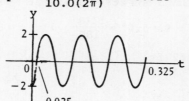

47. $y = A \sin 2\pi\left(\dfrac{t}{T} - \dfrac{x}{\lambda}\right)$

A = 2.00 cm, T = 0.100 s

λ = 20.0 cm, x = 5.00 cm

$y = 2.00 \sin 2\pi(10.0t - 0.250)$

ampl.: 2.00 ; per.: $\dfrac{2\pi}{2\pi(10.0)}$ =0.100

disp.: $-\dfrac{0.250(2\pi)}{10.0(2\pi)} = 0.025$

49. $y = 4 \sin 2t - 2 \cos 2t$

$y_1 = 4 \sin 2t$

amp. 4; per. π

disp. 0

$y_2 = -2 \cos 2t$

ampl.2;per.π;disp.0

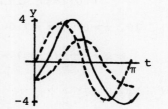

Exercises 6-2, p. 253

1. $\dfrac{\cot x}{\cos x} = \dfrac{\cos x/\sin x}{\cos x} = \dfrac{\cos x}{\sin x}\dfrac{1}{\cos x} = \dfrac{1}{\sin x} = \csc x$

3. $\tan y \csc y = \dfrac{\sin y}{\cos y}\dfrac{1}{\sin y} = \dfrac{1}{\cos y}$
$= \sec y$

5. $\cos^2 x - \sin^2 x = (1 - \sin^2 x) - \sin^2 x = 1 - 2 \sin^2 x$

7. $\sin x \tan x + \cos x = \sin x \dfrac{\sin x}{\cos x} + \cos x = \dfrac{\sin^2 x + \cos^2 x}{\cos x} = \dfrac{1}{\cos x} = \sec x$

9. $\tan x(\tan x + \cot x) = \tan^2 x + \tan x \cot x = \tan^2 x + 1 = \sec^2 x$

11. $\dfrac{\sec \theta}{\cos \theta} - \dfrac{\tan \theta}{\cot \theta} = \dfrac{\sec \theta}{1/\sec \theta} - \dfrac{\tan \theta}{1/\tan \theta} = \sec^2\theta - \tan^2\theta = 1$

13. $\sin(x+y)\sin(x-y) = (\sin x \cos y + \cos x \sin y)(\sin x \cos y - \cos x \sin y)$
$$= \sin^2 x \cos^2 y - \cos^2 x \sin^2 y = \sin^2 x(1 - \sin^2 y) - (1 - \sin^2 x)\sin^2 y$$
$$= \sin^2 x - \sin^2 x \sin^2 y - \sin^2 y + \sin^2 x \sin^2 y = \sin^2 x - \sin^2 y$$

15. $\cos(x-y) + \sin(x+y) = \cos x \cos y + \sin x \sin y + \sin x \cos y + \cos x \sin y$
$$= \cos x(\cos y + \sin y) + \sin x(\sin y + \cos y)$$
$$= (\cos x + \sin x)(\cos y + \sin y)$$

17. $2 \sin x + \sin 2x = 2 \sin x + 2 \sin x \cos x = 2 \sin x(1 + \cos x) = \dfrac{2 \sin x(1 + \cos x)(1 - \cos x)}{1 - \cos x}$
$$= \dfrac{2 \sin x(1 - \cos^2 x)}{1 - \cos x} = \dfrac{2 \sin x \sin^2 x}{1 - \cos x} = \dfrac{2 \sin^3 x}{1 - \cos x}$$

19. $\dfrac{\sin 3x}{\sin x} + \dfrac{\cos 3x}{\cos x} = \dfrac{\sin 3x \cos x + \cos 3x \sin x}{\sin x \cos x} = \dfrac{\sin(3x + x)}{\frac{1}{2}(2 \sin x \cos x)} = \dfrac{2 \sin 4x}{\sin 2x}$
$$= \dfrac{2(2 \sin 2x \cos 2x)}{\sin 2x} = 4 \cos 2x$$

21. $\dfrac{1 - \cos \alpha}{2 \sin(\alpha/2)} = \dfrac{1 - \cos \alpha}{2\sqrt{\dfrac{1 - \cos \alpha}{2}}} = \dfrac{\sqrt{1 - \cos \alpha}}{\sqrt{2}} = \sqrt{\dfrac{1 - \cos \alpha}{2}} = \sin\dfrac{\alpha}{2}$

23. $2 \sin^2\dfrac{x}{2} + \cos x = 2\left(\sqrt{\dfrac{1 - \cos x}{2}}\right)^2 + \cos x = 2\left(\dfrac{1 - \cos x}{2}\right) + \cos x = 1 - \cos x + \cos x = 1$

25. $y_1 + y_2 = A \sin 2\pi (\frac{t}{T} - \frac{x}{\lambda}) + A \sin 2\pi (\frac{t}{T} + \frac{x}{\lambda})$

$= A \sin \frac{2\pi t}{T} \cos \frac{2\pi x}{\lambda} - A \cos \frac{2\pi t}{T} \sin \frac{2\pi x}{\lambda} + A \sin \frac{2\pi t}{T} \cos \frac{2\pi x}{\lambda} + A \cos \frac{2\pi t}{T} \sin \frac{2\pi x}{\lambda}$

$= 2A \sin \frac{2\pi t}{T} \cos \frac{2\pi x}{\lambda}$

27. $y = A \sin 2t + B \cos 2t = C(\frac{A}{C} \sin 2t + \frac{B}{C} \cos 2t)$

$= C(\cos \alpha \sin 2t + \sin \alpha \cos 2t) = C \sin(2t + \alpha)$

29. $s = \theta r$; $r = 3.00$ ft ; $\theta = 5.0° = \frac{5.0\pi}{180}$ 31. $s = \theta r$; $r = 15.0$ ft

$s = \frac{5.0\pi}{180}(3.00) = 0.262$ ft

$\theta = 98.0° = \frac{98.0\pi}{180}$

$s = \frac{98.0\pi}{180}(15.0) = 25.7$ ft

33.

$\frac{x}{r} = \sin \frac{\theta}{2}$, $x = r \sin \frac{\theta}{2}$

$A = \frac{1}{2} r^2 \theta - \frac{1}{2}(2x)(r \cos \frac{\theta}{2}) = \frac{1}{2} r^2 \theta - (r \sin \frac{\theta}{2})(r \cos \frac{\theta}{2})$

$= \frac{1}{2} r^2 \theta - \frac{1}{2} r^2 (2 \sin \frac{\theta}{2} \cos \frac{\theta}{2}) = \frac{1}{2} r^2 \theta - \frac{1}{2} r^2 \sin \theta = \frac{1}{2} r^2 (\theta - \sin \theta)$

$r = 6.00$ cm ; $\cos \frac{\theta}{2} = \frac{3.00}{6.00} = 0.500$; $\frac{\theta}{2} = 60°$, $\theta = 120° = \frac{2}{3}\pi$

$A = \frac{1}{2}(6.00)^2(\frac{2}{3}\pi - \sin \frac{2}{3}\pi) = 22.11$ cm^2

$\frac{A}{\pi r^2} = \frac{22.11}{\pi(6.00)^2} = 0.196 = 19.6\%$

Exercises 6-3, p. 258

1. $y = \sin(x+2)$

$\frac{dy}{dx} = \cos(x+2)\frac{d(x+2)}{dx}$

$= [\cos(x+2)](1)$

$= \cos(x+2)$

3. $y = 2 \sin(2x-1)$

$\frac{dy}{dx} = 2 \cos(2x-1)\frac{d(2x-1)}{dx}$

$= [2 \cos(2x-1)](2)$

$= 4 \cos(2x-1)$

5. $y = 5 \cos 2x$

$\frac{dy}{dx} = 5(-\sin 2x)\frac{d(2x)}{dx}$

$= [-5 \sin 2x](2)$

$= -10 \sin 2x$

7. $y = 2 \cos(3x-1)$

$\frac{dy}{dx} = 2[-\sin(3x-1)]\frac{d(3x-1)}{dx}$

$= [-2 \sin(3x-1)](3)$

$= -6 \sin(3x-1)$

9. $y = \sin^2 4x$

$\frac{dy}{dx} = 2(\sin 4x)\frac{d(\sin 4x)}{dx}$

$= (2 \sin 4x)(\cos 4x)(4)$

$= 8 \sin 4x \cos 4x$

$= 4 \sin 8x$

11. $y = 3 \cos^3(5x+2)$

$\frac{dy}{dx} = 3[3\cos^2(5x+2)]\frac{d\cos(5x+2)}{dx}$

$= [9\cos^2(5x+2)][-\sin(5x+2)](5)$

$= -45 \cos^2(5x+2)\sin(5x+2)$

13. $y = x \sin 3x$

$\frac{dy}{dx} = x(3 \cos 3x) + (\sin 3x)(1)$

$= 3x \cos 3x + \sin 3x$

15. $y = 3x^3 \cos 5x$

$\frac{dy}{dx} = 3x^3(-5 \sin 5x) + (\cos 5x)(9x^2)$

$= 9x^2 \cos 5x - 15x^3 \sin 5x$

17. $y = \sin x^2 \cos 2x$

$\frac{dy}{dx} = (\sin x^2)[(-\sin 2x)(2)] + (\cos 2x)[(\cos x^2)(2x)] = 2x \cos x^2 \cos 2x - 2 \sin x^2 \sin 2x$

19. $y = 2 \sin 2x \cos x$

$\frac{dy}{dx} = (2 \sin 2x)(-\sin x) + (\cos x)(2 \cos 2x)(2) = -2 \sin 2x \sin x + 4 \cos 2x \cos x$

21. $y = \dfrac{\sin 3x}{x}$

$\dfrac{dy}{dx} = \dfrac{x(\cos 3x)(3) - (\sin 3x)(1)}{x^2}$

$= \dfrac{3x \cos 3x - \sin 3x}{x^2}$

23. $y = \dfrac{2 \cos x^2}{3x}$

$\dfrac{dy}{dx} = \dfrac{(3x)(-2 \sin x^2)(2x) - (2 \cos x^2)(3)}{(3x)^2}$

$= \dfrac{-4x^2 \sin x^2 - 2 \cos x^2}{3x^2}$

25. $y = 2 \sin^2 3x \cos 2x$

$\dfrac{dy}{dx} = (2 \sin^2 3x)(-\sin 2x)(2) + (\cos 2x)(2)(2 \sin 3x)(\cos 3x)(3)$

$= -4 \sin^2 3x \sin 2x + 12 \sin 3x \cos 2x \cos 3x = 4 \sin 3x(3 \cos 3x \cos 2x - \sin 3x \sin 2x)$

27. $y = \dfrac{\cos^2 3x}{1 + 2 \sin^2 2x}$

$\dfrac{dy}{dx} = \dfrac{(1 + 2 \sin^2 2x)(2 \cos 3x)(-\sin 3x)(3) - (\cos^2 3x)(2)(2 \sin 2x)(\cos 2x)(2)}{(1 + 2 \sin^2 2x)^2}$

$= \dfrac{-2 \cos 3x[3 \sin 3x(1 + 2 \sin^2 2x) + 4 \cos 3x \sin 2x \cos 2x]}{(1 + 2 \sin^2 2x)^2}$

29. $y = \sin^3 x - \cos 2x$

$\dfrac{dy}{dx} = 3 \sin^2 x \cos x - (-\sin 2x)(2)$

$= 3 \sin^2 x \cos x + 2 \sin 2x$

31. $y = x - \dfrac{1}{3}\sin^3 4x$

$\dfrac{dy}{dx} = 1 - \dfrac{1}{3}(3 \sin^2 4x)(\cos 4x)(4)$

$= 1 - 4 \sin^2 4x \cos 4x$

33.

θ	0.5	0.1	0.05	0.01	0.001
$\dfrac{\sin \theta}{\theta}$	0.9588511	0.9983342	0.9995834	0.9999833	0.9999998

35. (a) $\cos 1.0000 = 0.5403023$; $\dfrac{d \sin x}{dx}$ for $x = 1.0000$

slope of tangent line through $(1.0000, \sin 1.0000)$

(b) $\dfrac{\sin 1.0001 - \sin 1.0000}{0.0001} = 0.5402602$ (final digits vary with calculator)

slope of secant line through $(1.0000, \sin 1.0000)$ and $(1.0001, \sin 1.0001)$

37. $\dfrac{d \sin x}{dx} = \cos x$; $\dfrac{d^2 \sin x}{dx^2} = -\sin x$; $\dfrac{d^3 \sin x}{dx^3} = -\cos x$; $\dfrac{d^4 \sin x}{dx^4} = -(-\sin x) = \sin x$

39. $\cos 2x = 2 \cos^2 x - 1$

$(-\sin 2x)(2) = 2(2 \cos x)(-\sin x)$

$-2 \sin 2x = -4 \sin x \cos x$

$\sin 2x = 2 \sin x \cos x$

41. $y = 3 \sin 2x$; $x = \dfrac{\pi}{8}$

$\dfrac{dy}{dx} = 3(\cos 2x)(2) = 6 \cos 2x$

$\left.\dfrac{dy}{dx}\right|_{x=\pi/8} = 6 \cos 2(\dfrac{\pi}{8}) = 6 \cos \dfrac{\pi}{4} = 6(\dfrac{\sqrt{2}}{2}) = 3\sqrt{2}$

43. $y = x \cos 2x$; $x = 1.2$

$\dfrac{dy}{dx} = x(-\sin 2x)(2) + (\cos 2x)(1)$

$= -2x \sin 2x + \cos 2x$

$\left.\dfrac{dy}{dx}\right|_{x=1.2} = -2(1.2)\sin 2(1.2) + \cos 2(1.2)$

$= -2.4 \sin 2.4 + \cos 2.4$

$= -2.36 = m_{tan}$ for $x = 1.2$

45. $s = 2 \sin 3t + t^2$; $t = 2$ s

$v = \dfrac{ds}{dt} = 2(\cos 3t)(3) + 2t$

$= 6 \cos 3t + 2t$

$\left. v \right|_{t=2 s} = 6 \cos 3(2) + 2(2)$

$= 4 + 6 \cos 6 = 9.76$ ft/s

Exercises 6-4, p. 262

1. $y = \tan 5x$

$\dfrac{dy}{dx} = \sec^2 5x \dfrac{d(5x)}{dx}$

$\qquad = (\sec^2 5x)(5)$

$\qquad = 5\sec^2 5x$

3. $y = \cot(1-x)^2$

$\dfrac{dy}{dx} = -\csc^2(1-x)^2 \dfrac{d(1-x)^2}{dx}$

$\qquad = -\csc^2(1-x)^2[2(1-x)(-1)]$

$\qquad = 2(1-x)\csc^2(1-x)^2$

5. $y = 3\sec 2x$

$\dfrac{dy}{dx} = 3(\sec 2x \tan 2x)\dfrac{d(2x)}{dx}$

$\qquad = (3\sec 2x \tan 2x)(2)$

$\qquad = 6\sec 2x \tan 2x$

7. $y = -3\csc\sqrt{x} = -3\csc x^{1/2}$

$\dfrac{dy}{dx} = -3(-\csc x^{1/2}\cot^{1/2})\dfrac{d(x^{1/2})}{dx}$

$\qquad = (3\csc x^{1/2}\cot x^{1/2})(\tfrac{1}{2}x^{-1/2})$

$\qquad = \dfrac{3\csc\sqrt{x}\cot\sqrt{x}}{2\sqrt{x}}$

9. $y = 5\tan^2 3x$

$\dfrac{dy}{dx} = 5(2\tan 3x)\dfrac{d(\tan 3x)}{dx}$

$\qquad = (10\tan 3x)(\sec^2 3x)(3)$

$\qquad = 30\tan 3x \sec^2 3x$

11. $y = 2\cot^4 3x$

$\dfrac{dy}{dx} = 2(4\cot^3 3x)\dfrac{d(\cot 3x)}{dx}$

$\qquad = (8\cot^3 3x)(-\csc^2 3x)(3)$

$\qquad = -24\cot^3 3x \csc^2 3x$

13. $y = \sqrt{\sec 4x} = (\sec 4x)^{1/2}$

$\dfrac{dy}{dx} = \dfrac{1}{2}(\sec 4x)^{-1/2}(\sec 4x \tan 4x)(4)$

$\qquad = \dfrac{2\sec 4x \tan 4x}{(\sec 4x)^{1/2}} = 2\tan 4x\sqrt{\sec 4x}$

15. $y = 3\csc^4 7x$

$\dfrac{dy}{dx} = 3(4\csc^3 7x)(-\csc 7x \cot 7x)(7)$

$\qquad = -84\csc^4 7x \cot 7x$

17. $y = x^2\tan x$

$\dfrac{dy}{dx} = x^2(\sec^2 x) + (\tan x)(2x)$

$\qquad = x^2\sec^2 x + 2x\tan x$

19. $y = 4\cos x \csc x^2$

$\dfrac{dy}{dx} = (4\cos x)(-\csc x^2\cot x^2)(2x) + \csc x^2(-4\sin x) = -4\csc x^2(2x\cos x\cot x^2 + \sin x)$

21. $y = \dfrac{\csc x}{x}$

$\dfrac{dy}{dx} = \dfrac{x(-\csc x\cot x) - (\csc x)(1)}{x^2}$

$\qquad = -\dfrac{\csc x(x\cot x + 1)}{x^2}$

23. $y = \dfrac{\cos 4x}{1 + \cot 3x}$

$\dfrac{dy}{dx} = \dfrac{(1+\cot 3x)(-\sin 4x)(4) - (\cos 4x)(-\csc^2 3x)(3)}{(1 + \cot 3x)^2}$

$\qquad = \dfrac{-4\sin 4x - 4\sin 4x\cot 3x + 3\cos 4x\csc^2 3x}{(1 + \cot 3x)^2}$

25. $y = \dfrac{1}{3}\tan^3 x - \tan x$

$\dfrac{dy}{dx} = \dfrac{1}{3}(3\tan^2 x)(\sec^2 x) - \sec^2 x$

$\qquad = \sec^2 x(\tan^2 x - 1)$

27. $y = \tan 2x - \sec 2x$

$\dfrac{dy}{dx} = (\sec^2 2x)(2) - (\sec 2x \tan 2x)(2)$

$\qquad = 2\sec 2x(\sec 2x - \tan 2x)$

29. $y = 4\tan^2 3x$

$dy = 4(2\tan 3x)(\sec^2 3x)(3)dx$

$\qquad = 24\tan 3x \sec^2 3x\, dx$

31. $y = \tan 4x \sec 4x$

$dy = [\tan 4x(\sec 4x \tan 4x)(4) + \sec 4x(\sec^2 4x)(4)]dx$

$\qquad = 4\sec 4x(\tan^2 4x + \sec^2 4x)dx$

33. (a) $sec^2 1.0000 = 3.4255188$; $\dfrac{d\ tan\ x}{dx}$ for $x = 1.0000$

slope of tangent line through $(1.0000, tan\ 1.0000)$

(b) $\dfrac{tan\ 1.0001 - tan\ 1.0000}{0.0001} = 3.4260524$ (final digits vary with calculator)

slope of secant line through $(1.0000, tan\ 1.0000)$ and $(1.0001, tan\ 1.0001)$

35. $1 + tan^2 x = sec^2 x$

$2\,tan\ x(sec^2 x) = 2\ sec\ x(sec\ x\ tan\ x)$

$2\ tan\ x\ sec^2 x = 2\ sec^2 x\ tan\ x$

37. $y = 2\ cot\ 3x$; $x = \dfrac{\pi}{12}$

$\dfrac{dy}{dx} = (-2\ csc^2 3x)(3) = -6\ csc^2 3x$

$\dfrac{dy}{dx}\bigg|_{x=\pi/12} = -6\ csc^2(3)(\dfrac{\pi}{12}) = -6\ csc^2\dfrac{\pi}{4}$

$= -6(\sqrt{2})^2 = -12$

39. $s = 2t^2 + tan\ 2t$

$v = \dfrac{ds}{dt} = 4t + (sec^2 2t)(2)$

$= 4t + 2\ sec^2 2t$

$a = \dfrac{dv}{dt} = 4 + 2(2\ sec\ 2t)(sec\ 2t\ tan\ 2t)(2) = 4 + 8\ sec^2 2t\ tan\ 2t$

$a\big|_{t=0.375} = 4 + 8\ sec^2 0.750\ tan\ 0.750 = 17.92$

41. See Fig. 6-21 ; $h = 1000\ tan\ \dfrac{3t}{2t + 10}$

$\dfrac{dh}{dt} = 1000\ sec^2\dfrac{3t}{2t + 10}\left(\dfrac{(2t + 10)(3) - 3t(2)}{(2t + 10)^2}\right) = \dfrac{30000\ sec^2\dfrac{3t}{2t+10}}{(2t + 10)^2}$

$\dfrac{dh}{dt}\bigg|_{t=5\ s} = \dfrac{30000\ sec^2(15/20)}{(10 + 10)^2} = 140$ ft/s

Exercises 6-5, p. 267

1. $Arccos(\dfrac{1}{2}) = \dfrac{\pi}{3}$ since $cos\ \dfrac{\pi}{3} = \dfrac{1}{2}$ and $0 \le \dfrac{\pi}{3} \le \pi$

3. $Arcsin\ 0 = 0$ since $sin\ 0 = 0$ and $-\dfrac{\pi}{2} \le 0 \le \dfrac{\pi}{2}$

5. $Arctan(-\sqrt{3}) = -\dfrac{\pi}{3}$ since $tan(-\dfrac{\pi}{3}) = -\sqrt{3}$ and $-\dfrac{\pi}{2} < -\dfrac{\pi}{3} < \dfrac{\pi}{2}$

7. $Arcsec\ 2 = \dfrac{\pi}{3}$ since $sec\ \dfrac{\pi}{3} = 2$ and $0 \le \dfrac{\pi}{3} \le \pi$

9. $Arctan(\dfrac{\sqrt{3}}{3}) = \dfrac{\pi}{6}$ since $tan\ \dfrac{\pi}{6} = \dfrac{\sqrt{3}}{3}$ and $-\dfrac{\pi}{2} < \dfrac{\pi}{6} < \dfrac{\pi}{2}$

11. $Arcsin(-\dfrac{\sqrt{2}}{2}) = -\dfrac{\pi}{4}$ since $sin(-\dfrac{\pi}{4}) = -\dfrac{\sqrt{2}}{2}$ and $-\dfrac{\pi}{2} \le -\dfrac{\pi}{4} \le \dfrac{\pi}{2}$

13. $Arccsc\ \sqrt{2} = \dfrac{\pi}{4}$ since $csc\ \dfrac{\pi}{4} = \sqrt{2}$ and $-\dfrac{\pi}{2} \le \dfrac{\pi}{4} \le \dfrac{\pi}{2}$

15. $sin(Arctan\ \sqrt{3}) = sin\ \dfrac{\pi}{3} = \dfrac{1}{2}\sqrt{3}$ 17. $cos[Arctan(-1)] = cos(-\dfrac{\pi}{4}) = \dfrac{1}{2}\sqrt{2}$

19. $sec(2\ Arcsin\ 1) = sec[2(\dfrac{\pi}{2})] = sec\ \pi = -1$ 21. $Arctan(-3.7321) = -1.3090$

23. $Arcsin(-0.8326) = -0.9838$ 25. $Arccos\ 0.1291 = 1.4413$ 27. $Arctan\ 8.2614 = 1.4503$

29. $tan[Arccos(-0.6281)] = tan(2.2499) = -1.2389$

31. $sin[Arctan(-0.2297)] = sin(-0.2258) = -0.2239$

33. Let θ = Arcsin x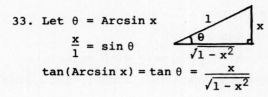

$$\frac{x}{1} = \sin\theta$$

$$\tan(\text{Arcsin } x) = \tan\theta = \frac{x}{\sqrt{1-x^2}}$$

35. Let θ = Arcsec x

$$\frac{x}{1} = \sec\theta$$

$$\cos(\text{Arcsec } x) = \cos\theta = \frac{1}{x}$$

37. Let θ = Arccsc 3x

$$\frac{3x}{1} = \csc\theta$$

$$\sec(\text{Arccsc } 3x) = \sec\theta = \frac{3x}{\sqrt{9x^2-1}}$$

39. Let θ = Arcsin x

$$\frac{x}{1} = \sin\theta$$

$$\sin(2\,\text{Arcsin } x) = \sin 2\theta = 2\sin\theta\cos\theta$$

$$= 2\left(\frac{x}{1}\right)\left(\frac{\sqrt{1-x^2}}{1}\right) = 2x\sqrt{1-x^2}$$

41. Let $\text{Arcsin}\frac{3}{5} + \text{Arcsin}\frac{5}{13} = \alpha + \beta$

$$\sin\alpha = \frac{3}{5} \;;\; \cos\alpha = \sqrt{1-\sin^2\alpha} = \sqrt{1-(3/5)^2} = \frac{4}{5} \;;\; \sin\beta = \frac{5}{13} \;;\; \cos\beta = \sqrt{1-(5/13)^2} = \frac{12}{13}$$

$$\sin(\alpha+\beta) = \sin\alpha\cos\beta + \cos\alpha\sin\beta = \frac{3}{5}\left(\frac{12}{13}\right) + \frac{4}{5}\left(\frac{5}{13}\right) = \frac{36}{65} + \frac{20}{65} = \frac{56}{65}$$

$$\sin(\alpha+\beta) = \frac{56}{65} \;;\; \alpha+\beta = \text{Arcsin}\frac{56}{65} \;;\; \text{Arcsin}\frac{3}{5} + \text{Arcsin}\frac{5}{13} = \text{Arcsin}\frac{56}{65}$$

43. $\text{Arcsin } 0.5 + \text{Arccos } 0.5 = \frac{\pi}{6} + \frac{\pi}{3} = \frac{\pi}{2}$

45. See Fig. 6-28.; $\sin A = \frac{a}{c}$; $A = \text{Arcsin}\left(\frac{a}{c}\right)$

47. See Fig. 6-30.; Let b = side opp. β

$$\tan\beta = \frac{b}{1} = b \;;\; \tan\alpha = \frac{a+b}{1} = a+b = a + \tan\beta$$

$$\alpha = \text{Arctan}(a + \tan\beta)$$

49. Let $\alpha = \text{Arctan}\frac{3}{6}$ (See Fig. 6-32)

$$\theta + \alpha = \text{Arctan}\frac{3}{6-x}$$

$$\theta = \text{Arctan}\frac{3}{6-x} - \text{Arctan}\frac{1}{2}$$

Exercises 6-6, p. 272

1. $y = \text{Arcsin}(x^2)$

$$\frac{dy}{dx} = \frac{1}{\sqrt{1-(x^2)^2}}\frac{dx^2}{dx}$$

$$= \frac{1}{\sqrt{1-x^4}}(2x)$$

$$= \frac{2x}{\sqrt{1-x^4}}$$

3. $y = 2\,\text{Arcsin } 3x^3$

$$\frac{dy}{dx} = 2\frac{1}{\sqrt{1-(3x^3)^2}}\frac{d\,3x^3}{dx}$$

$$= \frac{2}{\sqrt{1-9x^6}}(9x^2)$$

$$= \frac{18x^2}{\sqrt{1-9x^6}}$$

5. $y = \text{Arccos}\frac{1}{2}x$

$$\frac{dy}{dx} = -\frac{1}{\sqrt{1-(\frac{1}{2}x)^2}}\frac{d(\frac{1}{2}x)}{dx}$$

$$= -\frac{1}{\sqrt{1-\frac{1}{4}x^2}}\left(\frac{1}{2}\right)$$

$$= -\frac{1}{2\sqrt{\frac{4-x^2}{4}}} = -\frac{1}{\sqrt{4-x^2}}$$

7. $y = 2\,\text{Arccos}\sqrt{2-x} = 2\,\text{Arccos}(2-x)^{1/2}$

$$\frac{dy}{dx} = 2\left(-\frac{1}{\sqrt{1-[(2-x)^{1/2}]^2}}\right)\frac{d(2-x)^{1/2}}{dx}$$

$$= \frac{-2}{\sqrt{1-(2-x)}}\left[\frac{1}{2}(2-x)^{-1/2}(-1)\right]$$

$$= \frac{1}{\sqrt{x-1}\,(2-x)^{1/2}} = \frac{1}{\sqrt{(x-1)(2-x)}}$$

9. $y = \text{Arctan}\sqrt{x} = \text{Arctan } x^{1/2}$

$$\frac{dy}{dx} = \frac{1}{1+(x^{1/2})^2}\frac{dx^{1/2}}{dx}$$

$$= \frac{1}{1+x}\left(\frac{1}{2}x^{-1/2}\right)$$

$$= \frac{1}{2\sqrt{x}\,(1+x)}$$

11. $y = \text{Arctan}(\frac{1}{x}) = \text{Arctan } x^{-1}$

$\frac{dy}{dx} = \frac{1}{1 + (x^{-1})^2}\frac{d\,x^{-1}}{dx} = \frac{1}{1 + x^{-2}}(-x^{-2})$

$= -\frac{1}{1 + \frac{1}{x^2}}(\frac{1}{x^2}) = -\frac{1}{x^2 + 1}$

13. $y = x \text{ Arcsin } x$

$\frac{dy}{dx} = x(\frac{1}{\sqrt{1 - x^2}})(1) + (\text{Arcsin } x)(1)$

$= \frac{x}{\sqrt{1 - x^2}} + \text{Arcsin } x$

15. $y = 2x \text{ Arctan } 2x$

$\frac{dy}{dx} = 2x(\frac{1}{1 + (2x)^2})(2) + (\text{Arctan } 2x)(2)$

$= \frac{4x}{1 + 4x^2} + 2 \text{ Arctan } 2x$

17. $y = \frac{3x}{\text{Arcsin } 2x}$

$\frac{dy}{dx} = \frac{(\text{Arcsin } 2x)(3) - 3x(\frac{1}{\sqrt{1 - 4x^2}})(2)}{(\text{Arcsin } 2x)^2}$

$= \frac{3\sqrt{1 - 4x^2}\text{Arcsin } 2x - 6x}{\sqrt{1 - 4x^2}\text{Arcsin}^2 2x}$

19. $y = \frac{\text{Arcsin } 2x}{\text{Arccos } 2x}$

$\frac{dy}{dx} = \frac{(\text{Arccos } 2x)(\frac{1}{\sqrt{1-4x^2}})(2) - (\text{Arcsin } 2x)(\frac{-1}{\sqrt{1-4x^2}})(2)}{\text{Arccos}^2 2x} = \frac{2(\text{Arccos } 2x + \text{Arcsin } 2x)}{\sqrt{1 - 4x^2}\text{ Arccos}^2 2x}$

21. $y = 2 \text{ Arccos}^3 4x$

$\frac{dy}{dx} = 2(3 \text{ Arccos}^2 4x)(\frac{-1}{\sqrt{1 - 16x^2}})(4)$

$= \frac{-24 \text{ Arccos}^2 4x}{\sqrt{1 - 16x^2}}$

23. $y = \text{Arcsin}^2 4x$

$\frac{dy}{dx} = 2 \text{ Arcsin } 4x(\frac{1}{\sqrt{1 - 16x^2}})(4)$

$= \frac{8 \text{ Arcsin } 4x}{\sqrt{1 - 16x^2}}$

25. $y = 3 \text{ Arctan}^3 x$

$\frac{dy}{dx} = 3(3 \text{ Arctan}^2 x)(\frac{1}{1 + x^2})(1)$

$= \frac{9 \text{ Arctan}^2 x}{1 + x^2}$

27. $y = \frac{1}{1 + 4x^2} - \text{Arctan } 2x = (1 + 4x^2)^{-1} - \text{Arctan } 2x$

$\frac{dy}{dx} = -(1 + 4x^2)^{-2}(8x) - \frac{1}{1 + 4x^2}(2)$

$= -\frac{8x}{(1 + 4x^2)^2} - \frac{2}{1 + 4x^2} = \frac{-8x - 2(1 + 4x^2)}{(1 + 4x^2)^2}$

$= -\frac{8x^2 + 8x + 2}{(1 + 4x^2)^2} = \frac{-2(2x + 1)^2}{(1 + 4x^2)^2}$

29. (a) $\frac{1}{\sqrt{1 - 0.5^2}} = 1.1547005$

$\frac{d \text{ Arcsin } x}{dx}$ for $x = 0.5$; slope of tangent line through $(0.5, \text{Arcsin } 0.5)$

(b) $\frac{\text{Arcsin } 0.5001 - \text{Arcsin } 0.5000}{0.0001} = 1.1547390$ (final digits vary with calculator)

slope of secant line through $(0.5, \text{Arcsin } 0.5)$ and $(0.5001, \text{Arcsin } 0.5001)$

31. $y = \text{Arcsin}^3 x$

$dy = 3(\text{Arcsin}^2 x)(\frac{1}{\sqrt{1 - x^2}})(1)dx$

$= \frac{3 \text{ Arcsin}^2 x\, dx}{\sqrt{1 - x^2}}$

33. $y = \text{Arctan } 2x$

$\frac{dy}{dx} = \frac{1}{1 + 4x^2}(2) = \frac{2}{1 + 4x^2} = 2(1 + 4x^2)^{-1}$

$\frac{d^2 y}{dx^2} = -2(1 + 4x^2)^{-2}(8x) = \frac{-16x}{(1 + 4x^2)^2}$

35. $u = \sec y$; $\dfrac{du}{dx} = \sec y \tan y \dfrac{dy}{dx}$

$\dfrac{dy}{dx} = \dfrac{1}{\sec y \tan y} \dfrac{du}{dx} = \dfrac{1}{\sec y \sqrt{\sec^2 y - 1}} \dfrac{du}{dx}$

$= \dfrac{1}{u\sqrt{u^2 - 1}} \dfrac{du}{dx} = \dfrac{1}{\sqrt{u^2(u^2 - 1)}} \dfrac{du}{dx}$

37. $\theta = \text{Arccos} \dfrac{R_x}{R}$

$\dfrac{d\theta}{dt} = -\dfrac{1}{\sqrt{1 - (R_x/R)^2}} \left(\dfrac{1}{R}\dfrac{dR_x}{dt}\right)$

$= -\dfrac{R}{\sqrt{R^2 - R_x^2}}\left(\dfrac{1}{R}\dfrac{dR_x}{dt}\right) = -\dfrac{1}{\sqrt{R^2 - R_x^2}}\dfrac{dR_x}{dt}$

39. $\alpha = \text{Arccos} \dfrac{2f - r}{r}$

$\dfrac{d\alpha}{dr} = \dfrac{-1}{\sqrt{1 - (\frac{2f - r}{r})^2}}\left(\dfrac{r(-1) - (2f - r)(1)}{r^2}\right) = \dfrac{-r}{\sqrt{r^2 - (2f - r)^2}}\left(\dfrac{-2f}{r^2}\right) = \dfrac{2f}{r\sqrt{4fr - 4f^2}} = \dfrac{f}{r\sqrt{fr - f^2}}$

Exercises 6-7, p. 276

1. Let $y_1 = \sin x$, $y_2 = \cos x$
   $y_1' = \cos x$, $y_2' = -\sin x$
   At points of intersection
   $y_1 = y_2$, $y_1' = -y_2'$

3. $y = \text{Arctan } x$, $y' = \dfrac{1}{1 + x^2}$

   $\dfrac{1}{1 + x^2} > 0$ for all x

   y inc. for all x

5. $y = x - \tan x$

   Intercepts: $(0,0)$
   For $x=0$, $y=0-0=0$
   Symmetry: To origin
   $-y = -x - \tan(-x) = -x + \tan x$
   is the same as
   $y = x - \tan x$
   Vertical asymptotes:
   As $x \to -\pi/2$ from right,
   $\tan x \to -\infty$ and
   $x - \tan x \to -(-\infty) = +\infty$
   As $x \to \pi/2$ from left,
   $\tan x \to +\infty$ and
   $x - \tan x \to -(+\infty) = -\infty$
   Asymptotes: $x = -\dfrac{\pi}{2}, \dfrac{\pi}{2}$

   Derivatives:
   $y' = 1 - \sec^2 x$
   $y' < 0$ for all x ($x \neq 0$)
   since $\sec x > 1$ for all x ($x \neq 0$)
   y dec. for all x
   $y' = 0$ for $x=0$ since
   $1 - \sec^2 x = 0$, $\sec^2 x = 1$ for $x=0$.
   No change of sign means there
   is no rel. max. or min.
   $y'' = -2 \sec x(\sec x \tan x) = -2\sec^2 x \tan x$
   $y'' > 0$ for $-\dfrac{\pi}{2} < x < 0$, conc. up
   $y'' < 0$ for $0 < x < \dfrac{\pi}{2}$, conc. down   Infl. $(0,0)$

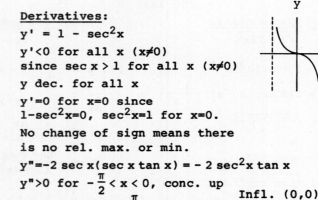

7. $y = \sin 2x$ at $x = \dfrac{5\pi}{8}$

   $y' = 2\cos 2x$
   $y'\big|_{x=5\pi/8} = 2\cos\dfrac{5\pi}{4} = 2\left(-\dfrac{\sqrt{2}}{2}\right) = -\sqrt{2}$
   $m = -\sqrt{2}$
   For $x = \dfrac{5\pi}{8}$, $y = \sin\dfrac{5\pi}{4} = -\dfrac{\sqrt{2}}{2}$

   $y - y_1 = m(x - x_1)$
   $y + \dfrac{\sqrt{2}}{2} = -\sqrt{2}\left(x - \dfrac{5\pi}{8}\right)$

   $8y + 4\sqrt{2} = -8\sqrt{2}x + 5\pi\sqrt{2}$
   $8\sqrt{2}x + 8y + 4\sqrt{2} - 5\pi\sqrt{2} = 0$

9. $x^2 - 4 \sin x = 0$

$f(x) = x^2 - 4 \sin x$

$f'(x) = 2x - 4 \cos x$

$f(0) = 0 - 0 = 0$

$f(1) = 1^2 - 4 \sin 1 = -2.37$

$f(2) = 2^2 - 4 \sin 2 = 0.36$

Let $x_1 = 1.8$

n	x_n	$f(x_n)$	$f'(x_n)$	$x_n - \dfrac{f(x_n)}{f'(x_n)}$
1	1.8	-0.6553905	4.5088084	1.9453578
2	1.9453578	0.0617444	5.3541734	1.9338258
3	1.9338258	0.0003809	5.2880831	1.9337538
4	1.9337538	1.54×10^{-8}	5.2876697	1.9337538
5	1.9337538			

$x_5 = x_4 = 1.9337538$

11. $y = 6 \cos x - 8 \sin x$

$y' = -6 \sin x - 8 \cos x$

$y' = 0, \; -6 \sin x - 8 \cos x = 0$

$\tan x = -\dfrac{4}{3}, \; x = 2.2143, \; 5.3559$

$y'' = -6 \cos x + 8 \sin x$

For $x = 2.2143$, $y'' > 0$ (min.)

For $x = 5.3559$, $y'' < 0$ (max.)

$\quad y = 6 \cos 5.3559 - 8 \sin 5.3559$

$\quad\quad = 10.0$ (max. value)

(Function has period 2π)

13. $s = \sin t + \cos 2t, \; t = \dfrac{\pi}{6}$ s

$v = \dfrac{ds}{dt} = \cos t - 2 \sin 2t$

$a = \dfrac{dv}{dt} = -\sin t - 4 \cos 2t$

For $t = \dfrac{\pi}{6}$ s

$v = \cos \dfrac{\pi}{6} - 2 \sin \dfrac{\pi}{3} = \dfrac{1}{2}\sqrt{3} - 2(\dfrac{1}{2}\sqrt{3}) = -\dfrac{1}{2}\sqrt{3}$ ft/s

$a = -\sin \dfrac{\pi}{6} - 4 \sin \dfrac{\pi}{3} = -\dfrac{1}{2} - 4(\dfrac{1}{2}) = -\dfrac{5}{2}$ ft/s^2

15. $A_y = 20.0 \sin \theta, \; \dfrac{d\theta}{dt} = 0.0551$ rad/s

$\dfrac{dA_y}{dt} = 20.0 \cos\theta \dfrac{d\theta}{dt}$

$\dfrac{dA_y}{dt}\bigg|_{\theta=0.323} = 20.0 \cos 0.323 \, (0.0551)$

$\quad\quad\quad = 1.05$ mm/s

17. $x = \sin 2\pi t, \; y = \cos 2\pi t, \; t = \dfrac{1}{12}$

$v_x = \dfrac{dx}{dt} = 2\pi \cos 2\pi t, \; v_y = \dfrac{dy}{dt} = -2\pi \sin 2\pi t$

For $t = \dfrac{1}{12}$: $v_x = 2\pi \cos \dfrac{\pi}{6} = 2\pi (\dfrac{\sqrt{3}}{2}) = \pi\sqrt{3}$

$\quad\quad v_y = -2\pi \sin \dfrac{\pi}{6} = -2\pi(\dfrac{1}{2}) = -\pi$

$v = \sqrt{(\pi\sqrt{3})^2 + (-\pi)^2} = 2\pi$

$\tan \theta = \dfrac{-\pi}{\pi\sqrt{3}} = -\dfrac{1}{\sqrt{3}}, \; \theta = 330°$ ($v_x > 0, v_y < 0$)

19. See Example E

$x = \cos 2t, \; y = \sin 2t$

$v_x = -2 \sin 2t, \; v_y = 2 \cos 2t$

$a_x = -4 \cos 2t, \; a_y = -4 \sin 2t$

For $t = \dfrac{\pi}{8}$: $a_x = -4 \cos \dfrac{\pi}{4} = -4(\dfrac{\sqrt{2}}{2}) = -2\sqrt{2}$

$\quad\quad a_y = -4 \sin \dfrac{\pi}{4} = -4(\dfrac{\sqrt{2}}{2}) = -2\sqrt{2}$

$\quad\quad a = \sqrt{(-2\sqrt{2})^2 + (-2\sqrt{2})^2} = 4$

$\quad \tan \theta = \dfrac{-2\sqrt{2}}{-2\sqrt{2}} = 1., \; \theta = 225°$ ($a_x < 0, a_y < 0$)

21. See Fig. 6-41

$s = 16t^2$

$\tan \theta = \dfrac{200 - s}{100} = \dfrac{200 - 16t^2}{100}$

$\quad\quad = 2 - 0.16t^2$

$\theta = \text{Arctan}(2 - 0.16t^2)$

$\dfrac{d\theta}{dt} = \dfrac{-0.32t}{1 + (2 - 0.16t^2)^2}$

$\dfrac{d\theta}{dt}\bigg|_{t=1 \text{ s}} = \dfrac{-0.32}{1 + (2 - 0.16)^2} = -0.0730$ rad/s

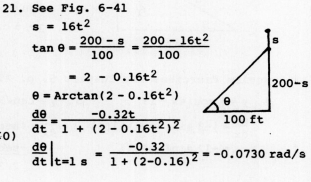

23. $\dfrac{d\theta}{dt} = 0.120$ rad/s

$\cos \theta = \dfrac{6.25}{x}$

$\theta = \text{Arccos} \dfrac{6.25}{x}, \; \dfrac{d\theta}{dt} = \dfrac{-1}{\sqrt{1 - (6.25/x)^2}}(-\dfrac{6.25}{x^2})\dfrac{dx}{dt} = \dfrac{6.25 \, dx/dt}{x\sqrt{x^2 - 6.25^2}}$

For $x = 10.5$: $0.120 = \dfrac{6.25 \, dx/dt}{10.5\sqrt{10.5^2 - 6.25^2}}, \; \dfrac{dx}{dt} = 1.70$ unit/s

25. $\mu = \tan \theta$, $\theta = 20° = 0.349$ **27.** $x^2 = 7.50^2 + 8.00^2 - 2(7.50)(8.00)\cos \theta$

$\quad d\theta = 1° = 0.0175$

$\quad d\mu = \sec^2\theta\, d\theta$

$\qquad = (\sec^2 0.349)(0.0175)$

$\qquad = 0.020$

$\quad 2x\, dx = -120(-\sin \theta\, d\theta)$

$\quad dx = \dfrac{60.0 \sin \theta\, d\theta}{x}$

For $\theta = 40.0° = 0.698$

$\quad d\theta = 0.5° = 0.00873$

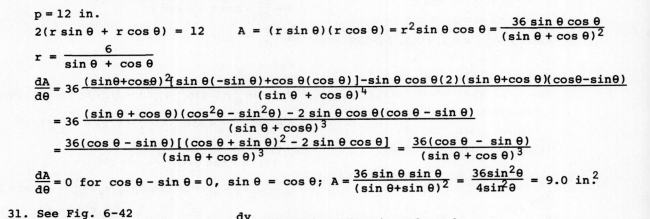

$x = \sqrt{7.50^2 + 8.00^2 - 120 \cos 0.698} = 5.32$

$dx = \dfrac{60.0(\sin 0.698)(0.00873)}{5.32} = 0.063$ cm

29.

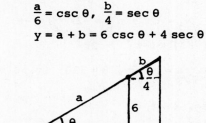

$p = 12$ in.

$2(r \sin \theta + r \cos \theta) = 12$ $\quad A = (r \sin \theta)(r \cos \theta) = r^2 \sin \theta \cos \theta = \dfrac{36 \sin \theta \cos \theta}{(\sin \theta + \cos \theta)^2}$

$r = \dfrac{6}{\sin \theta + \cos \theta}$

$\dfrac{dA}{d\theta} = 36\, \dfrac{(\sin\theta + \cos\theta)^2[\sin \theta(-\sin \theta) + \cos \theta(\cos \theta)] - \sin \theta \cos \theta(2)(\sin \theta + \cos \theta)(\cos \theta - \sin \theta)}{(\sin \theta + \cos \theta)^4}$

$\quad = 36\, \dfrac{(\sin \theta + \cos \theta)(\cos^2\theta - \sin^2\theta) - 2 \sin \theta \cos \theta(\cos \theta - \sin \theta)}{(\sin \theta + \cos \theta)^3}$

$\quad = \dfrac{36(\cos \theta - \sin \theta)[(\cos \theta + \sin \theta)^2 - 2 \sin \theta \cos \theta]}{(\sin \theta + \cos \theta)^3} = \dfrac{36(\cos \theta - \sin \theta)}{(\sin \theta + \cos \theta)^3}$

$\dfrac{dA}{d\theta} = 0$ for $\cos \theta - \sin \theta = 0$, $\sin \theta = \cos \theta$; $A = \dfrac{36 \sin \theta \sin \theta}{(\sin \theta + \sin \theta)^2} = \dfrac{36\sin^2\theta}{4\sin^2\theta} = 9.0$ in.^{2}

31. See Fig. 6-42

$\dfrac{a}{6} = \csc \theta$, $\dfrac{b}{4} = \sec \theta$

$y = a + b = 6 \csc \theta + 4 \sec \theta$

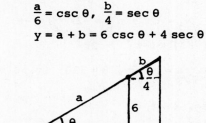

$\dfrac{dy}{d\theta} = -6 \csc \theta \cot \theta + 4 \sec \theta \tan \theta$

$-6 \csc \theta \cot \theta + 4 \sec \theta \tan \theta = 0$

$-6\, \dfrac{1}{\sin \theta}\, \dfrac{\cos \theta}{\sin \theta} + 4\, \dfrac{1}{\cos \theta}\, \dfrac{\sin \theta}{\cos \theta} = 0$

$\dfrac{-6 \cos^3\theta + 4 \sin^3\theta}{\sin^2\theta \cos^2\theta} = 0$, $\quad 4 \sin^3\theta = 6 \cos^3\theta$, $\quad \tan^3\theta = 1.5$

$\tan \theta = 1.1447$, $\theta = 0.8528$

$y = 6 \csc 0.8528 + 4 \sec 0.8528 = 14.05$ ft

Review Exercises for Chapter 6, p. 278

1. $y = 3 \cos(4x - 1)$

$\quad \dfrac{dy}{dx} = [-3 \sin(4x - 1)](4)$

$\qquad = -12 \sin(4x - 1)$

3. $y = \tan\sqrt{3 - x}$

$\quad \dfrac{dy}{dx} = (\sec^2\sqrt{3 - x})(\tfrac{1}{2})(3 - x)^{-1/2}(-1)$

$\qquad = \dfrac{-\sec^2\sqrt{3 - x}}{2\sqrt{3 - x}}$

5. $y = 3 \operatorname{Arctan}(\dfrac{x}{3})$

$\quad \dfrac{dy}{dx} = \dfrac{3}{1 + (x/3)^2}(\tfrac{1}{3})$

$\qquad = \dfrac{9}{9 + x^2}$

7. $y = \csc^2(3x + 2)$

$\quad \dfrac{dy}{dx} = [2 \csc(3x + 2)][-\csc(3x + 2)\cot(3x + 2)](3)$

$\qquad = -6 \csc^2(3x + 2)\cot(3x + 2)$

9. $y = 3 \cos^4 x^2$

$\quad \dfrac{dy}{dx} = 3(4 \cos^3 x^2)(-\sin x^2)(2x)$

$\qquad = -24x \cos^3 x^2 \sin x^2$

11. $y = \text{Arcsin}(\cos x)$

$$\frac{dy}{dx} = \frac{1}{\sqrt{1 - \cos^2 x}}(-\sin x)$$

$$= \frac{-\sin x}{\sqrt{\sin^2 x}} = -1$$

13. $y = \sqrt{\csc 4x + \cot 4x}$

$$\frac{dy}{dx} = \frac{1}{2}(\csc 4x + \cot 4x)^{-1/2}[(-\csc 4x \cot 4x)(4)$$
$$- (\csc^2 4x)(4)]$$

$$= \frac{-2(\csc 4x)(\cot 4x + \csc 4x)}{(\cos 4x + \cot 4x)^{1/2}}$$

$$= (-2 \csc 4x)\sqrt{\csc 4x + \cot 4x}$$

15. $y = x \sec 2x$

$$\frac{dy}{dx} = x(\sec 2x \tan 2x)(2) + (\sec 2x)(1)$$

$$= \sec 2x(2x \tan 2x + 1)$$

17. $y = \dfrac{x^2}{\text{Arctan } 2x}$

$$\frac{dy}{dx} = \frac{(\text{Arctan } 2x)(2x) - x^2\left(\dfrac{1}{1 + 4x^2}\right)(2)}{\text{Arctan}^2 2x}$$

$$= \frac{2x(1 + 4x^2)\text{Arctan } 2x - 2x^2}{(1 + 4x^2)\text{Arctan}^2 2x}$$

19. $y = x \text{ Arccos } x - \sqrt{1 - x^2}$

$$\frac{dy}{dx} = x\left(\frac{-1}{\sqrt{1-x^2}}\right)(1) + \text{Arccos } x - \frac{1}{2}(1-x^2)^{-1/2}(-2x)$$

$$= \text{Arccos } x$$

21. $y^2 \sin 2x + \tan x = 0$

$$y^2(\cos 2x)(2) + (\sin 2x)(2yy') + \sec^2 x = 0$$

$$y' = -\frac{2y^2 \cos 2x + \sec^2 x}{2y \sin 2x}$$

23. $x \sin 2y = y \cos 2x$

$$x(\cos 2y)(2y') + (\sin 2y)(1)$$
$$= y(-\sin 2x)(2) + (\cos 2x)(y')$$

$$y' = \frac{-2y \sin 2x - \sin 2y}{2x \cos 2y - \cos 2x}$$

$$= \frac{2y \sin 2x + \sin 2y}{\cos 2x - 2x \cos 2y}$$

25. $\sin^2 x + \cos^2 x = 1$

$$2 \sin x \cos x + 2 \cos x(-\sin x) = 0$$

$$0 = 0$$

27. $y = x - \cos x$

Intercepts: $(0, -1)$

For $x = 0$, $y = 0 - \cos 0 = -1$

Symmetry: None

Behavior as x becomes large:

Since $-1 \le \cos x \le 1$ for all x,

$y \to +\infty$ as $x \to +\infty$ and

$y \to -\infty$ as $x \to -\infty$

Derivatives: $y' = 1 + \sin x$

$y' = 0$ for $\sin x = -1$, $x = \dfrac{3\pi}{2}, \dfrac{7\pi}{2}, \cdots$

$y' > 0$ except for $\sin x = -1$

    since $-1 \le \sin x \le 1$ for all x.

y inc. except for $\sin x = -1$

no rel. max. or min.

$y'' = \cos x$

$-\dfrac{\pi}{2} < x < \dfrac{\pi}{2}$, $y'' > 0$, conc. up

$\dfrac{\pi}{2} < x < \dfrac{3\pi}{2}$, $y'' < 0$, conc. down

$\dfrac{3\pi}{2} < x < \dfrac{5\pi}{2}$, $y'' > 0$, conc. up

Infl. $\left(\dfrac{\pi}{2}, \dfrac{\pi}{2}\right)$

Infl. $\left(\dfrac{3\pi}{2}, \dfrac{3\pi}{2}\right)$

29. $y = 4 \cos^2 x^2$, $x = 1$

$$y' = (8 \cos x^2)(-\sin x^2)(2x)$$

$$= -16x \cos x^2 \sin x^2$$

For $x = 1$: $y = 4 \cos^2 1 = 1.17$, $y' = -16 \cos 1 \sin 1 = -7.27$

$y - y_1 = m(x - x_1)$; $x_1 = 1$, $y_1 = 1.17$, $m = -7.27$

$y - 1.17 = -7.27(x - 1)$, $7.27x + y - 8.44 = 0$

31. $\sin x = 1 - x$

$f(x) = 1 - x - \sin x$

$f'(x) = -1 - \cos x$

$f(0) = 1, f(1) = -0.84$

Let $x = 0.5$

n	x_n	$f(x_n)$	$f'(x_n)$	$x_n - \dfrac{f(x_n)}{f'(x_n)}$
1	0.5	0.0205745	-1.8775826	0.5109580
2	0.5109580	0.0000290	-1.8722765	0.5109734
3	0.5109734	1.76×10^{-10}	-1.8722689	0.5109734
4	0.5109734	$x_4 = x_3 = 0.5109734$		

33. $W = 10 \cos 2t$

$P = \dfrac{dW}{dt} = 10(-\sin 2t)(2)$

$\quad = -20 \sin 2t$

35. $x = \sin 2\pi t, \ y = 2 \cos^2 \pi t, \ t = \dfrac{13}{12}$

$v_x = 2\pi \cos 2\pi t, \ v_y = (4 \cos \pi t)(-\sin \pi t)(\pi) = -2\pi \sin 2\pi t$

For $t = \dfrac{13}{12}$: $v_x = 2\pi \cos \dfrac{13\pi}{6} = 2\pi(\dfrac{\sqrt{3}}{2}) = \pi\sqrt{3}$

$v_y = -2\pi \sin \dfrac{13\pi}{6} = -2\pi(\dfrac{1}{2}) = -\pi$

$v = \sqrt{(\pi\sqrt{3})^2 + (-\pi)^2} = 2\pi$

$\tan \theta = \dfrac{-\pi}{\pi\sqrt{3}} = -\dfrac{1}{\sqrt{3}}, \ \theta = 330° \ (v_x > 0, \ v_y < 0)$

37. $I = I_m \cos^2\theta, \ \dfrac{d\theta}{dt} = 0.03$ rad/s

$I = 8$ units when $\theta = 60°$

$8 = I_m \cos^2 60° = I_m(0.5)^2$

$I_m = 32, \ I = 32 \cos^2\theta$

$\dfrac{dI}{dt} = -64 \cos\theta \sin\theta \dfrac{d\theta}{dt}$

$\dfrac{dI}{dt}\Big|_{\theta=60°} = -64 \cos 60° \sin 60°(0.03) = -0.831$ units/s

39. $\omega = \sqrt{\dfrac{g}{\ell \cos\theta}} = \sqrt{\dfrac{g}{\ell}}(\cos\theta)^{-1/2}$

$g = 9.800$ m/s^2, $\ell = 0.6375$ m

$d\theta = 0.25° = 0.004363$

$d\omega = \sqrt{\dfrac{g}{\ell}}(-\dfrac{1}{2})(\cos\theta)^{-3/2}(-\sin\theta)d\theta$

$\quad = \sqrt{\dfrac{g}{\ell}}\dfrac{\sin\theta \, d\theta}{2(\cos\theta)^{3/2}}$

Evaluating:

$d\omega = \sqrt{\dfrac{9.800}{0.6375}}\dfrac{(\sin 32.5°)(0.004363)}{2(\cos 32.5°)^{3/2}}$

$\quad = 0.005933$ rad/s

41. See Fig. 6-45

$\dfrac{d\theta}{dt} = 6.00$ r/min

$\quad = 37.7$ rad/min

Find dx/dt

$\theta = \text{Arctan} \dfrac{x}{300}$

$\dfrac{d\theta}{dt} = \dfrac{1}{1 + (x/300)^2}\dfrac{dx/dt}{300} = \dfrac{300 \, dx/dt}{300^2 + x^2}$

Evaluating for $\theta = 45°$, $x = 300$ ft

$37.7 = \dfrac{300 \, dx/dt}{(300^2 + 300^2)}, \ \dfrac{dx}{dt} = 22{,}600$ ft/min

$\quad = 377$ ft/s

43. See Fig. 6-47

When taped →

$V = \dfrac{1}{3}\pi r^2 h$

Bottom circumference = arc (Fig. 6-47)

$2\pi r = 24.0(2\pi - \theta), \ r = \dfrac{12}{\pi}(2\pi - \theta)$

$h = \sqrt{24.0 - r^2} = \sqrt{576 - [(12/\pi)(2\pi - \theta)]^2}$

$\quad = \dfrac{12}{\pi}\sqrt{4\pi\theta - \theta^2}$

$V = \dfrac{1}{3}\pi(\dfrac{12^2}{\pi^2})(2\pi - \theta)^2(\dfrac{12}{\pi})\sqrt{4\pi\theta - \theta^2}$

$\quad = \dfrac{576}{\pi^2}(2\pi - \theta)^2(4\pi\theta - \theta^2)^{1/2}$

$\dfrac{dV}{d\theta} = \dfrac{576}{\pi^2}[(2\pi - \theta)^2(\dfrac{1}{2})(4\pi\theta - \theta^2)^{-1/2}(4\pi - 2\theta) + (4\pi\theta - \theta^2)^{1/2}(2)(2\pi - \theta)(-1)]$

$\quad = \dfrac{576}{\pi^2}[\dfrac{(2\pi - \theta)^3}{(4\pi\theta - \theta^2)^{1/2}} - 2(2\pi - \theta)(4\pi\theta - \theta^2)^{1/2}] = \dfrac{576(2\pi - \theta)[(2\pi - \theta)^2 - 2(4\pi\theta - \theta^2)]}{\pi^2(4\pi\theta - \theta^2)^{1/2}}$

$\quad = \dfrac{576}{\pi^2}\dfrac{(2\pi - \theta)(3\theta^2 - 12\pi\theta + 4\pi^2)}{(4\pi\theta - \theta^2)^{1/2}}; \ \theta = 2\pi$ gives $V = 0$

$3\theta^2 - 12\pi\theta + 4\pi^2 = 0, \ \theta = \dfrac{12\pi \pm \sqrt{144\pi^2 - 48\pi^2}}{6} = 2\pi \pm \dfrac{2}{3}\pi\sqrt{6}; \ \theta = 2\pi - \dfrac{2}{3}\pi\sqrt{6} = 1.153 \ (\theta < 2\pi)$

$V = \dfrac{576}{\pi^2}(2\pi - 1.153)^2[4\pi(1.153) - (1.153)^2]^{1/2} = 5570$ cm^3

45. See Fig. 6-48

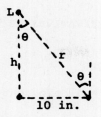

$$I = \frac{k \cos \theta}{r^2} \; , \; \frac{10}{r} = \sin \theta, \; \tan \theta = \frac{10}{h}$$

$$I = \frac{k \cos \theta}{(\frac{10}{\sin \theta})^2} = \frac{k}{100} \cos \theta \sin^2 \theta$$

$$\frac{dI}{d\theta} = \frac{k}{100} [\cos \theta (2 \sin \theta \cos \theta) + \sin^2 \theta (-\sin \theta)] = \frac{k}{100} \sin \theta (2 \cos^2 \theta - \sin^2 \theta)$$

$\sin \theta (2 \cos^2 \theta - \sin^2 \theta) = 0$; $\sin \theta = 0$, $\theta = 0$, L infinitely far, $I = 0$

$2 \cos^2 \theta - \sin^2 \theta = 0$, $\tan^2 \theta = 2$, $\tan \theta = 1.4142$ ($\theta = 54.7°$)

$$h = \frac{10}{\tan \theta} = \frac{10}{1.4142} = 7.07 \text{ in.}$$

Exercises 7-1, p. 286

1. $4^4 = 256$; base = 4, exponent(logarithm) = 4, number = 256; $\log_4 256 = 4$

3. $\log_8 16 = \frac{4}{3}$; base = 8, logarithm(exponent) = $\frac{4}{3}$, number = 16 ; $8^{4/3} = 16$

5. $8^{1/3} = 2$; base = 8, exponent(logarithm) = $\frac{1}{3}$, number = 2 ; $\log_8 2 = \frac{1}{3}$

7. $\log_{0.5} 16 = -4$; base = 0.5, logarithm(exponent) = -4, number = 16 ; $0.5^{-4} = 16$

9. $(12)^0 = 1$; base = 12, exponent(logarithm) = 0, number = 1 ; $\log_{12} 1 = 0$

11. $\log_8(\frac{1}{512}) = -3$; base = 8, logarithm(exponent) = -3, number = $\frac{1}{512}$; $8^{-3} = \frac{1}{512}$

13. $\log_7 y = 3$; $y = 7^3 = 343$ 15. $\log(10)^{0.2} = x$; $\log_{10}(10)^{0.2} = x$; $10^{0.2} = 10^x$; $x = 0.2$

17. $y = 3^x$

x	-3	-2	-1	0	1	2	3
y	$\frac{1}{27}$	$\frac{1}{9}$	$\frac{1}{3}$	1	3	9	27

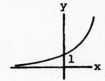

19. $y = \log_2 x$; $x = 2^y$

x	$\frac{1}{8}$	$\frac{1}{4}$	$\frac{1}{2}$	1	2	4	8
y	-3	-2	-1	0	1	2	3

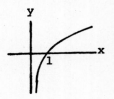

21. $\log_5 9 - \log_5 3 = \log_5 \frac{9}{3} = \log_5 3$ [Eq.(7-8)]

23. $\log_6 x^2 + \log_6 \sqrt{x} = \log_6(x^2)(\sqrt{x}) = \log_6(x^2 x^{1/2}) = \log_6 x^{5/2}$ [Eq.(7-7)]
$$= \frac{5}{2}\log_6 x \quad [Eq.(7-9)]$$

25. $2\log_b 2 + \log_b y = \log_b 2^2 + \log_b y = \log_b 4y$ [Eqs.(7-9),(7-7)]

27. $2\log_2 x + \log_2 3 - \log_2 5 = \log_2 x^2 + \log_2 3 - \log_2 5 = \log_2 3x^2 - \log_2 5 = \log_2(\frac{3}{5}x^2)$
   [Eqs.(7-9), (7-7), (7-8)]

29. $\ln 3 = \log_e 3 = \frac{\log 3}{\log e} = \frac{0.4771}{0.4343} = 1.099$ [Eq.(7-10)]

31. $\log_7 42 = \frac{\log 42}{\log 7} = \frac{1.6232}{0.8451} = 1.921$ [Eq.(7-10)]

33. $p = p_o e^{-kh}$

$\frac{p}{p_o} = e^{-kh}$

$\ln(\frac{p}{p_o}) = -kh$

$h = -\frac{1}{k}\ln(\frac{p}{p_o})$

$= \ln(\frac{p}{p_o})^{-1/k}$

35. $i = \frac{E}{R}(1 - e^{-Rt/L})$

$\frac{iR}{E} = 1 - e^{-Rt/L}$

$e^{-Rt/L} = 1 - \frac{iR}{E} = \frac{E - iR}{E}$

$-\frac{Rt}{L} = \ln(\frac{E - iR}{E})$

$t = -\frac{L}{R}\ln(\frac{E - iR}{E})$

37. (a) $\ln 6.700 = 1.9021$

(b) $\ln 67.00 = \ln(6.700)(10)$
$= \ln 6.700 + \ln 10$
$= 1.9021 + 2.3026$
$= 4.2047$

(c) $\ln 3400 = \ln(3.400)(1000)$
$= \ln 3.400 + 3 \ln 10$
$= 1.2238 + 6.9078$
$= 8.1316$

(d) $\ln 0.3400 = \ln \dfrac{3.400}{10} = \ln 3.400 - \ln 10$
$= 1.2238 - 2.3026 = -1.0788$

$\ln(6.700)(3400) = \ln 6.700 + \ln 3400$
$= 1.9021 + 8.1316$
$= 10.0337$

$10.0337 = 0.8233 + 4(2.3026)$

$(6.700)(3400) = 2.28 \times 10^4$

39. $\sinh 0.25 = \frac{1}{2}(e^{0.25} - e^{-0.25}) = 0.2526$; $\cosh 0.25 = \frac{1}{2}(e^{0.25} + e^{-0.25}) = 1.0314$

$1.0314^2 - 0.2526^2 = 1.0000$

Exercises 7-2 , p. 291

1. $y = \log x^2$
$u = x^2$; $\dfrac{du}{dx} = 2x$

$\dfrac{dy}{dx} = \dfrac{1}{x^2}(\log e)(2x)$

$= \dfrac{2 \log e}{x}$

3. $y = 2 \log_5(3x + 1)$
$u = 3x + 1$; $\dfrac{du}{dx} = 3$

$\dfrac{dy}{dx} = \dfrac{2}{3x + 1}(\log_5 e)(3)$

$= \dfrac{6 \log_5 e}{3x + 1}$

5. $y = \ln(1 - 3x)$
$u = 1 - 3x$; $\dfrac{du}{dx} = -3$

$\dfrac{dy}{dx} = \dfrac{1}{1 - 3x}(-3)$

$= \dfrac{-3}{1 - 3x}$

7. $y = 2 \ln \tan 2x$
$u = \tan 2x$; $\dfrac{du}{dx} = (\sec^2 2x)(2)$

$\dfrac{dy}{dx} = \dfrac{2}{\tan 2x}(2 \sec^2 2x)$

$= \dfrac{4 \sec^2 2x}{\tan 2x} = 4 \sec 2x \csc 2x$

9. $y = \ln\sqrt{x} = \ln x^{1/2} = \frac{1}{2}\ln x$
$u = x$; $\dfrac{du}{dx} = 1$

$\dfrac{dy}{dx} = \dfrac{1}{2}(\dfrac{1}{x})(1) = \dfrac{1}{2x}$

11. $y = \ln(x^2 + 2x)^3 = 3 \ln(x + 2x)$
$u = x^2 + 2x$; $\dfrac{du}{dx} = 2x + 2$

$\dfrac{dy}{dx} = \dfrac{3}{x^2 + 2x}(2x + 2) = \dfrac{6(x + 1)}{x^2 + 2x}$

13. $y = x \ln x$

$\dfrac{dy}{dx} = x[\dfrac{1}{x}(1)] + (\ln x)(1)$

$= 1 + \ln x$

15. $y = \dfrac{3x}{\ln(2x + 1)}$

$\dfrac{dy}{dx} = \dfrac{[\ln(2x + 1)](3) - 3x(\dfrac{1}{2x + 1})(2)}{[\ln(2x + 1)]^2}$

$= \dfrac{3(2x + 1)\ln(2x + 1) - 6x}{(2x + 1)\ln^2(2x + 1)}$

17. $y = \ln(\ln x)$

$\dfrac{dy}{dx} = \dfrac{1}{\ln x}(\dfrac{1}{x})(1)$

$= \dfrac{1}{x \ln x}$

19. $y = \ln\dfrac{x}{1 + x} = \ln x - \ln(1 + x)$

$\dfrac{dy}{dx} = \dfrac{1}{x}(1) - \dfrac{1}{1 + x}(1) = \dfrac{1}{x} - \dfrac{1}{x + 1}$

$= \dfrac{x + 1 - x}{x(x + 1)} = \dfrac{1}{x^2 + x}$

21. $y = \sin \ln x$

$\dfrac{dy}{dx} = (\cos \ln x)[\dfrac{1}{x}(1)]$

$= \dfrac{\cos \ln x}{x}$

23. $y = (\ln 2x)^2$

$\dfrac{dy}{dx} = 2(\ln 2x)[\dfrac{1}{2x}(2)]$

$= \dfrac{2 \ln 2x}{x}$

25. $y = \ln(x \tan x) = \ln x + \ln \tan x$

$\dfrac{dy}{dx} = \dfrac{1}{x}(1) + \dfrac{1}{\tan x}(\sec^2 x)$

$= \dfrac{1}{x} + \dfrac{\sec^2 x}{\tan x} = \dfrac{\tan x + x \sec^2 x}{x \tan x}$

27. $y = \ln\dfrac{x^2}{x+2} = \ln x^2 - \ln(x+2)$

$\dfrac{dy}{dx} = \dfrac{1}{x^2}(2x) - \dfrac{1}{x+2}(1)$

$= \dfrac{2}{x} - \dfrac{1}{x+2} = \dfrac{2(x+2) - x}{x(x+2)}$

$= \dfrac{x+4}{x(x+2)}$

29. $\dfrac{\ln 2.0001 - \ln 2.0000}{0.0001} = 0.4999875$

(final digits vary with calculator)
slope of secant line through
(2.0000, ln 2.0000) and
(2.0001, ln 2.0001)

$0.5 = \dfrac{d \ln x}{dx}$ for $x = 2$ and is slope of
tangent line through
(2.0000, ln 2.0000)

31.

x	0.1	0.01	0.001	0.0001
$(1+x)^{1/x}$	2.5937	2.7048	2.7169	2.71815

33. $y = \text{Arctan } 2x + \ln(4x^2 + 1)$; $x = 0.625$

$\dfrac{dy}{dx} = \dfrac{1}{1+4x^2}(2) + \dfrac{1}{4x^2+1}(8x) = \dfrac{8x+2}{1+4x^2}$

$\dfrac{dy}{dx}\Big|_{x=0.625} = \dfrac{8(0.625)+2}{1+4(0.625)^2} = 2.73$

35. $y = \ln \cos x$; $x = \dfrac{\pi}{4}$

$\dfrac{dy}{dx} = \dfrac{1}{\cos x}(-\sin x) = -\tan x$

$\dfrac{dy}{dx}\Big|_{x=\pi/4} = -\tan\dfrac{\pi}{4} = -1$

$= m_{\tan}$ for $x = \pi/4$

37. $y = x^x$

$\ln y = \ln x^x = x \, \text{l} x \, x$

$\dfrac{1}{y}\dfrac{dy}{dx} = x(\dfrac{1}{x})(1) + (\ln x)(1) = 1 + \ln x$

$\dfrac{dy}{dx} = y(1 + \ln x) = x^x(1 + \ln x)$

39. $b = 10 \log(I/I_o)$

$\dfrac{db}{dt} = 10 \log e (\dfrac{1}{I/I_o})(\dfrac{d \, I/I_o}{dt})$

$= \dfrac{10 \log e}{I/I_o} \dfrac{1}{I_o}\dfrac{dI}{dt} = \dfrac{10 \log e}{I}\dfrac{dI}{dt}$

Exercises 7-3, p. 295

1. $y = 3^{2x}$; $b = 3$, $u = 2x$, $\dfrac{du}{dx} = 2$

$\dfrac{dy}{dx} = \dfrac{1}{\log_3 e}(3^{2x})(2) = \dfrac{2(3^{2x})}{\log_3 e}$

3. $y = 4^{6x}$; $b = 4$, $u = 6x$, $\dfrac{du}{dx} = 6$

$\dfrac{dy}{dx} = \dfrac{1}{\log_4 e}(4^{6x})(6) = \dfrac{6(4^{6x})}{\log_4 e}$

5. $y = e^{6x}$; $b = e$, $u = 6x$, $\dfrac{du}{dx} = 6$

$\dfrac{dy}{dx} = e^{6x}(6) = 6e^{6x}$

7. $y = e^{\sqrt{x}}$; $b = e$, $u = \sqrt{x} = x^{1/2}$, $\dfrac{du}{dx} = \dfrac{1}{2}x^{-1/2}$

$\dfrac{dy}{dx} = e^{\sqrt{x}}(\dfrac{1}{2}x^{-1/2}) = \dfrac{e^{\sqrt{x}}}{2\sqrt{x}}$

9. $y = xe^x$

$\dfrac{dy}{dx} = x(e^x)(1) + e^x(1) = e^x(x+1)$

11. $y = xe^{\sin x}$

$\dfrac{dy}{dx} = x(e^{\sin x})(\cos x)(1) + e^{\sin x}(1)$

$= e^{\sin x}(x \cos x + 1)$

13. $y = \dfrac{3e^{2x}}{x+1}$

$\dfrac{dy}{dx} = \dfrac{(x+1)(3e^{2x})(2) - 3e^{2x}(1)}{(x+1)^2}$

$\qquad = \dfrac{3e^{2x}(2x+1)}{(x+1)^2}$

15. $y = e^{-3x}\sin 4x$

$\dfrac{dy}{dx} = e^{-3x}(\cos 4x)(4) + (\sin 4x)(e^{-3x})(-3)$

$\qquad = e^{-3x}(4\cos 4x - 3\sin 4x)$

17. $y = \dfrac{2e^{3x}}{4x+3}$

$\dfrac{dy}{dx} = \dfrac{(4x+3)(2e^{3x})(3) - 2e^{3x}(4)}{(4x+3)^2}$

$\qquad = \dfrac{2e^{3x}(12x+5)}{(4x+3)^2}$

19. $y = \ln(e^{x^2} + 4)$

$\dfrac{dy}{dx} = \dfrac{1}{e^{x^2}+4}(e^{x^2})(2x) = \dfrac{2xe^{x^2}}{e^{x^2}+4}$

21. $y = (2e^{2x})^3\sin x^2 = 8e^{6x}\sin x^2$

$\dfrac{dy}{dx} = 8e^{6x}(\cos x^2)(2x) + (\sin x^2)(8e^{6x})(6)$

$\qquad = 16e^{6x}(x\cos x^2 + 3\sin x^2)$

23. $y = (\ln 2x + e^{2x})^2$

$\dfrac{dy}{dx} = 2(\ln 2x + e^{2x})[\dfrac{1}{2x}(2) + e^{2x}(2)]$

$\qquad = 2(\ln 2x + e^{2x})(\dfrac{1}{x} + 2e^{2x})$

25. $y = e^{2x} - e^{-2x}$

$\dfrac{dy}{dx} = e^{2x}(2) - e^{-2x}(-2)$

$\qquad = 2(e^{2x} + e^{-2x})$

27. $y = e^{2x}\ln x$

$\dfrac{dy}{dx} = e^{2x}(\dfrac{1}{x})(1) + (\ln x)(e^{2x})(2)$

$\qquad = \dfrac{e^{2x}}{x} + 2e^{2x}\ln x$

29. $y = \ln \sin 2e^{6x}$

$\dfrac{dy}{dx} = \dfrac{(\cos 2e^{6x})(2e^{6x})(6)}{\sin 2e^{6x}}$

$\qquad = 12e^{6x}\cot 2e^{6x}$

31. $y = 2\,\text{Arcsin}\, e^{2x}$

$\dfrac{dy}{dx} = \dfrac{2}{\sqrt{1-(e^{2x})^2}}(e^{2x})(2)$

$\qquad = \dfrac{4e^{2x}}{\sqrt{1-e^{4x}}}$

33. (a) $e = 2.7182818$; $\dfrac{d\,e^x}{dx}$ for $x=1$

slope of tangent line through $(1,e)$

(b) $\dfrac{e^{1.0001} - e^{1.0000}}{0.0001} = 2.7184178$

(final digits vary with calculator)
slope of secant line through
$(1.0001, e^{1.0001})$ and $(1.0000, e^{1.0000})$

35. $y = e^{-2x}\cos 2x$; $x = 0.625$

$\dfrac{dy}{dx} = e^{-2x}(-\sin 2x)(2) + (\cos 2x)(e^{-2x})(-2)$

$\qquad = -2e^{-2x}(\sin 2x + \cos 2x)$

$\dfrac{dy}{dx}\Big|_{x=0.625} = -2e^{-2(0.625)}[\sin 2(0.625) + \cos 2(0.625)]$

$\qquad = -2e^{-1.25}(\sin 1.25 + \cos 1.25) = -0.724$

37. $y = xe^{-x}$

$\dfrac{dy}{dx} = x(e^{-x})(-1) + e^{-x}(1)$

$\qquad = e^{-x}(1-x)$

$e^{-x}(1-x) + xe^{-x} = e^{-x}$

$\qquad\qquad e^{-x} = e^{-x}$

39. $y = e^x$

$\dfrac{dy}{dx} = e^x$

$y\big|_{x=a} = e^a$

$\dfrac{dy}{dx}\Big|_{x=a} = e^a$

41. $V = 100(1 - e^{-0.1t})$

$\dfrac{dV}{dt} = 100(-e^{-0.1t})(-0.1)$

$\qquad = 10e^{-0.1t}$

43. $\sinh u = \dfrac{1}{2}(e^u - e^{-u})$

$\dfrac{d}{dx}\sinh u = \dfrac{1}{2}[e^u\dfrac{du}{dx} - e^{-u}(-\dfrac{du}{dx})]$

$\qquad = \dfrac{1}{2}(e^u + e^{-u})\dfrac{du}{dx}$

$\qquad = \cosh u\,\dfrac{du}{dx}$

Exercises 7-4, p. 298

1. $y = \ln \cos x$

Intercepts: $(0,0)$, $(2\pi,0)$,...

For $x=2n\pi$, n an integer,
$y = \ln 1 = 0$

For $y=0$, $x=0$

Symmetry: To y-axis

$y = \ln \cos(-x) = \ln \cos x$

Behavior as x becomes large:

$\ln x$ defined only if $x>0$,
therefore y defined if $\cos x > 0$,
or $-\dfrac{\pi}{2} < x < \dfrac{\pi}{2}$, $\dfrac{3\pi}{2} < x < \dfrac{5\pi}{2}$, ...

Vertical asymptotes: $x = -\dfrac{\pi}{2}, \dfrac{\pi}{2}$,...

As $x \to \dfrac{\pi}{2}$, $\cos x \to 0$, $y \to -\infty$

Same for other asymptotes

Derivatives: $y' = \dfrac{1}{\cos x}(-\sin x) = -\tan x$

$\tan x = 0$ for $x = 0, \pi, 2\pi, \ldots$,
where $x = 0, 2\pi, 4\pi, \ldots$ in defined ranges

$y'' = -\sec^2 x$ is negative for all x

$(0,0)$, $(2\pi,0)$, ... are rel. max.

y concave down for all defined values

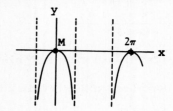

3. $y = x e^{-x}$

Intercepts: $(0,0)$

Symmetry: None

Behavior as x becomes large:

As $x \to +\infty$, $y \to 0$ (due to e^{-x})

As $x \to -\infty$, $y \to -\infty$

$y = 0$ is an asymptote

Derivatives:

$y' = -xe^{-x} + e^{-x} = e^{-x}(1-x)$

$y' = 0$ for $1-x=0$, $x=1$

$x<1$, $y'>0$, y inc.

$x>1$, $y'<0$, y dec.

$(1, 1/e)$ rel. max.

$y'' = -e^{-x}(1-x) - e^{-x} = -e^{-x}(x-2)$

$x<2$, $y''>0$, conc. up

$x>2$, $y''<0$, conc. down

Infl. $(2, 2/e^2)$

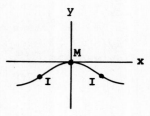

5. $y = \ln \dfrac{1}{x^2 + 1} = -\ln(x^2 + 1)$

Intercepts: $(0,0)$

For $x=0$, $y = -\ln 1 = 0$

Symmetry: To y-axis

$y = -\ln[(-x)^2 + 1] = -\ln(x^2 + 1)$

Behavior as x becomes large:

As $x \to +\infty$, $y \to -\infty$

As $x \to -\infty$, $y \to -\infty$

Derivatives: $y' = -\dfrac{2x}{x^2+1}$

$x<0$, $y'>0$, y inc.

$x>0$, $y'<0$, y dec. $(0,0)$ rel. max

$y'' = -\dfrac{(x^2+1)(2) - 2x(2x)}{(x^2+1)^2}$

$= \dfrac{2x^2-2}{(x^2+1)^2} = \dfrac{2(x+1)(x-1)}{(x^2+1)^2}$

$y'' = 0$ for $x = -1, 1$

$x<-1$, $y''>0$, conc. up

$-1<x<1$, $y''<0$, conc. down

$x>1$, $y''>0$, conc. up

Infl. $(-1, -\ln 2)$, $(1, -\ln 2)$

7. $y = e^{-x^2}$

Intercepts: $(0,1)$
For $x=0$, $y=e^0=1$
Symmetry: To y-axis
$y = e^{(-x)^2} = e^{x^2}$
Behavior as x becomes large:
As $x \to +\infty$, $y \to 0$
As $x \to -\infty$, $y \to 0$
$y=0$ is an asymptote

Derivatives:
$y' = e^{-x^2}(-2x) = -2xe^{-x^2}$
$x<0$, $y'>0$, y inc.
$x>0$, $y'<0$, y dec.
$(0,1)$ rel. max.

$y'' = -2xe^{-x^2}(-2x) - 2e^{-x^2}$
$\quad = 2e^{-x^2}(2x^2 - 1)$
$y''=0$ for $x = \pm\sqrt{1/2}$
$x < -\sqrt{1/2}$, $y''>0$, conc. up
$-\sqrt{1/2}<x<\sqrt{1/2}$, $y''<0$, conc. down
$x>\sqrt{1/2}$, $y''>0$, conc. up
Infl.: $(\pm\frac{1}{2}\sqrt{2}, \frac{1}{e}\sqrt{e})$

9. $y = \ln x - x$

Intercepts: None
$x=0$, $\ln 0$ not defined
$y=0$, $\ln x \neq x$ for any x
Symmetry: None
Behavior as x becomes large:
Only values of $x>0$ valid
since $\ln x$ not defined for $x\leq 0$
As $x \to +\infty$, $y \to -\infty$ ($-x$ term)

Vertical asymptotes: $x = 0$
As $x \to 0$, $\ln x \to -\infty$
Derivatives: $y' = \frac{1}{x} - 1 = \frac{1-x}{x}$
$0<x<1$, $y'>0$, y inc.
$x>1$, $y'<0$, y dec.
$(1,-1)$ rel. max.
$y'' = -\frac{1}{x^2}$, conc. down $x>0$

11. $y = \frac{1}{2}(e^x - e^{-x})$

Intercepts: $(0,0)$
For $x=0$, $y = \frac{1}{2}(e^0 - e^0)=0$
For $y=0$, $e^x = e^{-x}$ only for $x=0$
Symmetry: To origin
$-y = \frac{1}{2}(e^{-x}-e^x) = \frac{1}{2}(e^x - e^{-x})$
Behavior as x becomes large:
As $x \to +\infty$, $e^x \to +\infty$, $y \to +\infty$
As $x \to -\infty$, $e^{-x} \to +\infty$, $y \to -\infty$

Derivatives:
$y' = \frac{1}{2}(e^x + e^{-x})$
All x, $y'>0$, y inc.
$y'' = \frac{1}{2}(e^x - e^{-x})$
$x<0$, $e^{-x}>e^x$, $y''<0$, conc. down
$x>0$, $e^x>e^{-x}$, $y''>0$, conc. up
Infl. $(0,0)$

13. $y = x^2 \ln x$ at $(1,0)$
$y' = x^2(\frac{1}{x}) + 2x \ln x = x + 2x \ln x$
$y'|_{x=1} = 1 + 2\ln 1 = 1$
$m=1$, $(x_1,y_1) = (1,0)$
$y - 0 = 1(x - 1)$, $y = x - 1$

15. $y = 2 \sin \frac{1}{2} x$, $x = \frac{3}{2}\pi$
$y' = 2(\cos\frac{1}{2}x)(\frac{1}{2}) = \cos\frac{1}{2}x$
$y'|_{x=3\pi/2} = \cos\frac{3\pi}{4} = -\frac{1}{\sqrt{2}}$, $m = \sqrt{2}$, $x_1 = \frac{3}{2}\pi$
$y_1 = 2\sin\frac{3\pi}{4} = 2(\frac{\sqrt{2}}{2}) = \sqrt{2}$
$y - \sqrt{2} = \sqrt{2}(x - \frac{3}{2}\pi)$
$2\sqrt{2}\, x - 2y + 2\sqrt{2} - 3\pi\sqrt{2} = 0$

17. $x^2 - 2 + \ln x = 0$
$f(x) = x^2 - 2 + \ln x$
$f'(x) = 2x + \frac{1}{x}$
$f(0)$ not defined
$f(1) = -1$, $f(2) = 2.7$
Let $x_1 = 1.3$

n	x_n	$f(x_n)$	$f'(x_n)$	$x_n - \dfrac{f(x_n)}{f'(x_n)}$
1	1.3	-0.0476357	3.3692308	1.3141385
2	1.3141385	0.0001412	3.3892317	1.3140968
3	1.3140968			

$x_3 = x_2 = 1.3141$ (to four decimal places)

19. $P = 100e^{-0.005t}$, $t = 100$ d

$$\frac{dP}{dt} = 100e^{-0.005t}(-0.005) = -0.5e^{-0.005t}$$

$$\frac{dP}{dt}\Big|_{t=100} = -0.5e^{-0.5} = -0.303 \text{ W/d}$$

21. $v = 40(1 - e^{-0.05t})$, $t = 10$

$$a = \frac{dv}{dt} = 40(-e^{-0.05t})(-0.05)$$

$$= 2e^{-0.05t}$$

$$a\Big|_{t=10} = 2e^{-0.5} = 1.21$$

23. $x = ae^{kt} + be^{-kt}$

$$v = \frac{dx}{dt} = ake^{kt} - bke^{-kt}$$

$$a = \frac{dv}{dt} = ak^2e^{kt} + bk^2e^{-kt}$$

$$= k^2(ae^{kt} + be^{-kt}) = k^2x$$

$$F = k_1a = k_1k^2x = k_2x$$

$$(k_2 = k_1k^2)$$

25. $y = e^{-0.5t}(0.4 \cos 6t - 0.2 \sin 6t)$, $t = \frac{\pi}{12}$ s

$$v = \frac{dy}{dt} = e^{-0.5t}(-2.4 \sin 6t - 1.2 \cos 6t)$$

$$-0.5e^{-0.5t}(0.4 \cos 6t - 0.2 \sin 6t)$$

$$= -e^{-0.5t}(2.3 \sin 6t + 1.4 \cos 6t)$$

$$v\Big|_{t=\pi/12} = -e^{-\pi/24}(2.3 \sin \frac{\pi}{2} + 1.4 \cos \frac{\pi}{2})$$

$$= -e^{-0.131}[2.3(1) + 1.4(0)]$$

$$= -2.3e^{-0.131} = -2.02 \text{ cm/s}$$

27. $\dfrac{T_2}{T_1} = e^{k/\sin(\theta/2)} = e^{k \csc(\theta/2)}$

$$d(\frac{T_2}{T_1}) = e^{k \csc(\theta/2)}(-k \csc \frac{\theta}{2})(\cot \frac{\theta}{2})(\frac{1}{2})d\theta = -\frac{k}{2}e^{k/\sin(\theta/2)}\csc \frac{\theta}{2} \cot \frac{\theta}{2} \, d\theta$$

Review Exercises for Chapter 7, p. 299

1. $y = 3 \ln(x^2 + 1)$

$$\frac{dy}{dx} = \frac{3}{x^2 + 1}(2x)$$

$$= \frac{6x}{x^2 + 1}$$

3. $y = (e^{x-3})^2 = e^{2(x-3)}$

$$\frac{dy}{dx} = e^{2(x-3)}(2)$$

$$= 2e^{2(x-3)}$$

5. $y = \ln \csc x^2$

$$\frac{dy}{dx} = \frac{1}{\csc x^2}(-\csc x^2 \cot x^2)(2x)$$

$$= -2x \cot x^2$$

7. $y = [\ln(3 + \sin x)]^2$

$$\frac{dy}{dx} = 2[\ln(3 + \sin x)](\frac{1}{3 + \sin x})(\cos x)$$

$$= \frac{2 \cos x \ln(3 + \sin x)}{3 + \sin x}$$

9. $y = e^{\sin 2x}$

$$\frac{dy}{dx} = e^{\sin 2x}(\cos 2x)(2)$$

$$= 2 \cos 2x \, e^{\sin 2x}$$

11. $y = x^3e^x$

$$\frac{dy}{dx} = x^3e^x + e^x(3x^2)$$

$$= x^2(x + 3)e^x$$

13. $y = \ln(x - e^{-x})^2 = 2 \ln(x - e^{-x})$

$$\frac{dy}{dx} = \frac{2}{x - e^{-x}}[1 - e^{-x}(-1)] = \frac{2(1 + e^{-x})}{x - e^{-x}}$$

15. $y = \dfrac{\cos^2x}{e^{3x} + 1}$

$$\frac{dy}{dx} = \frac{(e^{3x} + 1)(2 \cos x)(-\sin x) - (\cos^2x)(e^{3x})(3)}{(e^{3x} + 1)^2}$$

$$= \frac{-\cos x[2(e^{3x} + 1)\sin x + 3e^{3x}\cos x]}{(e^{3x} + 1)^2}$$

$$= \frac{-\cos x(2e^{3x}\sin x + 3e^{3x}\cos x + 2 \sin x)}{(e^{3x} + 1)^2}$$

17. $y = x^2 \ln x$

$$\frac{dy}{dx} = x^2(\frac{1}{x}) + (\ln x)(2x)$$

$$= x + 2x \ln x$$

$$= x(1 + 2 \ln x)$$

19. $y = \dfrac{e^{2x}}{x^2 + 1}$

$\dfrac{dy}{dx} = \dfrac{(x^2+1)(e^{2x})(2) - e^{2x}(2x)}{(x^2+1)^2}$

$= \dfrac{2e^{2x}(x^2 - x + 1)}{(x^2+1)^2}$

21. $y = e^{-x} \sec x$

$\dfrac{dy}{dx} = e^{-x}(\sec x \tan x) + (\sec x)(e^{-x})(-1)$

$= e^{-x} \sec x (\tan x - 1)$

23. $x^2 \ln y = y + x$

$x^2 \left(\dfrac{1}{y}\right)\dfrac{dy}{dx} + (\ln y)(2x) = \dfrac{dy}{dx} + 1$

$x^2 \dfrac{dy}{dx} - y \dfrac{dy}{dx} = y - 2xy \ln y$

$\dfrac{dy}{dx} = \dfrac{y(1 - 2x \ln y)}{x^2 - y}$

25. $y = \ln \cos x, \quad x = \dfrac{\pi}{6}$

$y' = \dfrac{1}{\cos x}(-\sin x) = -\tan x$

$y' \big|_{x=\pi/6} = -\tan \dfrac{\pi}{6} = -\dfrac{1}{3}\sqrt{3}, \quad x_1 = \dfrac{\pi}{6}$

$y_1 = \ln \cos \dfrac{\pi}{6} = \ln \dfrac{1}{2}\sqrt{3} = -0.144$

$y + 0.144 = -\dfrac{1}{3}\sqrt{3}\left(x - \dfrac{\pi}{6}\right)$

$3y + 0.432 = -\sqrt{3}x + \dfrac{1}{6}\pi\sqrt{3}$

$3y + 0.432 = -1.732x + 0.907$

$1.73x + 3y - 0.48 = 0$

27. $y = \ln(1 + x)$

<u>Intercepts:</u> $(0,0)$

For $x=0$, $y=\ln 1 = 0$

<u>Symmetry:</u> None

<u>Behavior as x becomes large:</u>

As $x \to +\infty$, $\ln(1+x) \to +\infty$, $y \to +\infty$

$\ln(x+1)$ not defined for $x \le -1$

<u>Vertical asymptotes:</u>

As $x \to -1$, $\ln(x+1) \to -\infty$, $y \to -\infty$

$x = -1$ is an asymptote

<u>Derivatives:</u> $y' = \dfrac{1}{1+x}$

$x > -1$, $y' > 0$, y inc.

$y'' = -\dfrac{1}{(1+x)^2}$

$x > -1$, $y'' < 0$, conc. down

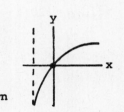

29. $e^x - x^2 = 0$

$f(x) = e^x - x^2$

$f'(x) = e^x - 2x$

$f(0) = 1$, $f(-1) = -0.63$

Let $x_1 = -0.7$

n	x_n	$f(x_n)$	$f'(x_n)$	$x_n - \dfrac{f(x_n)}{f'(x_n)}$
1	-0.7	0.0065853	1.8965853	-0.7034722
2	-0.7034722	-0.0000091	1.9018084	-0.7034674
3	-0.7034674			

$x_3 = x_2 = -0.70347$ (to 5 decimal places)

31. $n = 3(e^{-0.05t} + 1)$

$\dfrac{dn}{dt} = 3e^{-0.05t}(-0.05)$

$= -0.05(3e^{-0.05t})$

$= -0.05(n-3) = 0.05(3-n)$

33. $E = 170e^{-0.015t}$, $t = 0.100$ s, $dt = 0.005$ s

$dE = 170e^{-0.015t}(-0.015)dt$

$= -2.55e^{-0.015t}dt$

Evaluating: $dE = -2.55e^{-0.0015}(0.005) = -0.0127$ V

35. $i = i_0 e^{-t/RC}$, $i_0 = 10$ A

$R = 2 \ \Omega$, $C = 50 \ \mu F = 5 \times 10^{-5}$ F

$t = 200 \ \mu s = 2 \times 10^{-4}$ s, $dt = 10 \ \mu s = 10^{-5}$ s

$di = i_0 e^{-t/RC}\left(-\dfrac{1}{RC}\right)dt = -\dfrac{i_0 e^{-t/RC}dt}{RC}$

Evaluating:

$di = -\dfrac{10(e^{-2 \times 10^{-4}/10^{-4}})(10^{-5})}{2(5 \times 10^{-5})}$

$= -e^{-2} = -0.135$ A

37. $V = 100(0.95)^t$, $t = 3$ years

$dt = 1$ year

$dV = 100 \ \dfrac{1}{\log_{0.95} e}(0.95)^t(1)dt$

Evaluating: $(1/\log_{0.95} e = \ln 0.95)$

$dV = 100(\ln 0.95)(0.95)^3(1)$

$= -\$4.40$

39. $t = a \ln \dfrac{x}{b-x} - c = a \ln x - a \ln(b-x) - c$

$\dfrac{dt}{dx} = \dfrac{a}{x} - \dfrac{a}{b-x}(-1) = \dfrac{a}{x} + \dfrac{a}{b-x} = \dfrac{ab}{x(b-x)}$

43. $W = k \ln(P_1/P_2) = k \ln P_1 - k \ln P_2$

$k = 2400 \text{ J/mol}, \quad P_1 = 200 \text{ kPa} = 2\times10^5 \text{ Pa}$

$dP_2/dt = 2 \text{ kPa/min} = 2\times10^3 \text{ Pa/min}$

$\dfrac{dW}{dt} = 0 - k \dfrac{1}{P_2} \dfrac{dP_2}{dt}$

Evaluating for $P_2 = 300 \text{ kPa} = 3\times10^5 \text{ Pa}$

$\dfrac{dW}{dt} = -\dfrac{2400}{3\times10^5}(2\times10^3) = -16.0 \text{ J/min}$

45. $y = 320(t + 10e^{-0.1t} - 10)$

$v = \dfrac{dy}{dt} = 320 + 3200e^{-0.1t}(-0.1)$

$\quad = 320(1 - e^{-0.1t})$

$v\big|_{t=10} = 320(1 - e^{-1}) = 202 \text{ ft/s}$

41. See Fig. 7-9

$y = e^{-x^2}$

$A = xy = xe^{-x^2}$

$\dfrac{dA}{dx} = xe^{-x^2}(-2x) + e^{-x^2}(1)$

$\quad = e^{-x^2}(1 - 2x^2)$

$\dfrac{dA}{dx} = 0, \quad 1-2x^2=0, \quad x = \sqrt{\dfrac{1}{2}} = \dfrac{1}{2}\sqrt{2}$

$A = \dfrac{1}{2}\sqrt{2}\, e^{-1/2} = 0.429$

47. $x = e^{-t}\sin 2t, \quad y = e^{-t}\cos 2t$

$v_x = e^{-t}(2 \cos 2t - \sin 2t)$

$v_y = e^{-t}(-2 \sin 2t - \cos 2t)$

$a_x = e^{-t}(-4 \sin 2t - 2 \cos 2t - 2 \cos 2t + \sin 2t)$

$\quad = e^{-t}(-3 \sin 2t - 4 \cos 2t)$

$a_y = e^{-t}(-4 \cos 2t + 2 \sin 2t + 2 \sin 2t + \cos 2t)$

$\quad = e^{-t}(4 \sin 2t - 3 \cos 2t)$

For $t = \pi/4$:

$a_x = e^{-\pi/4}[-3(1)-4(0)] = -3e^{-\pi/4} = -1.368$

$a_y = e^{-\pi/4}[4(1)-3(0)] = 4e^{-\pi/4} = 1.824$

$a = \sqrt{(-1.368)^2 + (1.824)^2} = 2.28$

$\tan\theta = \dfrac{1.824}{-1.368} = -1.333, \quad \theta = 126.9°$

$\qquad\qquad (a_x<0, \ a_y>0)$

Exercises 8-1, p. 304

1. $\int \sin^4 x \cos x \, dx$; $u = \sin x$, $du = \cos x \, dx$; $n = 4$, $n+1 = 5$

$\int \sin^4 x \cos x \, dx = \dfrac{\sin^5 x}{5} + C = \dfrac{1}{5}\sin^5 x + C$

3. $\int \sqrt{\cos x} \sin x \, dx$; $u = \cos x$, $du = -\sin x \, dx$; $n = \dfrac{1}{2}$, $n+1 = \dfrac{3}{2}$

$\int \sqrt{\cos x} \sin x \, dx = -\int (\cos x)^{1/2}(-\sin x \, dx) = -\dfrac{(\cos x)^{3/2}}{3/2} + C = -\dfrac{2}{3}(\cos x)^{3/2} + C$

5. $\int \tan^2 x \sec^2 x \, dx$; $u = \tan x$, $du = \sec^2 x \, dx$; $n = 2$, $n+1 = 3$

$\int \tan^2 x \sec^2 x \, dx = \dfrac{\tan^3 x}{3} + C = \dfrac{1}{3}\tan^3 x + C$

7. $\int_0^{\pi/8} \cos 2x \sin 2x \, dx$; $u = \sin 2x$, $du = 2\cos 2x \, dx$; $n = 1$, $n+1 = 2$
   (could also let $u = \cos 2x$, $du = -2\sin 2x \, dx$)

$\int_0^{\pi/8} \sin 2x \cos 2x \, dx = \dfrac{1}{2}\int_0^{\pi/8} \sin 2x (2\cos 2x \, dx) = \dfrac{1}{2}\dfrac{\sin^2 2x}{2}\Big|_0^{\pi/8} = \dfrac{1}{4}\sin^2 2x \Big|_0^{\pi/8}$

$= \dfrac{1}{4}\sin^2 2(\dfrac{\pi}{8}) - \dfrac{1}{4}\sin^2 2(0) = \dfrac{1}{4}\sin^2 \dfrac{\pi}{4} - 0 = \dfrac{1}{4}(\dfrac{\sqrt{2}}{2})^2 = \dfrac{1}{8}$

9. $\int (\text{Arcsin } x)^3 (\dfrac{dx}{\sqrt{1-x^2}})$; $u = \text{Arcsin } x$, $du = \dfrac{dx}{\sqrt{1-x^2}}$; $n = 3$, $n+1 = 4$

$\int (\text{Arcsin } x)^3 (\dfrac{dx}{\sqrt{1-x^2}}) = \dfrac{(\text{Arcsin } x)^4}{4} + C = \dfrac{1}{4}\text{Arcsin}^4 x + C$

11. $\int \dfrac{\text{Arctan } 5x}{1 + 25x^2}dx$; $u = \text{Arctan } 5x$, $du = \dfrac{5\,dx}{1 + 25x^2}$; $n = 1$, $n+1 = 2$

$\int \dfrac{\text{Arctan } 5x}{1 + 25x^2}dx = \dfrac{1}{5}\int (\text{Arctan } 5x)(\dfrac{5\,dx}{1 + 25x^2}) = \dfrac{1}{5}\dfrac{\text{Arctan}^2 5x}{2} + C = \dfrac{1}{10}\text{Arctan}^2 5x + C$

13. $\int [\ln(x+1)]^2 \dfrac{dx}{x+1}$; $u = \ln(x+1)$, $du = \dfrac{dx}{x+1}$; $n = 2$, $n+1 = 3$

$\int [\ln(x+1)]^2 \dfrac{dx}{x+1} = \dfrac{[\ln(x+1)]^3}{3} + C = \dfrac{1}{3}\ln^3(x+1) + C$

15. $\int_0^{1/2} \dfrac{\ln(2x+3)dx}{2x+3}$; $u = \ln(2x+3)$, $du = \dfrac{2\,dx}{2x+3}$; $n = 1$, $n+1 = 2$

$\int_0^{1/2} \dfrac{\ln(2x+3)dx}{2x+3} = \dfrac{1}{2}\int_0^{1/2} \ln(2x+3)\dfrac{2\,dx}{2x+3} = \dfrac{1}{2}\dfrac{\ln^2(2x+3)}{2}\Big|_0^{1/2}$

$= \dfrac{1}{4}\ln^2[2(\dfrac{1}{2})+3] - \dfrac{1}{4}\ln^2[2(0)+3] = \dfrac{1}{4}(\ln^2 4 - \ln^2 3) = 0.179$

17. $\int (4 + e^x)^3 e^x dx$; $u = 4 + e^x$, $du = e^x dx$; $n = 3$, $n+1 = 4$

$\int (4 + e^x)^3 e^x dx = \dfrac{(4 + e^x)^4}{4} + C = \dfrac{1}{4}(4 + e^x)^4 + C$

19. $\int (2e^{2x} - 1)^{1/3} e^{2x} dx$; $u = 2e^{2x} - 1$, $du = 4e^{2x} dx$; $n = \dfrac{1}{3}$, $n+1 = \dfrac{4}{3}$

$\int (2e^{2x} - 1)^{1/3} e^{2x} dx = \dfrac{1}{4}\int (2e^{2x} - 1)^{1/3}(4e^{2x} dx) = \dfrac{1}{4}\dfrac{(2e^{2x} - 1)^{4/3}}{4/3} + C = \dfrac{3}{16}(2e^{2x} - 1)^{4/3} + C$

21. $\int(1+\sec^2 x)^4(\sec^2 x \tan x\, dx)$; $u = 1+\sec^2 x$, $du = 2\sec x(\sec x \tan x)dx = 2\sec^2 x \tan x\, dx$;

$\int(1+\sec^2 x)^4(\sec^2 x \tan x\, dx) = \frac{1}{2}\int(1+\sec^2 x)^4(2\sec^2 x \tan x\, dx)$ $n=4$, $n+1=5$

$$= \frac{1}{2}\frac{(1+\sec^2 x)^5}{5} + C = \frac{1}{10}(1+\sec^2 x)^5 + C$$

23. $\int_0^{\pi/6}\frac{\tan x}{\cos^2 x}dx$; $u = \tan x$, $du = \sec^2 x\, dx$; $n = 1$, $n+1 = 2$

$\int_0^{\pi/6}\frac{\tan x}{\cos^2 x}dx = \int_0^{\pi/6}\tan x(\sec^2 x\, dx) = \frac{\tan^2 x}{2}\Big|_0^{\pi/6} = \frac{1}{2}(\tan^2\frac{\pi}{6} - \tan^2 0) = \frac{1}{2}[(\frac{\sqrt{3}}{3})^2 - 0] = \frac{1}{6}$

25. $A = \int_0^{\pi/2}y\, dx = \int_0^{\pi/2}\sin x \cos x\, dx = \frac{1}{2}\sin^2 x\Big|_0^{\pi/2} = \frac{1}{2}(\sin^2\frac{\pi}{2} - \sin^2 0)$

(Can also let $u = \cos x$, $du = -\sin x\, dx$ $= \frac{1}{2}(1 - 0) = \frac{1}{2}$
or $u = \sin 2x$, $du = 2\cos 2x\, dx$.)

27. $\frac{dy}{dx} = \frac{(\ln x)^2}{x}$; $dy = (\ln x)^2\frac{dx}{x}$

$\int dy = \int(\ln x)^2\frac{dx}{x}$; $y = \frac{1}{3}(\ln x)^3 + C$

Curve passes through (1,2)

$2 = \frac{1}{3}(\ln 1)^3 + C$; $2 = 0 + C$; $C = 2$

$y = \frac{1}{3}(\ln x)^3 + 2$

29. $P = mnv^2\int_0^{\pi/2}\sin\theta \cos^2\theta\, d\theta$

$= -mnv^2\int_0^{\pi/2}\cos^2\theta(-\sin\theta\, d\theta)$

$= -mnv^2(\frac{\cos^3\theta}{3})\Big|_0^{\pi/2} = -\frac{mnv^2}{3}(\cos^3\frac{\pi}{2} - \cos^3 0)$

$= -\frac{mnv^2}{3}(0 - 1) = \frac{1}{3}mnv^2$

Exercises 8-2, p. 307

1. $\int\frac{dx}{1 + 4x}$; $u = 1+4x$, $du = 4dx$

$\int\frac{dx}{1 + 4x} = \frac{1}{4}\int\frac{4dx}{1 + 4x} = \frac{1}{4}\ln|1 + 4x| + C$

3. $\int\frac{xdx}{4 - 3x^2}$; $u = 4 - 3x^2$, $du = -6xdx$

$\int\frac{xdx}{4 - 3x^2} = -\frac{1}{6}\int\frac{-6xdx}{4 - 3x^2} = -\frac{1}{6}\ln|4 - 3x^2| + C$

5. $\int\frac{\csc^2 2x}{\cot 2x}dx$; $u = \cot 2x$, $du = -2\csc^2 2x\, dx$

$\int\frac{\csc^2 2x}{\cot 2x}dx = -\frac{1}{2}\int\frac{-2\csc^2 2x\, dx}{\cot 2x}$

$= -\frac{1}{2}|\cot 2x| + C$

7. $\int_0^{\pi/2}\frac{\cos x\, dx}{1 + \sin x}$; $u = 1 + \sin x$, $du = \cos x\, dx$

$\int_0^{\pi/2}\frac{\cos x\, dx}{1 + \sin x} = \ln|1 + \sin x|\Big|_0^{\pi/2}$

$= \ln|1 + \sin\frac{\pi}{2}| - \ln|1 + 0|$

$= \ln 2 - 0 = 0.693$

9. $\int\frac{e^{-x}}{1 - e^{-x}}dx$; $u = 1 - e^{-x}$, $du = e^{-x}dx$

$\int\frac{e^{-x}}{1 - e^{-x}}dx = \int\frac{e^{-x}dx}{1 - e^{-x}} = \ln|1 - e^{-x}| + C$

11. $\int\frac{1 + e^x}{x + e^x}dx$; $u = x + e^x$, $du = (1 + e^x)dx$

$\int\frac{1 + e^x}{x + e^x}dx = \int\frac{(1 + e^x)dx}{x + e^x} = \ln|x + e^x| + C$

13. $\int\frac{\sec x \tan x\, dx}{1 + 4\sec x}$; $u = 1 + 4\sec x$, $du = 4\sec x \tan x\, dx$

$\int\frac{\sec x \tan x\, dx}{1 + 4\sec x} = \frac{1}{4}\int\frac{4\sec x \tan x\, dx}{1 + 4\sec x} = \frac{1}{4}\ln|1 + 4\sec x| + C$

15. $\int_1^3\frac{1 + x}{4x + 2x^2}dx$; $u = 4x + 2x^2$, $du = (4 + 4x)dx = 4(1 + x)dx$

$\int_1^3\frac{1 + x}{4x + 2x^2}dx = \frac{1}{4}\int_1^3\frac{4(1 + x)dx}{4x + 2x^2} = \frac{1}{4}\ln|4x + 2x^2|\Big|_1^3 = \frac{1}{4}\ln|4(3)+2(3)^2| - \frac{1}{4}\ln|4(1)+2(1)^2|$

$= \frac{1}{4}\ln 30 - \frac{1}{4}\ln 6 = \frac{1}{4}\ln 5 = 0.402$

17. $\int \dfrac{dx}{x \ln x}$; $u = \ln x$, $du = \dfrac{dx}{x}$

$\int \dfrac{dx}{x \ln x} = \int \dfrac{dx/x}{\ln x} = \ln |\ln x| + C$

19. $\int \dfrac{2 + \sec^2 x}{2x + \tan x} dx$; $u = 2x + \tan x$, $du = (2 + \sec^2 x) dx$

$\int \dfrac{2 + \sec^2 x}{2x + \tan x} dx = \int \dfrac{(2 + \sec^2 x) dx}{2x + \tan x} = \ln |2x + \tan x| + C$

21. $\int \dfrac{dx}{\sqrt{1 - 2x}}$; $u = 1 - 2x$, $du = -2dx$; $n = -\dfrac{1}{2}$, $n+1 = \dfrac{1}{2}$; NOT logarithmic form

$\int \dfrac{dx}{\sqrt{1 - 2x}} = -\dfrac{1}{2} \int (1 - 2x)^{-1/2} (-2x\,dx) = -\dfrac{1}{2} \dfrac{(1 - 2x)^{1/2}}{1/2} + C = -(1 - 2x)^{1/2} + C$

23. $\int_0^{\pi/12} \dfrac{\sec^2 3x}{4 + \tan 3x} dx$; $u = 4 + \tan 3x$, $du = 3 \sec^2 3x\,dx$

$\int_0^{\pi/12} \dfrac{\sec^2 3x}{4 + \tan 3x} dx = \dfrac{1}{3} \int_0^{\pi/12} \dfrac{3 \sec^2 3x\,dx}{4 + \tan 3x} = \dfrac{1}{3} \ln |4 + \tan 3x| \Big|_0^{\pi/12}$

$= \dfrac{1}{3} \ln |4 + \tan 3(\dfrac{\pi}{12})| - \dfrac{1}{3} \ln |4 + \tan 3(0)| = \dfrac{1}{3} \ln 5 - \dfrac{1}{3} \ln 4 = \dfrac{1}{3} \ln \dfrac{5}{4} = 0.0744$

25. $A = \int_0^2 y\,dx = \int_0^2 \dfrac{dx}{x + 1} = \ln |x + 1| \Big|_0^2$ See Fig. 8-4

$= \ln 3 - \ln 1 = \ln 3 = 1.10$

27. $V = 2\pi \int_0^1 xy\,dx = 2\pi \int_0^1 \dfrac{x\,dx}{x^2 + 1} = \pi \int_0^1 \dfrac{2x\,dx}{x^2 + 1} = \pi \ln |x^2 + 1| \Big|_0^1$

$= \pi \ln |1^2 + 1| - \pi \ln |0^2 + 1| = \pi \ln 2 - 0 = 2.18$

29. $\dfrac{dy}{dx} = \dfrac{\sin x}{3 + \cos x}$; $dy = \dfrac{\sin x\,dx}{3 + \cos x}$

$\int dy = \int \dfrac{\sin x\,dx}{3 + \cos x} = -\int \dfrac{-\sin x\,dx}{3 + \cos x}$

$y = -\ln |3 + \cos x| + C$

Curve passes through $(\dfrac{\pi}{3}, 2)$

$2 = -\ln |3 + \cos \dfrac{\pi}{3}| + C$

$2 = -\ln \dfrac{7}{2} + C$, $C = 2 + \ln \dfrac{7}{2}$

$y = -\ln |3 + \cos x| + \ln \dfrac{7}{2} + 2$

$= \ln \dfrac{3.5}{3 + \cos x} + 2$

31. $i = \dfrac{t}{1 + 2t^2}$; $\dfrac{dq}{dt} = \dfrac{t}{1 + 2t^2}$

$dq = \dfrac{t\,dt}{1 + 2t^2}$; $\int dq = \dfrac{1}{4} \int \dfrac{4t\,dt}{1 + 2t^2}$

$q = \dfrac{1}{4} \ln |1 + 2t^2| + C$; $q = 0$ when $t = 0$

$0 = \dfrac{1}{4} \ln |1 + 2(0)^2| + C$, $C = 0$

$q = \dfrac{1}{4} \ln (1 + 2t^2)$

$q \big|_{t=2\,s} = \dfrac{1}{4} \ln [1 + 2(2)^2] = \dfrac{1}{4} \ln 9 = 0.549$ C

33. $F = \dfrac{1}{1 + 3x}$; $W = \int F\,dx$

$W = \int_0^5 \dfrac{dx}{1 + 3x} = \dfrac{1}{3} \int_0^5 \dfrac{3\,dx}{1 + 3x} = \dfrac{1}{3} \ln |1 + 3x| \Big|_0^5 = \dfrac{1}{3} \ln |1 + 3(5)| - \dfrac{1}{3} \ln |1 + 3(0)| = \dfrac{1}{3} \ln 16 = 0.924$ ft·lb

Exercises 8-3, p. 311

1. $\int e^{7x} (7\,dx)$; $u = 7x$, $du = 7\,dx$

$\int e^{7x} (7\,dx) = e^{7x} + C$

3. $\int e^{2x+5} dx$; $u = 2x + 5$, $du = 2\,dx$

$\int e^{2x+5} dx = \dfrac{1}{2} \int e^{2x+5} (2\,dx) = \dfrac{1}{2} e^{2x+5} + C$

5. $\int_0^2 e^{x/2} dx$; $u = \dfrac{x}{2}$, $du = \dfrac{1}{2} dx$

$\int_0^2 e^{x/2} dx = 2 \int_0^2 e^{x/2} (\dfrac{1}{2} dx) = 2 e^{x/2} \Big|_0^2 = 2e - 2e^0 = 2(e - 1) = 3.44$

7. $\int x^2 e^{x^3} dx$; $u = x^3$, $du = 3x^2 dx$

$\int x^2 e^{x^3} dx = \frac{1}{3}\int e^{x^3}(3x^2 dx) = \frac{1}{3}e^{x^3} + C$

9. $\int \frac{e^{\sqrt{x}}}{\sqrt{x}} dx$; $u = \sqrt{x} = x^{1/2}$, $du = \frac{1}{2}x^{-1/2} = \frac{dx}{2\sqrt{x}}$

$\int \frac{e^{\sqrt{x}}}{\sqrt{x}} = 2\int e^{\sqrt{x}}(\frac{dx}{2\sqrt{x}}) = 2e^{\sqrt{x}} + C$

11. $\int (\sec x \tan x)e^{2\sec x} dx$; $u = 2\sec x$, $du = 2\sec x \tan x\, dx$

$\int (\sec x \tan x)e^{2\sec x} dx = \frac{1}{2}\int e^{2\sec x}(2\sec x \tan x\, dx) = \frac{1}{2}e^{2\sec x} + C$

13. $\int \frac{(3 - e^x)dx}{e^{2x}} = 3\int e^{-2x}dx - \int e^{-x}dx = 3(-\frac{1}{2})\int e^{-2x}(-2dx) + \int e^{-x}(-dx)$

$= -\frac{3}{2}e^{-2x} + e^{-x} + C = \frac{2e^x - 3}{2e^{2x}} + C$

15. $\int_1^3 3e^{2x}(e^{-2x} - 1)dx = 3\int_1^3 dx - \frac{3}{2}\int_1^3 e^{2x}(2dx) = 3x - \frac{3}{2}e^{2x}\Big|_1^3 = 3(3) - \frac{3}{2}e^{2(3)} - [3(1) - \frac{3}{2}e^{2(1)}]$

$= 9 - \frac{3}{2}e^6 - 3 + \frac{3}{2}e^2 = 6 - \frac{3(e^6 - e^2)}{2} = -588.06$

17. $\int \frac{2dx}{\sqrt{x}\, e^{\sqrt{x}}} = \int \frac{2e^{-\sqrt{x}}\, dx}{\sqrt{x}}$; $u = -\sqrt{x} = -x^{1/2}$, $du = -\frac{1}{2}x^{-1/2}dx = -\frac{dx}{2\sqrt{x}}$

$\int \frac{2dx}{\sqrt{x}\, e^{\sqrt{x}}} = -2(2)\int e^{-\sqrt{x}}(-\frac{dx}{2\sqrt{x}}) = -4e^{-\sqrt{x}} + C = -\frac{4}{e^{\sqrt{x}}} + C$

19. $\int \frac{e^{\text{Arctan } x}}{x^2 + 1} dx$; $u = \text{Arctan } x$, $du = \frac{dx}{x^2 + 1}$

$\int \frac{e^{\text{Arctan } x}}{x^2 + 1} dx = \int e^{\text{Arctan } x}(\frac{dx}{x^2 + 1}) = e^{\text{Arctan } x} + C$

21. $\int \frac{e^{\cos 3x}\, dx}{\csc 3x}$; $u = \cos 3x$, $du = -3\sin 3x\, dx$

$\int \frac{e^{\cos 3x}\, dx}{\csc 3x} = \int e^{\cos 3x}\sin 3x\, dx = -\frac{1}{3}\int e^{\cos 3x}(-3\sin 3x\, dx) = -\frac{1}{3}e^{\cos 3x} + C$

23. $\int_0^\pi (\sin 2x)e^{\cos^2 x} dx$; $u = \cos^2 x$, $du = 2\cos x(-\sin x\, dx) = -\sin 2x\, dx$

$\int_0^\pi (\sin 2x)e^{\cos^2 x} dx = -\int_0^\pi e^{\cos^2 x}(-\sin 2x\, dx) = -e^{\cos^2 x}\Big|_0^\pi = -e^{\cos^2 \pi} - (-e^{\cos^2 0})$

$= -e + e = 0$

25. $A = \int_0^2 y\, dx = \int_0^2 e^x dx = e^x\Big|_0^2$

$= e^2 - e^0 = e^2 - 1 = 6.389$

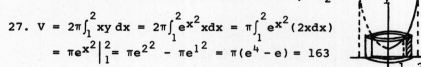

27. $V = 2\pi\int_1^2 xy\, dx = 2\pi\int_1^2 e^{x^2}x\, dx = \pi\int_1^2 e^{x^2}(2x\, dx)$ See Fig. 8-5

$= \pi e^{x^2}\Big|_1^2 = \pi e^{2^2} - \pi e^{1^2} = \pi(e^4 - e) = 163$

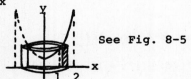

29. $y_{av} = \frac{\int_0^4 e^{2x} dx}{4 - 0} = \frac{\frac{1}{2}\int_0^4 e^{2x}(2dx)}{4} = \frac{e^{2x}\Big|_0^4}{8} = \frac{1}{8}(e^8 - 1) = 372$

31. $v = \frac{1}{C}\int i\, dt = \frac{1}{C}\int 2e^{-30t} dt = \frac{2}{C}(-\frac{1}{30})\int e^{-30t}(-30dt) = -\frac{1}{15C}e^{-30t} + C_1$

$v = 60$ V when $t = 0$, $C = 250 \times 10^{-6}$F $= 2.5 \times 10^{-4}$F ; $60 = -\frac{1}{15(2.5 \times 10^{-4})}e^0 + C_1$

$C_1 = 60 + \frac{800}{3} = \frac{980}{3}$ V ; $v = -\frac{800}{3}e^{-30t} + \frac{980}{3} = \frac{1}{3}(980 - 800e^{-30t})$

Exercises 8-4 , p. 315

1. $\int \cos 2x\, dx$; $u = 2x$, $du = 2dx$

$\int \cos 2x\, dx = \frac{1}{2}\int \cos 2x(2dx) = \frac{1}{2}\sin 2x + C$

3. $\int \sec^2 3x\, dx$; $u = 3x$, $du = 3dx$

$\int \sec^2 3x\, dx = \frac{1}{3}\int \sec^2 3x(3dx) = \frac{1}{3}\tan 3x + C$

5. $\int \sec \frac{1}{2}x \tan \frac{1}{2}x\, dx$; $u = \frac{1}{2}x$, $du = \frac{1}{2}dx$

$\int \sec \frac{1}{2}x \tan \frac{1}{2}x\, dx = 2\int \sec \frac{1}{2}x \tan \frac{1}{2}x(\frac{1}{2}dx) = 2 \sec \frac{1}{2}x + C$

7. $\int x^2 \cot x^3 dx$; $u = x^3$, $du = 3x^2 dx$

$\int x^2 \cot x^3 dx = \frac{1}{3}\int \cot x^3(3x^2 dx)$

$= \frac{1}{3}\ln|\sin x^3| + C$

9. $\int x \sec x^2 dx$; $u = x^2$, $du = 2xdx$

$\int x \sec x^2 dx = \frac{1}{2}\int \sec x^2(2xdx)$

$= \frac{1}{2}\ln|\sec x^2 + \tan x^2| + C$

11. $\int \frac{\sin(1/x)}{x^2} dx$; $u = \frac{1}{x}$, $du = -\frac{dx}{x^2}$

$\int \frac{\sin(1/x)}{x^2} dx = -\int \sin \frac{1}{x}(-\frac{dx}{x^2}) = -(-\cos \frac{1}{x}) + C = \cos \frac{1}{x} + C$

13. $\int_0^{\pi/6} \frac{dx}{\cos^2 2x} = \int_0^{\pi/6} \sec^2 2x\, dx$; $u = 2x$, $du = 2dx$

$\int_0^{\pi/6} \sec^2 2x\, dx = \frac{1}{2}\int_0^{\pi/6} \sec^2 2x(2dx) = \frac{1}{2}\tan 2x\Big|_0^{\pi/6} = \frac{1}{2}\tan 2(\frac{\pi}{6}) - \frac{1}{2}\tan 2(0) = \frac{1}{2}\tan \frac{\pi}{3} = \frac{1}{2}\sqrt{3}$

15. $\int \frac{\sec 5x}{\cot 5x} dx = \int \sec 5x \tan 5x\, dx$; $u = 5x$, $du = 5dx$

$\int \sec 5x \tan 5x\, dx = \frac{1}{5}\int \sec 5x \tan 5x(5dx) = \frac{1}{5}\sec 5x + C$

17. $\int \sqrt{\tan^2 2x + 1}\, dx = \int \sqrt{\sec^2 2x}\, dx = \int \sec 2x\, dx$; $u = 2x$, $du = 2dx$

$\int \sec 2x\, dx = \frac{1}{2}\int \sec 2x(2dx) = \frac{1}{2}\ln|\sec 2x + \tan 2x| + C$

19. $\int_0^{\pi/9} \sin 3x(\csc 3x + \sec 3x)dx = \int_0^{\pi/9} \sin 3x \csc 3x\, dx + \int_0^{\pi/9} \frac{\sin 3x}{\cos 3x} dx$

$= \int_0^{\pi/9} dx + \int_0^{\pi/9} \tan 3x\, dx = \int_0^{\pi/9} dx + \frac{1}{3}\int_0^{\pi/9} \tan 3x(3dx) = x + \frac{1}{3}(-\ln|\cos 3x|)\Big|_0^{\pi/9}$

$= \frac{\pi}{9} - \frac{1}{3}\ln \cos 3(\frac{\pi}{9}) - [0 - \ln \cos 3(0)] = \frac{\pi}{9} - \frac{1}{3}\ln \cos \frac{\pi}{3} + \ln 1 = \frac{\pi}{9} - \frac{1}{3}\ln \frac{1}{2} = \frac{\pi}{9} + \frac{1}{3}\ln 2 = 0.580$

21. $\int \frac{1 - \sin x}{1 + \cos x} dx = \int \frac{(1-\sin x)(1-\cos x)}{(1+\cos x)(1-\cos x)} dx = \int \frac{1 - \sin x - \cos x + \sin x \cos x}{1 - \cos^2 x} dx$

$= \int \frac{1 - \sin x - \cos x + \sin x \cos x}{\sin^2 x} dx = \int \csc^2 x\, dx - \int \csc x\, dx - \int \cot x \csc x\, dx$

$+ \int \cot x\, dx = -\cot x - \ln|\csc x - \cot x| + \csc x + \ln|\sin x| + C$

23. $\int \dfrac{1 + \sin 2x}{\tan 2x}\,dx = \int \cot 2x\,dx + \int \sin 2x(\dfrac{\cos 2x}{\sin 2x})dx = \dfrac{1}{2}\int \cot 2x\,(2dx) + \dfrac{1}{2}\int \cos 2x\,(2dx)$

$\qquad = \dfrac{1}{2}\ln|\sin 2x| + \dfrac{1}{2}\sin 2x + C = \dfrac{1}{2}[\ln|\sin 2x| + \sin 2x] + C$

25. $A = \int_0^{\pi/4} y\,dx = \int_0^{\pi/4} \tan x\,dx = -\ln|\cos x|\Big|_0^{\pi/4}$

$\qquad = -\ln \cos \dfrac{\pi}{4} - (-\ln \cos 0) = -\ln\dfrac{1}{2}\sqrt{2} + 0$

$\qquad = \ln\sqrt{2} = \dfrac{1}{2}\ln 2 = 0.347$

27. $V = \pi\int_0^{\pi/3} y^2\,dx = \pi\int_0^{\pi/3} \sec^2 x\,dx = \pi\tan x\Big|_0^{\pi/3}$

$\qquad = \pi\tan\dfrac{\pi}{3} - \pi\tan 0 = \pi\sqrt{3} - 0 = \pi\sqrt{3} = 5.44$

29. $v = \dfrac{ds}{dt} = 4\cos 3t$

$\qquad ds = 4\cos 3t\,dt$

$\qquad \int ds = \dfrac{4}{3}\int \cos 3t\,(3dt)$

$\qquad s = \dfrac{4}{3}\sin 3t + C$

$\qquad s = 0$ when $t = 0$; $0 = \dfrac{4}{3}\sin 3(0) + C$

$\qquad C = 0$; $s = \dfrac{4}{3}\sin 3t$

31. $\bar{x} = \dfrac{\int_0^1 xy\,dx}{0.3103} = \dfrac{\int_0^1 x\sin(x^2)dx}{0.3103}$ See Fig. 8-6

$\qquad = \dfrac{\dfrac{1}{2}\int_0^1 \sin(x^2)(2x\,dx)}{0.3103}$

$\qquad = \dfrac{-\dfrac{1}{2}\cos(x^2)\Big|_0^1}{0.3103} = \dfrac{-\dfrac{1}{2}\cos 1^2 - (-\dfrac{1}{2}\cos 0^2)}{0.3103}$

$\qquad = \dfrac{-0.2702 + 0.5000}{0.3103} = 0.741$

Exercises 8-5 , p. 320

1. $\int \sin^2 x\cos x\,dx = \int \sin^2 x(\cos x\,dx) = \dfrac{1}{3}\sin^3 x + C$

3. $\int \sin^3 2x\,dx = \int \sin^2 2x(\sin 2x\,dx) = \int(1 - \cos^2 2x)(\sin 2x\,dx)$

$\qquad = -\dfrac{1}{2}\int(-\sin 2x\,dx)(2\,dx) - [-\dfrac{1}{2}\int \cos^2 2x(-\sin 2x)(2dx)]$

$\qquad = -\dfrac{1}{2}\cos 2x + \dfrac{1}{2}\dfrac{\cos^3 2x}{3} + C = -\dfrac{1}{2}\cos 2x + \dfrac{1}{6}\cos^3 2x + C$

5. $\int \sin^2 x\cos^3 x\,dx = \int \sin^2 x(\cos^2 x)(\cos x\,dx) = \int \sin^2 x(1 - \sin^2 x)(\cos x\,dx)$

$\qquad = \int \sin^2 x(\cos x\,dx) - \int \sin^4 x(\cos x\,dx) = \dfrac{1}{3}\sin^3 x - \dfrac{1}{5}\sin^5 x + C$

7. $\int_0^{\pi/4}\sin^5 x\,dx = \int_0^{\pi/4}\sin^4 x(\sin x\,dx) = \int_0^{\pi/4}(1 - \cos^2 x)^2(\sin x\,dx)$

$\qquad = \int_0^{\pi/4}\sin x\,dx - 2\int_0^{\pi/4}\cos^2 x(\sin x\,dx) + \int_0^{\pi/4}\cos^4 x(\sin x\,dx)$

$\qquad = \int_0^{\pi/4}\sin x\,dx + 2\int_0^{\pi/4}\cos^2 x(-\sin x\,dx) - \int_0^{\pi/4}\cos^4 x(-\sin x\,dx)$

$\qquad = -\cos x + \dfrac{2}{3}\cos^3 x - \dfrac{1}{5}\cos^5 x\Big|_0^{\pi/4} = -\cos\dfrac{\pi}{4} + \dfrac{2}{3}\cos^3\dfrac{\pi}{4} - \dfrac{1}{5}\cos^5\dfrac{\pi}{4}$

$\qquad -[-\cos 0 + \dfrac{2}{3}\cos^3 0 - \dfrac{1}{5}\cos^5 0] = -\dfrac{1}{2}\sqrt{2} + \dfrac{2}{3}(\dfrac{1}{2}\sqrt{2})^3 - \dfrac{1}{5}(\dfrac{1}{2}\sqrt{2})^5 + 1 - \dfrac{2}{3} + \dfrac{1}{5}$

$\qquad = -\dfrac{1}{2}\sqrt{2} + \dfrac{1}{6}\sqrt{2} - \dfrac{1}{40}\sqrt{2} + \dfrac{1}{3} + \dfrac{1}{5} = \dfrac{-60\sqrt{2} + 20\sqrt{2} - 3\sqrt{2} + 40 + 24}{120} = \dfrac{64 - 43\sqrt{2}}{120}$

$\qquad = 0.0266$

9. $\int \sin^2 x \, dx = \int [\frac{1}{2}(1 - \cos 2x)] dx = \frac{1}{2}\int dx - \frac{1}{2}\int \cos 2x \, dx = \frac{1}{2}\int dx - \frac{1}{2}(\frac{1}{2})\int \cos 2x (2dx)$

$\qquad = \frac{1}{2} x - \frac{1}{4} \sin 2x + C$

11. $\int \cos^2 3x \, dx = \int \frac{1}{2}(1 + \cos 6x) dx = \frac{1}{2}\int dx + \frac{1}{2}(\frac{1}{6})\int \cos 6x \, (6dx) = \frac{1}{2} x + \frac{1}{12}\sin 6x + C$

13. $\int \tan^3 x \, dx = \int \tan^2 x \tan x \, dx = \int (\sec^2 x - 1)\tan x \, dx = \int \tan x (\sec^2 x \, dx) - \int \tan x \, dx$

$\qquad = \frac{1}{2}\tan^2 x - (-\ln|\cos x|) + C = \frac{1}{2}\tan^2 x + \ln|\cos x| + C$

15. $\int \tan x \sec^4 x \, dx = \int \sec^3 x (\sec x \tan x \, dx) = \frac{1}{4}\sec^4 x + C$

17. $\int \tan^4 2x \, dx = \int \tan^2 2x \tan^2 2x \, dx = \int \tan^2 2x (\sec^2 2x - 1) dx = \int \tan^2 2x \sec^2 2x \, dx - \int \tan^2 2x \, dx$

$\qquad = \int \tan^2 2x \sec^2 2x \, dx - \int (\sec^2 2x - 1) dx = \frac{1}{2}\int \tan^2 2x (\sec^2 2x \, 2dx)$

$\qquad - \frac{1}{2}\int \sec^2 2x (2dx) + \int dx = \frac{1}{2}\frac{1}{3}\tan^3 2x - \frac{1}{2}\tan 2x + x + C = \frac{1}{6}\tan^3 2x - \frac{1}{2}\tan 2x + x + C$

19. $\int \tan^3 3x \sec^3 3x \, dx = \int \tan^2 3x \tan 3x \sec^3 3x \, dx = \int (\sec^2 3x - 1)\tan 3x \sec^2 3x \sec 3x \, dx$

$\qquad = \int \sec^4 3x (\sec 3x \tan 3x \, dx) - \int \sec^2 3x (\sec 3x \tan 3x \, dx)$

$\qquad = \frac{1}{3}\int \sec^4 3x (\sec 3x \tan 3x \, 3dx) - \frac{1}{3}\int \sec^2 3x (\sec 3x \tan 3x \, 3dx)$

$\qquad = \frac{1}{3}\frac{1}{5}\sec^5 3x - \frac{1}{3}\frac{1}{3}\sec^3 3x + C = \frac{1}{15}\sec^5 3x - \frac{1}{9}\sec^3 3x + C$

21. $\int (\sin x + \cos x)^2 dx = \int (\sin^2 x + 2\sin x \cos x + \cos^2 x) dx = \int (1 + \sin 2x) dx$

$\qquad = \int dx + \frac{1}{2}\int \sin 2x \, (2dx) = x - \frac{1}{2}\cos 2x + C$

23. $\int \frac{1 - \cot x}{\sin^4 x} dx = \int (\csc^4 x - \csc^4 x \cot x) dx = \int [\csc^2 x (\cot^2 x + 1) - \csc^2 x (\cot^2 x + 1)\cot x] dx$

$\qquad = -\int \cot^2 x (-\csc^2 x \, dx) + \int \csc^2 x \, dx + \int \cot^3 x (-\csc^2 x \, dx) + \int \cot x (-\csc^2 x \, dx)$

$\qquad = -\frac{1}{3}\cot^3 x - \cot x + \frac{1}{4}\cot^4 x + \frac{1}{2}\cot^2 x + C$

$\qquad = \frac{1}{4}\cot^4 x - \frac{1}{3}\cot^3 x + \frac{1}{2}\cot^2 x - \cot x + C$

25. $\int_{\pi/6}^{\pi/4} \cot^5 x \, dx = \int_{\pi/6}^{\pi/4} \cot^4 x \cot x \, dx = \int_{\pi/6}^{\pi/4} (\csc^2 x - 1)^2 \cot x \, dx$

$\qquad = \int_{\pi/6}^{\pi/4} (\csc^4 x - 2\csc^2 x + 1)\cot x \, dx = -\int_{\pi/6}^{\pi/4} \csc^3 x (-\csc x \cot x \, dx)$

$\qquad + 2\int_{\pi/6}^{\pi/4} \cot x (-\csc^2 x \, dx) + \int_{\pi/6}^{\pi/4} \cot x \, dx = -\frac{1}{4}\csc^4 x + \cot^2 x + \ln|\sin x| \Big|_{\pi/6}^{\pi/4}$

$\qquad = -\frac{1}{4}\csc^4 \frac{\pi}{4} + \cot^2 \frac{\pi}{4} + \ln \sin \frac{\pi}{4} - (-\frac{1}{4}\csc^4 \frac{\pi}{6} + \cot^2 \frac{\pi}{6} + \ln \sin \frac{\pi}{6})$

$\qquad = -\frac{1}{4}(\sqrt{2})^4 + 1^2 + \ln \frac{1}{2}\sqrt{2} + \frac{1}{4}(2)^4 - (\sqrt{3})^2 - \ln \frac{1}{2}$

$\qquad = -1 + 1 - \frac{1}{2}\ln 2 + 4 - 3 + \ln 2 = 1 + \frac{1}{2}\ln 2 = 1.347$

27. $\int \sec^6 x \, dx = \int \sec^4 x \sec^2 x \, dx = \int (\tan^2 x + 1)^2 \sec^2 x \, dx = \int (\tan^4 x + 2\tan^2 x + 1)\sec^2 x \, dx$

$\qquad = \frac{1}{5}\tan^5 x + \frac{2}{3}\tan^3 x + \tan x + C$

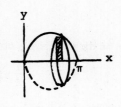

29. $V = \pi\int_0^\pi y^2 dx = \pi\int_0^\pi \sin^2 x\, dx = \frac{\pi}{2}\int_0^\pi (1 - \cos 2x)dx$

$= \frac{\pi}{2}\int_0^\pi dx - \frac{\pi}{4}\int_0^\pi \cos 2x(2dx) = \frac{\pi}{2}x - \frac{\pi}{4}\sin 2x\Big|_0^\pi$

$= \frac{\pi}{2}(\pi) - \frac{\pi}{4}\sin 2\pi - [\frac{\pi}{2}(0) - \frac{\pi}{4}\sin 2(0)] = \frac{1}{2}\pi^2 = 4.935$

31. $\int_0^\pi \sin^3\theta d\theta = \int_0^\pi \sin^2\theta\sin\theta\, d\theta = \int_0^\pi (1 - \cos^2\theta)\sin\theta\, d\theta = \int_0^\pi \sin\theta\, d\theta + \int_0^\pi \cos^2\theta(-\sin\theta\, d\theta)$

$= -\cos\theta + \frac{1}{3}\cos^3\theta\Big|_0^\pi = -\cos\pi + \frac{1}{3}\cos^3\pi - (-\cos 0 + \frac{1}{3}\cos^3 0)$

$= -(-1) + \frac{1}{3}(-1)^3 + 1 - \frac{1}{3} = 1 - \frac{1}{3} + 1 - \frac{1}{3} = \frac{4}{3}$

33. $i = 2\sin t$; period $= \frac{2\pi}{1} = 2\pi$; $i_{rms} = \sqrt{\frac{1}{2\pi}\int_0^{2\pi}(2\sin t)^2 dt}$

$\frac{1}{2\pi}\int_0^{2\pi}(2\sin t)^2 dt = \frac{2}{\pi}\int_0^{2\pi}\sin^2 t\, dt = \frac{1}{\pi}\int_0^{2\pi}(1 - \cos 2t)dt = \frac{1}{\pi}\int_0^{2\pi}dt - \frac{1}{2\pi}\int_0^{2\pi}\cos 2t(2dt)$

$= \frac{1}{\pi}t - \frac{1}{2\pi}\sin 2t\Big|_0^{2\pi} = \frac{1}{\pi}(2\pi) - \frac{1}{2\pi}\sin 4\pi - [\frac{1}{\pi}(0) - \frac{1}{2\pi}\sin 0] = 2$

$i_{rms} = \sqrt{2}$

35. $i = i_o\sin\omega t$; period $= \frac{2\pi}{\omega}$; $i_{rms} = \sqrt{\frac{\omega}{2\pi}\int_0^{2\pi/\omega}(i_o\sin\omega t)^2 dt}$

$\frac{\omega}{2\pi}\int_0^{2\pi/\omega}(i_o\sin\omega t)^2 dt = \frac{\omega i_o^2}{2\pi}\int_0^{2\pi/\omega}\sin^2\omega t\, dt = \frac{\omega i_o^2}{4\pi}\int_0^{2\pi/\omega}(1 - \cos 2\omega t)dt$

$= \frac{\omega i_o^2}{4\pi}[t - \frac{1}{2\omega}\sin 2\omega t]\Big|_0^{2\pi/\omega} = \frac{\omega i_o^2}{4\pi}[(\frac{2\pi}{\omega}) - \frac{1}{2\omega}\sin 2\omega(\frac{2\pi}{\omega})] - 0$

$= \frac{\omega i_o^2}{4\pi}(\frac{2\pi}{\omega}) = \frac{i_o^2}{2}$; $i_{rms} = i_o/\sqrt{2}$

Exercises 8-6, p. 324

1. $\int \frac{dx}{\sqrt{4 - x^2}}$; $u = x$, $du = dx$, $a = 2$

$\int \frac{dx}{\sqrt{4 - x^2}} = \text{Arcsin } \frac{x}{2} + C$

3. $\int \frac{dx}{64 + x^2}$; $u = x$, $du = dx$, $a = 8$

$\int \frac{dx}{64 + x^2} = \frac{1}{8}\text{Arctan } \frac{x}{8} + C$

5. $\int \frac{dx}{\sqrt{1 - 16x^2}}$; $u = 4x$, $du = 4dx$, $a = 1$

$\int \frac{dx}{\sqrt{1 - 16x^2}} = \frac{1}{4}\int \frac{4dx}{\sqrt{1 - 16x^2}} = \frac{1}{4}\text{Arcsin } \frac{4x}{1} + C = \frac{1}{4}\text{Arcsin } 4x + C$

7. $\int_0^2 \frac{dx}{1 + 9x^2}$; $u = 3x$, $du = 3dx$, $a = 1$

$\int_0^2 \frac{dx}{1 + 9x^2} = \frac{1}{3}\int_0^2 \frac{3dx}{1 + 9x^2} = \frac{1}{3}\frac{1}{1}\text{Arctan } \frac{3x}{1}\Big|_0^2 = \frac{1}{3}\text{Arctan } 3x\Big|_0^2 = \frac{1}{3}\text{Arctan } 6 - \frac{1}{3}\text{Arctan } 0$

$= \frac{1}{3}\text{Arctan } 6 = 0.469$

9. $\int \frac{2dx}{\sqrt{4 - 5x^2}}$; $u = \sqrt{5}x$, $du = \sqrt{5}dx$, $a = 2$

$\int \frac{2dx}{\sqrt{4 - 5x^2}} = \frac{2}{\sqrt{5}}\int \frac{\sqrt{5}dx}{\sqrt{4 - 5x^2}} = \frac{2}{\sqrt{5}}\text{Arcsin } \frac{\sqrt{5}x}{2} + C = \frac{2}{5}\sqrt{5}\text{ Arcsin } \frac{1}{2}\sqrt{5}x + C$

11. $\int \dfrac{8x\,dx}{9x^2+16}$; $u=9x^2+16$, $du=18x\,dx$; NOT inverse tangent form

$\int \dfrac{8x\,dx}{9x^2+16} = \dfrac{8}{18}\int \dfrac{18x\,dx}{9x^2+16} = \dfrac{4}{9}\ln|9x^2+16| + C$

13. $\int_1^2 \dfrac{dx}{5x^2+7}$; $u=\sqrt{5}x$, $du=\sqrt{5}\,dx$, $a=\sqrt{7}$

$\int_1^2 \dfrac{dx}{5x^2+7} = \dfrac{1}{\sqrt{5}}\int_1^2 \dfrac{\sqrt{5}\,dx}{7+5x^2} = \dfrac{1}{\sqrt{5}}\dfrac{1}{\sqrt{7}}\operatorname{Arctan}\dfrac{1}{\sqrt{7}}\sqrt{5}x\ \Big|_1^2 = \dfrac{\sqrt{35}}{35}[\operatorname{Arctan}\dfrac{\sqrt{35}}{7}(2)-\operatorname{Arctan}\dfrac{\sqrt{35}}{7}(1)]$

$\qquad = \dfrac{1}{35}\sqrt{35}(\operatorname{Arctan}\dfrac{2}{7}\sqrt{35} - \operatorname{Arctan}\dfrac{1}{7}\sqrt{35}) = 0.0566$

15. $\int \dfrac{e^x\,dx}{\sqrt{1-e^{2x}}}$; $u=e^x$, $du=e^x\,dx$, $a=1$

$\int \dfrac{e^x\,dx}{\sqrt{1-e^{2x}}} = \int \dfrac{e^x\,dx}{\sqrt{1-(e^x)^2}} = \operatorname{Arcsin}\dfrac{e^x}{1} + C = \operatorname{Arcsin} e^x + C$

17. $\int \dfrac{dx}{x^2+2x+2} = \int \dfrac{dx}{(x+1)^2+1}$; $u=x+1$, $du=dx$, $a=1$

$\int \dfrac{dx}{(x+1)^2+1} = \dfrac{1}{1}\operatorname{Arctan}\dfrac{x+1}{1} + C = \operatorname{Arctan}(x+1) + C$

19. $\int \dfrac{4\,dx}{\sqrt{-4x-x^2}} = \int \dfrac{4\,dx}{\sqrt{4-4-4x-x^2}} = \int \dfrac{4\,dx}{\sqrt{4-(2+x)^2}}$; $u=2+x$, $du=dx$, $a=2$

$\int \dfrac{4\,dx}{\sqrt{4-(2+x)^2}} = 4\int \dfrac{dx}{\sqrt{4-(2+x)^2}} = 4\operatorname{Arcsin}\dfrac{x+2}{2} + C = 4\operatorname{Arcsin}\dfrac{1}{2}(x+2) + C$

21. $\int_{\pi/6}^{\pi/2} \dfrac{\cos 2x}{1+\sin^2 2x}\,dx$; $u=\sin 2x$, $du=2\cos 2x\,dx$, $a=1$

$\int_{\pi/6}^{\pi/2} \dfrac{\cos 2x}{1+\sin^2 2x}\,dx = \dfrac{1}{2}\int_{\pi/6}^{\pi/2}\dfrac{2\cos 2x\,dx}{1+\sin^2 2x} = \dfrac{1}{2}(\dfrac{1}{1})\operatorname{Arctan}\dfrac{\sin 2x}{1}\Big|_{\pi/6}^{\pi/2} = \dfrac{1}{2}\operatorname{Arctan}(\sin 2x)\Big|_{\pi/6}^{\pi/2}$

$\qquad = \dfrac{1}{2}\operatorname{Arctan}(\sin \pi) - \dfrac{1}{2}\operatorname{Arctan}(\sin\dfrac{\pi}{3}) = \dfrac{1}{2}\operatorname{Arctan}0 - \dfrac{1}{2}\operatorname{Arctan}\dfrac{1}{2}\sqrt{3} = -0.357$

23. $\int \dfrac{(2-x)\,dx}{\sqrt{4-x^2}} = \int \dfrac{2\,dx}{\sqrt{4-x^2}} - \int \dfrac{x\,dx}{\sqrt{4-x^2}} = 2\int \dfrac{dx}{\sqrt{4-x^2}} + \dfrac{1}{2}\int\dfrac{-2x\,dx}{\sqrt{4-x^2}}$

$\qquad = 2\operatorname{Arcsin}\dfrac{1}{2}x + \dfrac{1}{2}\dfrac{(4-x^2)^{1/2}}{1/2} + C = 2\operatorname{Arcsin}\dfrac{1}{2}x + \sqrt{4-x^2} + C$

25. (a) $\int \dfrac{2\,dx}{4+9x^2} = \dfrac{2}{3}\int\dfrac{3\,dx}{4+9x^2}$; Inverse tangent form; $u=3x$

(b) $\int \dfrac{2\,dx}{4+9x} = \dfrac{2}{9}\int\dfrac{9\,dx}{4+9x}$; Logarithmic form ; $u=4+9x$

(c) $\int \dfrac{2x\,dx}{\sqrt{4+9x^2}} = \dfrac{2}{18}\int(4+9x^2)^{-1/2}(18x\,dx)$; General power form; $u=4+9x^2$

27. (a) $\int \dfrac{2x\,dx}{\sqrt{4-9x^2}} = \dfrac{2}{-18}\int(4-9x^2)^{-1/2}(-18x\,dx)$; General power form; $u=4-9x^2$

(b) $\int \dfrac{2\,dx}{\sqrt{4-9x^2}} = \dfrac{2}{3}\int\dfrac{3\,dx}{\sqrt{4-9x^2}}$; Inverse sine form; $u=3x$

(c) $\int \dfrac{2\,dx}{4-9x} = \dfrac{2}{-9}\int\dfrac{-9\,dx}{4-9x}$; Logarithmic form; $u=4-9x$

29. $A = \int_0^2 y\,dx = \int_0^2 \dfrac{dx}{1+x^2} = $ Arctan $x\ \Big|_0^2 = $ Arctan $2 = 1.11$

31. $V = \pi \int_0^3 y^2 dx = \pi \int_0^3 (\dfrac{1}{\sqrt{16+x^2}})^2 dx = \pi \int_0^3 \dfrac{dx}{16+x^2}$

$= \dfrac{\pi}{4}$ Arctan $\dfrac{x}{4}\ \Big|_0^3 = \dfrac{\pi}{4}($Arctan $\dfrac{3}{4} - $ Arctan $0)$

$= \dfrac{\pi}{4}$ Arctan $\dfrac{3}{4} = 0.505$

33. $\dfrac{dx}{\sqrt{A^2-x^2}} = \sqrt{\dfrac{k}{m}}\,dt$; $\int \dfrac{dx}{\sqrt{A^2-x^2}} = \int \sqrt{\dfrac{k}{m}}\,dt$

Arcsin $\dfrac{x}{A} = \sqrt{\dfrac{k}{m}}\,t + C$

$x = x_0$ when $t = 0$, Arcsin $\dfrac{x_0}{A} = \sqrt{\dfrac{k}{m}}(0) + C$

$C = $ Arcsin $\dfrac{x_0}{A}$; Arcsin $\dfrac{x}{A} = \sqrt{\dfrac{k}{m}}\,t + $ Arcsin $\dfrac{x_0}{A}$

35. $I_y = \int_1^2 x^2 y\,dx$ $(k=1)$

$= \int_1^2 \dfrac{x^2 dx}{1+x^6} = \dfrac{1}{3}\int_1^2 \dfrac{3x^2 dx}{1+(x^3)^2}$

$= \dfrac{1}{3}\dfrac{1}{1}$ Arctan $\dfrac{x^3}{1}\ \Big|_1^2 = \dfrac{1}{3}$ Arctan $x^3\ \Big|_1^2$

$= \dfrac{1}{3}$ Arctan $8 - \dfrac{1}{3}$ Arctan $1 = 0.22$

Review Exercises for Chapter 8, p. 325

1. $\int e^{-2x}\,dx = -\dfrac{1}{2}\int e^{-2x}(-2dx) = -\dfrac{1}{2}e^{-2x} + C$

3. $\int \dfrac{dx}{x(\ln 2x)^2} = \int (\ln 2x)^{-2}(\dfrac{dx}{x}) = \dfrac{(\ln 2x)^{-1}}{-1} + C$

$= -\dfrac{1}{\ln 2x} + C$

5. $\int \dfrac{\cos x\,dx}{1+\sin x} = \ln|1+\sin x| + C$

7. $\int \dfrac{2dx}{25+49x^2} = \dfrac{2}{7}\int \dfrac{7dx}{25+49x^2} = \dfrac{2}{7}(\dfrac{1}{5})$ Arctan $\dfrac{7x}{5} + C = \dfrac{2}{35}$ Arctan $\dfrac{7}{5}x + C$

9. $\int_0^{\pi/2} \cos^3 2x\,dx = \int_0^{\pi/2} \cos^2 2x(\cos 2x\,dx) = \int_0^{\pi/2}(1-\sin^2 2x)(\cos 2x\,dx)$

$= \dfrac{1}{2}\int_0^{\pi/2}\cos 2x(2dx) - \dfrac{1}{2}\int_0^{\pi/2}\sin^2 2x(\cos 2x\,2dx) = \dfrac{1}{2}\sin 2x - \dfrac{1}{2}\dfrac{\sin^3 2x}{3}\ \Big|_0^{\pi/2}$

$= \dfrac{1}{2}\sin 2x - \dfrac{1}{6}\sin^3 2x\ \Big|_0^{\pi/2} = \dfrac{1}{2}\sin \pi - \dfrac{1}{6}\sin^3 \pi - (\dfrac{1}{2}\sin 0 - \dfrac{1}{6}\sin^3 0) = 0$

11. $\int_0^2 \dfrac{x\,dx}{4+x^2} = \dfrac{1}{2}\int_0^2 \dfrac{2x\,dx}{4+x^2} = \dfrac{1}{2}\ln|4+x^2|\ \Big|_0^2 = \dfrac{1}{2}\ln 8 - \dfrac{1}{2}\ln 4 = \dfrac{1}{2}\ln 2 = 0.3466$

13. $\int \sec^4 3x \tan 3x\,dx = \dfrac{1}{3}\int \sec^3 3x(\sec 3x \tan 3x\,3dx) = \dfrac{1}{3}\dfrac{\sec^4 3x}{4} + C = \dfrac{1}{12}\sec^4 3x + C$

15. $\int \tan 3x\,dx = \dfrac{1}{3}\int \tan 3x(3\,dx) = -\dfrac{1}{3}\ln|\cos 3x| + C$

17. $\int \sec^4 3x\,dx = \int \sec^2 3x \sec^2 3x\,dx = \int (\tan^2 3x + 1)\sec^2 3x\,dx$

$= \dfrac{1}{3}\int \tan^2 3x(\sec^2 3x\,3dx) + \dfrac{1}{3}\int \sec^2 3x(3\,dx) = \dfrac{1}{3}\dfrac{\tan^3 3x}{3} + \dfrac{1}{3}\tan 3x + C$

$= \dfrac{1}{9}\tan^3 3x + \dfrac{1}{3}\tan 3x + C$

19. $\int_0^{0.5} \dfrac{2e^{2x} - 3e^x}{e^{2x}}\,dx = \int_0^{0.5}(2 - 3e^{-x})\,dx = 2\int_0^{0.5} dx + 3\int_0^{0.5} e^{-x}(-dx) = 2x + 3e^{-x}\ \Big|_0^{0.5}$

$= 2(0.5) + 3e^{-0.5} - [2(0) + 3e^0] = 1 + \dfrac{3}{\sqrt{e}} - 3 = \dfrac{3}{\sqrt{e}} - 2 = -0.1804$

21. $\int \frac{3x\,dx}{4+x^4} = \frac{3}{2}\int \frac{2x\,dx}{4+(x^2)^2} = \frac{3}{2}\frac{1}{2}\text{Arctan}\frac{x^2}{2} + C = \frac{3}{4}\text{Arctan}\frac{1}{2}x^2 + C$

23. $\int \frac{e^x dx}{\sqrt{9-e^{2x}}} = \int \frac{e^x dx}{\sqrt{9-(e^x)^2}} = \text{Arcsin}\frac{e^x}{3} + C = \text{Arcsin}\frac{1}{3}e^x + C$

25. $\int \frac{e^{2x}dx}{\sqrt{e^{2x}+1}} = \frac{1}{2}\int (e^{2x}+1)^{-1/2}(2e^{2x}\,dx) = \frac{1}{2}\frac{(e^{2x}+1)^{1/2}}{1/2} + C = \sqrt{e^{2x}+1} + C$

27. $\int \sin^2 3x\,dx = \frac{1}{2}\int(1-\cos 6x)dx = \frac{1}{2}\int dx - \frac{1}{12}\int \cos 6x\,(6dx) = \frac{1}{2}x - \frac{1}{12}\sin 6x + C$

$= \frac{1}{2}x - \frac{1}{12}(2\sin 3x\cos 3x) + C = \frac{1}{6}(3x - \sin 3x\cos 3x) + C$

29. $\int \frac{\tan^2 x\,dx}{\tan x - x} = \int \frac{(\sec^2 x - 1)dx}{\tan x - x} = \ln|\tan x - x| + C$

31. $\int e^{2x}\cos e^{2x}dx = \frac{1}{2}\int \cos e^{2x}(2e^{2x}dx) = \frac{1}{2}\sin e^{2x} + C$

33. $\int_1^e \frac{(\ln x)^2 dx}{x} = \int_1^e \ln^2 x(\frac{dx}{x}) = \frac{\ln^3 x}{3}\Big|_1^e = \frac{1}{3}(\ln^3 e - \ln^3 1) = \frac{1}{3}\ln^3 e = \frac{1}{3}$

35. $\int \frac{x^2-1}{x+2}dx = \int(x-2+\frac{3}{x+2})dx = \frac{1}{2}x^2 - 2x + 3\ln|x+2| + C$ (First divide x^2-1 by $x+2$.)

37.(a) $\int e^x(e^x+1)^2 dx = \int (e^x+1)^2(e^x dx) = \frac{1}{3}(e^x+1)^3 + C_1 = \frac{1}{3}(e^{3x}+3e^{2x}+3e^x+1) + C_1$

$= \frac{1}{3}e^{3x} + e^{2x} + e^x + \frac{1}{3} + C_1 \qquad\qquad C_2 = C_1 + \frac{1}{3}$

(b) $\int e^x(e^x+1)^2 dx = \int e^x(e^{2x}+2e^x+1)dx = \int(e^{3x}+2e^{2x}+e^x)dx = \frac{1}{3}e^{3x} + e^{2x} + e^x + C_2$

39. $\Delta S = \int \frac{c_v}{T}dT = \int \frac{a+bT+cT^2}{T}dT = a\int \frac{dT}{T} + b\int dT + c\int T\,dT = a\ln T + bT + \frac{1}{2}cT^2 + C$

41. $\frac{dv}{32-0.1v} = dt$; $-10\int \frac{-0.1dv}{32-0.1v} = \int dt$

$-10\ln(32-0.1v) = t + C$

$v = 0$ when $t = 0$; $-10\ln 32 = C$

$-10\ln(32-0.1v) = t - 10\ln 32$

$-10\ln\frac{32-0.1v}{32} = t$

$\frac{32-0.1v}{32} = -0.1t$

$32 - 0.1v = 32e^{-0.1t}$

$32 - 32e^{-0.1t} = 0.1v$

$v = 320(1 - e^{-t/10})$

43. $A = \int_0^{1.5} y\,dx = \int_0^{1.5} 4e^{2x}dx$

$= 2\int_0^{1.5} e^{2x}(2dx)$

$= 2e^{2x}\Big|_0^{1.5}$

$= 2(e^3 - e^0) = 38.17$

45. $\dfrac{dy}{dx} = \sec^4 x$

$dy = \sec^4 x \, dx$

$\int dy = \int \sec^2 x \sec^2 x \, dx$

$\quad = \int (\tan^2 x + 1)\sec^2 x \, dx$

$\quad = \int \tan^2 x (\sec^2 x \, dx) + \int \sec^2 x \, dx$

$y = \dfrac{1}{3}\tan^3 x + \tan x + C$

$y = 0$ when $x = 0$

$0 = \dfrac{1}{3}(0)^3 + 0 + C, \quad C = 0$

$y = \dfrac{1}{3}\tan^3 x + \tan x$

47. $V = \pi \int_2^4 y^2 dx = \pi \int_2^4 (e^{-0.1x})^2 dx$

$\quad = \pi \int_2^4 e^{-0.2x} dx = -5\pi \int_2^4 e^{-0.2x}(-0.2dx)$

$\quad = -5\pi \, e^{-0.2x} \Big|_2^4 = -5\pi(e^{-0.8} - e^{-0.4})$

$\quad = 3.47 \text{ cm}^3$

See
Fig. 8-10

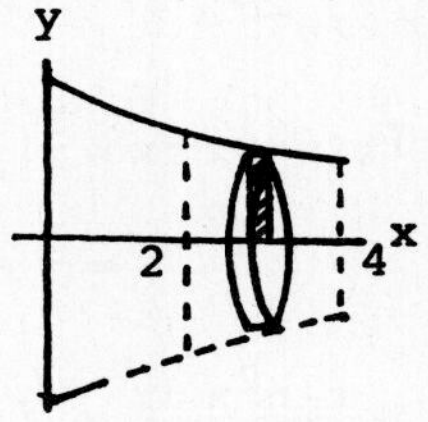

49. $y = \ln \sin x$; $y' = \dfrac{\cos x}{\sin x} = \cot x$

$s = \int_{\pi/3}^{2\pi/3} \sqrt{1 + \cot^2 x} \, dx = \int_{\pi/3}^{2\pi/3} \sqrt{\csc^2 x} \, dx = \int_{\pi/3}^{2\pi/3} \csc x \, dx = \ln|\csc x - \cot x| \Big|_{\pi/3}^{2\pi/3}$

$\quad = \ln\left|\csc \dfrac{2\pi}{3} - \cot \dfrac{2\pi}{3}\right| - \ln\left|\csc \dfrac{\pi}{3} - \cot \dfrac{\pi}{3}\right| = \ln\left|\dfrac{2}{3}\sqrt{3} + \dfrac{1}{3}\sqrt{3}\right| - \ln\left|\dfrac{2}{3}\sqrt{3} - \dfrac{1}{3}\sqrt{3}\right|$

$\quad = \ln \sqrt{3} - \ln \dfrac{1}{3}\sqrt{3} = \ln 3 = 1.099$

CHAPTER 9   **METHODS OF INTEGRATION**

Exercises 9-1 , p. 330

1. $\int x \cos x \, dx$; $u = x$, $du = dx$, $dv = \cos x \, dx$, $v = \int \cos x \, dx = \sin x$

   $\int x \cos x \, dx = x \sin x - \int \sin x \, dx = x \sin x + \cos x + C$

3. $\int x \, e^{2x} dx$; $u = x$, $du = dx$, $dv = e^{2x} dx$, $v = \int e^{2x} dx = \frac{1}{2}\int e^{2x}(2dx) = \frac{1}{2} e^{2x}$

   $\int x \, e^{2x} dx = x(\frac{1}{2} e^{2x}) - \int (\frac{1}{2} e^{2x}) dx = \frac{1}{2} x \, e^{2x} - \frac{1}{4}\int e^{2x}(2dx) = \frac{1}{2} x \, e^{2x} - \frac{1}{4} e^{2x} + C$

5. $\int x \sec^2 x \, dx$; $u = x$, $du = dx$, $dv = \sec^2 x \, dx$, $v = \int \sec^2 x \, dx = \tan x$

   $\int x \sec^2 x \, dx = x \tan x - \int \tan x \, dx = x \tan x + \ln|\cos x| + C$

7. $\int \text{Arctan } x \, dx$; $u = \text{Arctan } x$, $du = \frac{dx}{1 + x^2}$, $dv = dx$, $v = \int dx = x$

   $\int \text{Arctan } x \, dx = x \, \text{Arctan } x - \int \frac{x \, dx}{1 + x^2} = x \, \text{Arctan } x - \frac{1}{2}\int \frac{2x \, dx}{1 + x^2}$

   $= x \, \text{Arctan } x - \frac{1}{2}\ln|1 + x^2| + C = x \, \text{Arctan } x - \ln\sqrt{1 + x^2} + C$

9. $\int \frac{x \, dx}{\sqrt{1 - x}}$; $u = x$, $du = dx$, $dv = \frac{dx}{\sqrt{1 - x}}$, $v = -\int (1 - x)^{-1/2}(-dx) = -2(1 - x)^{1/2}$

   $\int \frac{x \, dx}{\sqrt{1 - x}} = x[-2(1 - x)^{1/2}] - \int [-2(1 - x)^{1/2}] dx = -2x(1 - x)^{1/2} - 2\int (1 - x)^{1/2}(-dx)$

   $= -2x(1 - x)^{1/2} - 2\frac{(1 - x)^{3/2}}{3/2} + C = -2x(1 - x)^{1/2} - \frac{4}{3}(1 - x)^{3/2} + C$

11. $\int x \ln x \, dx$; $u = \ln x$, $du = \frac{dx}{x}$, $dv = x \, dx$, $v = \int x \, dx = \frac{1}{2} x^2$

    $\int x \ln x \, dx = \frac{1}{2} x^2 \ln x - \int \frac{1}{2} x^2 (\frac{dx}{x}) = \frac{1}{2} x^2 \ln x - \frac{1}{2}\int x \, dx = \frac{1}{2} x^2 \ln x - \frac{1}{4} x^2 + C$

13. $\int x^2 \sin 2x \, dx$; $u = x^2$, $du = 2x \, dx$, $dv = \sin 2x \, dx$, $v = \frac{1}{2}\int \sin 2x(2dx) = -\frac{1}{2}\cos 2x$

    $\int x^2 \sin 2x \, dx = x^2(-\frac{1}{2}\cos 2x) - \int (-\frac{1}{2}\cos 2x)(2x \, dx) = -\frac{1}{2} x^2 \cos 2x + \int x \cos 2x \, dx$

    For $\int x \cos 2x \, dx$; $u = x$, $du = dx$, $dv = \cos 2x \, dx$, $v = \frac{1}{2}\int \cos 2x(2 \, dx) = \frac{1}{2}\sin 2x$

    $\int x \cos 2x \, dx = x(\frac{1}{2}\sin 2x) - \int \frac{1}{2}\sin 2x \, dx = \frac{1}{2} x \sin 2x - \frac{1}{4}\int \sin 2x(2dx)$

    $= \frac{1}{2} x \sin 2x + \frac{1}{4}\cos 2x + C$

    $\int x^2 \sin 2x \, dx = -\frac{1}{2} x^2 \cos 2x + [\frac{1}{2} x \sin 2x + \frac{1}{4}\cos 2x] + C$

    $= \frac{1}{2} x \sin 2x - \frac{1}{4}(2x^2 - 1)\cos 2x + C$

119

15. $\int e^x \cos x \, dx$; $u = \cos x$, $du = -\sin x \, dx$, $dv = e^x dx$, $v = \int e^x dx = e^x$

$\int e^x \cos x \, dx = e^x \cos x - \int e^x(-\sin x \, dx) = e^x \cos x + \int e^x \sin x \, dx$

For $\int e^x \sin x \, dx$; $u = \sin x$, $du = \cos x \, dx$, $dv = e^x dx$, $v = \int e^x dx = e^x$

$\int e^x \sin x \, dx = e^x \sin x - \int e^x \cos x \, dx$

$\int e^x \cos x \, dx = e^x \cos x + e^x \sin x - \int e^x \cos x \, dx$; $2\int e^x \cos x \, dx = e^x(\cos x + \sin x) + 2C$

$\int e^x \cos x \, dx = \frac{1}{2} e^x(\cos x + \sin x) + C$

$\int_0^{\pi/2} e^x \cos x \, dx = \frac{1}{2} e^x(\cos x + \sin x)\Big|_0^{\pi/2} = \frac{1}{2} e^{\pi/2}(\cos \frac{\pi}{2} + \sin \frac{\pi}{2}) - \frac{1}{2} e^0(\cos 0 + \sin 0)$

$= \frac{1}{2} e^{\pi/2} - \frac{1}{2} = \frac{1}{2}(e^{\pi/2} - 1) = 1.91$

17. $A = \int_0^2 y \, dx = \int_0^2 xe^{-x} dx$; $u = x$, $du = dx$, $dv = e^{-x} dx$, $v = \int e^{-x} dx = -e^{-x}$ See Fig. 9-1

$= -xe^{-x}\Big|_0^2 - \int_0^2(-e^{-x})dx = -xe^{-x} - e^{-x}\Big|_0^2 = -2e^{-2} - e^{-2} - (0 - 1)$

$= 1 - \frac{3}{e^2} = 0.594$

19. $\bar{x} = \dfrac{\int_0^{\pi/2} xy \, dx}{\int_0^{\pi/2} y \, dx} = \dfrac{\int_0^{\pi/2} x \cos x \, dx}{\int_0^{\pi/2} \cos x \, dx}$ For $\int x \cos x \, dx$, see solution for #1.

$= \dfrac{x \sin x + \cos x\Big|_0^{\pi/2}}{\sin x\Big|_0^{\pi/2}} = \dfrac{\frac{\pi}{2} \sin \frac{\pi}{2} + \cos \frac{\pi}{2} - (0 \sin 0 + \cos 0)}{\sin \frac{\pi}{2} - \sin 0} = \dfrac{\frac{\pi}{2} - 1}{1}$

$= 0.571$

21. $y_{rms} = \sqrt{\dfrac{1}{1 - 0} \int_0^1 (\sqrt{\text{Arcsin } x})^2 dx} = \sqrt{\int_0^1 \text{Arcsin } x \, dx}$

$\int \text{Arcsin } x \, dx$; $u = \text{Arcsin } x$, $du = \dfrac{dx}{\sqrt{1 - x^2}}$, $dv = dx$, $v = \int dx = x$

$\int \text{Arcsin } x \, dx = x \, \text{Arcsin } x - \int \dfrac{x \, dx}{\sqrt{1 - x^2}} = x \, \text{Arcsin } x + \frac{1}{2}\int(1 - x^2)^{-1/2}(-2x \, dx)$

$= x \, \text{Arcsin } x + \frac{1}{2}\dfrac{(1 - x^2)^{1/2}}{1/2} + C = x \, \text{Arcsin } x + \sqrt{1 - x^2} + C$

$\int_0^1 \text{Arcsin } x \, dx = x \, \text{Arcsin } x + \sqrt{1 - x^2}\Big|_0^1 = 1 \, \text{Arcsin } 1 + \sqrt{1 - 1} - (0 \, \text{Arcsin } 0 - \sqrt{1 - 0}) = \frac{\pi}{2} - 1$

$\sqrt{\int_0^1 \text{Arcsin } x \, dx} = \sqrt{\frac{\pi}{2} - 1} = 0.756$

23. $q = \int i \, dt = \int e^{-2t} \cos t \, dt$; $u = \cos t$, $du = -\sin t \, dt$, $dv = e^{-2t} dt$, $v = \int e^{-2t} dt = -\frac{1}{2}e^{-2t}$

$\int e^{-2t} \cos t \, dt = -\frac{1}{2}e^{-2t} \cos t - \int(-\frac{1}{2}e^{-2t})(-\sin t \, dt) = -\frac{1}{2}e^{-2t} \cos t - \frac{1}{2}\int e^{-2t} \sin t \, dt$

For $\int e^{-2t} \sin t \, dt$; $u = \sin t$, $du = \cos t \, dt$, $dv = e^{-2t} dt$, $v = \int e^{-2t} dt = -\frac{1}{2}e^{-2t}$

$\int e^{-2t} \sin t \, dt = -\frac{1}{2}e^{-2t} \sin t - \int(-\frac{1}{2}e^{-2t})\cos t \, dt = -\frac{1}{2}e^{-2t} + \frac{1}{2}\int e^{-2t} \cos t \, dt$

$\int e^{-2t} \cos t \, dt = -\frac{1}{2}e^{-2t} \cos t - \frac{1}{2}[-\frac{1}{2}e^{-2t} \sin t + \frac{1}{2}\int e^{-2t} \cos t \, dt]$

$\frac{5}{4}\int e^{-2t} \cos t \, dt = -\frac{1}{2}e^{-2t} \cos t + \frac{1}{4}e^{-2t} \sin t + \frac{5}{4}C$; $q = \frac{1}{5}e^{-2t}(\sin t - 2 \cos t) + C$

$q = 0$ when $t = 0$; $0 = \frac{1}{5}e^0(\sin 0 - 2 \cos 0) + C$; $C = \frac{2}{5}$

$q = \frac{1}{5}[e^{-2t}(\sin t - 2 \cos t) + 2]$

Exercises 9-2, p. 335

1. $\int x\sqrt{x+1}\,dx$; $u = \sqrt{x+1}$, $u^2 = x+1$, $x = u^2 - 1$, $dx = 2u\,du$

$\int x\sqrt{x+1}\,dx = \int (u^2-1)(u)(2u\,du) = 2\int (u^4 - u^2)du = \frac{2}{5}u^5 - \frac{2}{3}u^3 + C$

$\quad = \frac{2}{5}(x+1)^{5/2} - \frac{2}{3}(x+1)^{3/2} + C = (x+1)^{3/2}[\frac{2}{5}(x+1) - \frac{2}{3}] + C = \frac{2}{15}(3x-2)(x+1)^{3/2}+C$

3. $\int x\sqrt{2x+1}\,dx$; $u = \sqrt{2x+1}$, $u^2 = 2x+1$, $x = \frac{1}{2}(u^2-1)$, $dx = u\,du$

$\int x\sqrt{2x+1}\,dx = \int \frac{1}{2}(u^2-1)(u)(u\,du) = \frac{1}{2}\int (u^4-u^2)du = \frac{1}{10}u^5 - \frac{1}{6}u^3 + C$

$\quad = \frac{1}{10}(2x+1)^{5/2} - \frac{1}{6}(2x+1)^{3/2} + C = (2x+1)^{3/2}[\frac{1}{10}(2x+1) - \frac{1}{6}] + C$

$\quad = \frac{1}{15}(3x-1)(2x+1)^{3/2} + C$

5. $\int \frac{x\,dx}{\sqrt{x+3}}$; $u = \sqrt{x+3}$, $u^2 = x+3$, $x = u^2 - 3$, $dx = 2u\,du$

$\int \frac{x\,dx}{\sqrt{x+3}} = \int \frac{(u^2-3)(2u\,du)}{u} = 2\int (u^2-3)du = \frac{2}{3}u^3 - 6u + C$

$\quad = \frac{2}{3}(x+3)^{3/2} - 6(x+3)^{1/2} + C = (x+3)^{1/2}[\frac{2}{3}(x+3) - 6] + C = \frac{2}{3}(x-6)(x+3)^{1/2} + C$

7. $\int \frac{x^2\,dx}{\sqrt{x-2}}$; $u = \sqrt{x-2}$, $u^2 = x-2$, $x = u^2 + 2$, $dx = 2u\,du$

$\int \frac{x^2\,dx}{\sqrt{x-2}} = \int \frac{(u^2+2)^2(2u\,du)}{u} = 2\int (u^2+2)^2 du = 2\int (u^4+4u^2+4)du = \frac{2}{5}u^5 + \frac{8}{3}u^3 + 8u + C$

$\quad = \frac{2}{5}(x-2)^{5/2} + \frac{8}{3}(x-2)^{3/2} + 8(x-2)^{1/2} + C = (x-2)^{1/2}[\frac{2}{5}(x-2)^2 + \frac{8}{3}(x-2) + 8] + C$

$\quad = 2(x-2)^{1/2}[\frac{3(x-2)^2 + 20(x-2)+60}{15}]+C = \frac{2}{15}(3x^2 + 8x + 32)(x-2)^{1/2} + C$

9. $\int x^2\sqrt{1+x}\,dx$; $u = \sqrt{1+x}$, $u^2 = 1+x$, $x = u^2 - 1$, $dx = 2u\,du$

$\int x^2\sqrt{1+x}\,dx = \int (u^2-1)^2(u)(2u\,du) = 2\int (u^6 - 2u^4 + u^2)du = \frac{2}{7}u^7 - \frac{4}{5}u^5 + \frac{2}{3}u^3 + C$

$\quad = \frac{2}{7}(1+x)^{7/2} - \frac{4}{5}(1+x)^{5/2} + \frac{2}{3}(1+x)^{3/2} + C = 2(1+x)^{3/2}[\frac{1}{7}(1+x)^2 - \frac{2}{5}(1+x) + \frac{1}{3}]+C$

$\quad = 2(1+x)^{3/2}[\frac{15(1+x)^2 - 42(1+x) + 35}{105}] + C = \frac{2}{105}(15x^2 - 12x + 8)(1+x)^{3/2} + C$

$\int_0^3 x^2\sqrt{1+x}\,dx = \frac{2}{105}(15x^2 - 12x + 8)(1+x)^{3/2}\Big|_0^3 = \frac{2}{105}[(135-36+8)(8) - (0-0+8)(1)] = \frac{1696}{105}$

Alternate evaluation: When x=0, u=1, and when x=3, u=2. Therefore,

$\int_0^3 x^2\sqrt{1+x}\,dx = 2\int_1^2 (u^6 - 2u^4 + u^2)du = \frac{2}{7}u^7 - \frac{4}{5}u^5 + \frac{2}{3}u^3\Big|_1^2 = \frac{256}{7} - \frac{128}{5} + \frac{16}{3} - (\frac{2}{7} - \frac{4}{5} + \frac{2}{3}) = \frac{1696}{105}$

11. $\int x\sqrt[3]{x-1}\,dx$; $u = \sqrt[3]{x-1}$, $u^3 = x-1$, $x = u^3 + 1$, $dx = 3u^2du$

$\int x\sqrt[3]{x-1}\,dx = \int (u^3 + 1)(u)(3u^2du) = 3\int (u^6 + u^3)du = \frac{3}{7}u^7 + \frac{3}{4}u^4 + C$

$\quad = \frac{3}{7}(x-1)^{7/3} + \frac{3}{4}(x-1)^{4/3} + C = 3(x-1)^{4/3}[\frac{1}{7}(x-1) + \frac{1}{4}]+C = \frac{3}{28}(4x+3)(x-1)^{4/3}+C$

13. $\int \frac{x\,dx}{\sqrt[4]{2x+3}}$; $u = \sqrt[4]{2x+3}$, $u^4 = 2x+3$, $x = \frac{1}{2}(u^4 - 3)$, $dx = 2u^3du$

$\int \frac{x\,dx}{\sqrt[4]{2x+3}} = \frac{1}{2}\int \frac{(u^4 - 3)(2u^3du)}{u} = \int (u^6 - 3u^2)du = \frac{1}{7}u^7 - u^3 + C$

$\quad = \frac{1}{7}(2x+3)^{7/4} - (2x+3)^{3/4}+C = (2x+3)^{3/4}[\frac{1}{7}(2x+3)-1]+C = \frac{2}{7}(x-2)(2x+3)^{3/4} + C$

15. $\int x(x+1)^{2/3}dx$; $u = (x+1)^{1/3}$, $u^3 = x+1$, $x = u^3 - 1$, $dx = 3u^2du$

$\int x(x+1)^{2/3}dx = \int (u^3-1)(u^2)(3u^2du) = 3\int (u^7 - u^4)du = \frac{3}{8}u^8 - \frac{3}{5}u^5 + C$

$\qquad = \frac{3}{8}(x+1)^{8/3} - \frac{3}{5}(x+1)^{5/3} + C = 3(x+1)^{5/3}[\frac{1}{8}(x+1)-\frac{1}{5}]+C = \frac{3}{40}(5x-3)(x+1)^{5/3}+C$

$\int_0^7 x(x+1)^{2/3}dx = \frac{3}{40}(5x-3)(x+1)^{5/3}\Big|_0^7 = \frac{3}{40}[32(8)^{5/3}-(-3)(1)^{5/3}]=\frac{3}{40}(1024+3) = \frac{3081}{40}$

Alternate evaluation: When x=0, u=1, and when x=7, u=2. Therefore,

$\int_0^7 x(x+1)^{2/3}dx = 3\int_1^2 (u^7 - u^4)du = \frac{3}{8}u^8 - \frac{3}{5}u^5\Big|_1^2 = \frac{3}{8}(256) - \frac{3}{5}(32) -(\frac{3}{8} - \frac{3}{5}) = \frac{3081}{40}$

17. $A = \int_1^2 ydx = \int_1^2 x^3\sqrt{x-1}\,dx$; $u = \sqrt{x-1}$, $x = u^2 + 1$, $dx = 2u\,du$
When x=1, u=0, and when x=2, u=1.

$A = \int_0^1 (u^2+1)^3(u)(2udu) = 2\int_0^1 (u^8+3u^6+3u^4+u^2)du$

$\qquad = 2(\frac{1}{9}u^9+\frac{3}{7}u^7+\frac{3}{5}u^5+\frac{1}{3}u^3)\Big|_0^1 = 2(\frac{1}{9} + \frac{3}{7} + \frac{3}{5} + \frac{1}{3})-2(0) = \frac{928}{315} = 2.946$

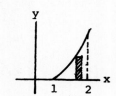

19. $V = 2\pi\int_1^9 xydx = 2\pi\int_1^9 x^2\sqrt[3]{x-1}\,dx$; $u = \sqrt[3]{x-1}$, $x = u^3 + 1$, $dx = 3u^2du$
When x=1, u=0, and when x=9, u=2.

$V = 2\pi\int_0^2 (u^3+1)^2(u)(3u^2du) = 6\pi\int_0^2 (u^9+2u^6+u^3)du$

$\qquad = 6\pi(\frac{1}{10}u^{10} + \frac{2}{7}u^7 + \frac{1}{4}u^4)\Big|_0^2 = 6\pi[\frac{1}{10}(1024) + \frac{2}{7}(128) + \frac{1}{4}(16)]-6\pi(0)$

$\qquad = 6\pi(\frac{512}{5} + \frac{256}{7} + 4) = \frac{30024\pi}{35} = 2695$

21. $W = \int_0^{2.50} 4s\sqrt{4s+3}\,ds$; $u = \sqrt{4s+3}$, $s = \frac{1}{4}(u^2-3)$, $ds = \frac{1}{2}u\,du$; When s=0, u=$\sqrt{3}$, and when s=2.50, u=$\sqrt{13.0}$.

$W = \int_0^{2.50} 4s\sqrt{4s+3}\,ds = \int_{\sqrt{3}}^{\sqrt{13}} 4(\frac{1}{4})(u^2-3)(u)(\frac{1}{2}u\,du) = \frac{1}{2}\int_{\sqrt{3}}^{\sqrt{13}} (u^4 - 3u^2)du$

$\qquad = \frac{1}{10}u^5 - \frac{1}{2}u^3\Big|_{\sqrt{3}}^{\sqrt{13}} = \frac{1}{10}(\sqrt{13})^5 - \frac{1}{2}(\sqrt{13})^3 - [\frac{1}{10}(\sqrt{3})^5 - \frac{1}{2}(\sqrt{3})^3] = 38.5 \text{ ft·lb}$

23. $\int x\sqrt{1-x}\,dx$; $v = 1-x$, $x = 1-v$, $dx = -dv$

$\int x\sqrt{1-x}\,dx = \int (1-v)(v^{1/2})(-dv) = -\int (v^{1/2} - v^{3/2})dv = \int (v^{3/2} - v^{1/2})dv = \frac{2}{5}v^{5/2} - \frac{2}{3}v^{3/2} +C$

$\qquad = \frac{2}{5}(1-x)^{5/2} - \frac{2}{3}(1-x)^{3/2}+ C = 2(1-x)^{3/2}[\frac{1}{5}(1-x) - \frac{1}{3}]+C = -\frac{2}{15}(1-x)^{3/2}(2+3x) + C$

Exercises 9-3, p. 339

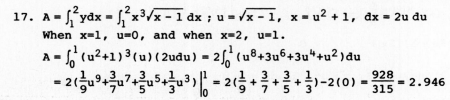

1. $\int \frac{\sqrt{1-x^2}}{x^2}\,dx$; Let $x = \sin\theta$, $dx = \cos\theta\,d\theta$

$\int \frac{\sqrt{1-x^2}}{x^2}\,dx = \int \frac{\sqrt{1-\sin^2\theta}}{\sin^2\theta}\cos\theta\,d\theta = \int \frac{\sqrt{\cos^2\theta}\cos\theta\,d\theta}{\sin^2\theta} = \int \frac{\cos^2\theta\,d\theta}{\sin^2\theta} = \int\cot^2\theta\,d\theta$

$\qquad = \int(\csc^2\theta - 1)d\theta = -\cot\theta - \theta + C = -\frac{\sqrt{1-x^2}}{x} - \text{Arcsin } x + C$

3. $\int \dfrac{dx}{\sqrt{x^2-4}}$; Let $x = 2\sec\theta$, $dx = 2\sec\theta\tan\theta\,d\theta$

$\int \dfrac{dx}{\sqrt{x^2-4}} = \int \dfrac{2\sec\theta\tan\theta\,d\theta}{\sqrt{4\sec^2\theta-4}} = \int \dfrac{\sec\theta\tan\theta\,d\theta}{\sqrt{\tan^2\theta}} = \int \sec\theta\,d\theta$

$= \ln|\sec\theta + \tan\theta| + C = \ln\left|\dfrac{x}{2} + \dfrac{\sqrt{x^2-4}}{2}\right| + C = \ln|x + \sqrt{x^2-4}| + C_1$; $C_1 = C - \ln 2$

5. $\int \dfrac{dx}{x^2\sqrt{x^2+9}}$; Let $x = 3\tan\theta$, $dx = 3\sec^2\theta\,d\theta$

$\int \dfrac{dx}{x^2\sqrt{x^2+9}} = \int \dfrac{3\sec^2\theta\,d\theta}{9\tan^2\theta\sqrt{9\tan^2\theta+9}} = \dfrac{1}{9}\int \dfrac{\sec^2\theta\,d\theta}{\tan^2\theta\sqrt{\tan^2\theta+1}} = \dfrac{1}{9}\int \dfrac{\sec\theta\,d\theta}{\tan^2\theta}$

$= \dfrac{1}{9}\int \cot\theta\csc\theta\,d\theta = -\dfrac{1}{9}\csc\theta + C = -\dfrac{1}{9}\dfrac{\sqrt{x^2+9}}{x} + C = -\dfrac{\sqrt{x^2+9}}{9x} + C$

7. $\int \dfrac{dx}{(4-x^2)^{3/2}}$; Let $x = 2\sin\theta$, $dx = 2\cos\theta\,d\theta$

$\int \dfrac{dx}{(4-x^2)^{3/2}} = \int \dfrac{2\cos\theta\,d\theta}{(4-4\sin^2\theta)^{3/2}} = \dfrac{1}{4}\int \dfrac{\cos\theta\,d\theta}{(1-\sin^2\theta)^{3/2}} = \dfrac{1}{4}\int \dfrac{\cos\theta\,d\theta}{\cos^3\theta}$

$= \dfrac{1}{4}\int \sec^2\theta\,d\theta = \dfrac{1}{4}\tan\theta + C = \dfrac{1}{4}\dfrac{x}{\sqrt{4-x^2}} + C = \dfrac{x}{4\sqrt{4-x^2}} + C$

9. $\int \dfrac{x^3\,dx}{\sqrt{1-x^2}}$; Let $x = \sin\theta$, $dx = \cos\theta\,d\theta$

$\int \dfrac{x^3\,dx}{\sqrt{1-x^2}} = \int \dfrac{\sin^3\theta\cos\theta\,d\theta}{\sqrt{1-\sin^2\theta}} = \int \sin^3\theta\,d\theta = \int (1-\cos^2\theta)\sin\theta\,d\theta$

$= -\cos\theta + \dfrac{1}{3}\cos^3\theta + C = -(1-x^2)^{1/2} + \dfrac{1}{3}(1-x^2)^{3/2} + C$

$\int_0^{0.5} \dfrac{x^3\,dx}{\sqrt{1-x^2}} = -(1-x^2)^{1/2} + \dfrac{1}{3}(1-x^2)^{3/2}\Big|_0^{0.5} = -\sqrt{\dfrac{3}{4}} + \dfrac{1}{3}\left(\sqrt{\dfrac{3}{4}}\right)^3 - \left(-1 + \dfrac{1}{3}\right)$

$= -\dfrac{\sqrt{3}}{2} + \dfrac{1}{3}\left(\sqrt{\dfrac{3}{4}}\right)^3 + \dfrac{2}{3} = -\dfrac{\sqrt{3}}{2} + \dfrac{\sqrt{3}}{8} + \dfrac{2}{3} = \dfrac{16-9\sqrt{3}}{24} = 0.017$

11. $\int \dfrac{dx}{\sqrt{x^2+2x+2}} = \int \dfrac{dx}{\sqrt{(x+1)^2+1}}$; Let $x+1 = \tan\theta$, $dx = \sec^2\theta\,d\theta$

$\int \dfrac{dx}{\sqrt{(x+1)^2+1}} = \int \dfrac{\sec^2\theta\,d\theta}{\sqrt{\tan^2\theta+1}} = \int \sec\theta\,d\theta = \ln|\sec\theta+\tan\theta| + C = \ln|\sqrt{x^2+2x+2}+x+1| + C$

13. $\int \dfrac{dx}{x\sqrt{4x^2-9}}$; Let $x = \dfrac{3}{2}\sec\theta$, $dx = \dfrac{3}{2}\sec\theta\tan\theta\,d\theta$

$\int \dfrac{(3/2)\sec\theta\tan\theta\,d\theta}{(3/2)\sec\theta\sqrt{4(9/4)\sec^2\theta-9}} = \int \dfrac{\tan\theta\,d\theta}{3\sqrt{\sec^2\theta-1}} = \dfrac{1}{3}\int d\theta = \dfrac{1}{3}\theta + C = \dfrac{1}{3}\text{Arcsec}\dfrac{2}{3}x + C$

15. $\int \dfrac{dx}{\sqrt{e^{2x}-1}}$; Let $e^x = \sec\theta$, $e^x\,dx = \sec\theta\tan\theta\,d\theta$, $dx = \tan\theta\,d\theta$

$\int \dfrac{dx}{\sqrt{e^{2x}-1}} = \int \dfrac{\tan\theta\,d\theta}{\sqrt{\sec^2\theta-1}} = \int d\theta = \theta + C = \text{Arcsec}\,e^x + C$

17. $A = 4\int_0^1 y\,dx = 4\int_0^1 \sqrt{1-x^2}\,dx$; Let $x = \sin\theta$, $dx = \cos\theta\,d\theta$

$\int\sqrt{1-x^2}\,dx = \int\sqrt{1-\sin^2\theta}\cos\theta\,d\theta = \int\cos^2\theta\,d\theta = \frac{1}{2}\int(1+\cos 2\theta)\,d\theta$

$\qquad = \frac{1}{2}\theta + \frac{1}{4}\sin 2\theta + C = \frac{1}{2}\theta + \frac{1}{2}\sin\theta\cos\theta + C$

$\qquad = \frac{1}{2}\text{Arcsin } x + \frac{1}{2}x\sqrt{1-x^2} + C$

$A = 4\int_0^1\sqrt{1-x^2}\,dx = 4(\frac{1}{2}\text{Arcsin } x + \frac{1}{2}x\sqrt{1-x^2})\Big|_0^1$

$\qquad = 2\text{ Arcsin }1 + 2(1)\sqrt{0} - [2\text{ Arcsin }0 + 2(0)] = 2\text{ Arcsin }1 = 2(\frac{\pi}{2}) = \pi$

19. $I_y = \int_0^a x^2 y\,dx = \int_0^a x^2\sqrt{a^2-x^2}\,dx$; Let $x = a\sin\theta$, $dx = a\cos\theta\,d\theta$ (k=1)

$\int x^2\sqrt{a^2-x^2}\,dx = \int a^2\sin^2\theta\sqrt{a^2-a^2\sin^2\theta}(a\cos\theta\,d\theta) = a^4\int\sin^2\theta\cos^2\theta\,d\theta$

$\qquad = a^4\int(\frac{\sin 2\theta}{2})^2 d\theta = \frac{a^4}{4}\int\sin^2 2\theta\,d\theta = \frac{a^4}{8}\int(1-\cos 4\theta)\,d\theta$

$\qquad = \frac{a^4}{8}(\theta - \frac{1}{4}\sin 4\theta) + C = \frac{a^4}{8}[\theta - \frac{1}{4}(2\sin 2\theta\cos 2\theta)] + C$

$\qquad = \frac{a^4}{8}[\theta - \sin\theta\cos\theta(\cos^2\theta - \sin^2\theta)] + C$

$\qquad = \frac{a^4}{8}[\text{Arcsin }\frac{x}{a} - (\frac{x}{a})(\frac{\sqrt{a^2-x^2}}{a})(\frac{a^2-x^2}{a^2} - \frac{x^2}{a^2})] + C$

$I_y = \int_0^a x^2\sqrt{a^2-x^2}\,dx = \frac{a^4}{8}[\text{Arcsin}\frac{x}{a} - \frac{x\sqrt{a^2-x^2}(a^2-2x^2)}{a^4}]\Big|_0^a$

$\qquad = \frac{a^4}{8}[\text{Arcsin }1 - 0 - (\text{Arcsin }0 - 0)] = \frac{a^4}{8}\text{Arcsin }1 = \frac{a^4}{8}(\frac{\pi}{2}) = \frac{\pi a^4}{16}$; $m = \frac{1}{4}\pi a^2$; $I_y = \frac{1}{4}ma^2$

21. $V = 2\pi\int_4^5 xy\,dx = 2\pi\int_4^5\frac{\sqrt{x^2-16}}{x}\,dx$; Let $x = 4\sec\theta$, $dx = 4\tan\theta\sec\theta\,d\theta$

$\int\frac{\sqrt{x^2-16}}{x}\,dx = \int\frac{\sqrt{16\sec^2\theta - 16}}{4\sec\theta}(4\sec\theta\tan\theta\,d\theta) = 4\int\tan^2\theta\,d\theta$

$\qquad = 4\int(\sec^2\theta - 1)\,d\theta = 4(\tan\theta - \theta) + C = 4\frac{\sqrt{x^2-16}}{4} - 4\text{ Arcsec }\frac{x}{4} + C$

$V = 2\pi\int_4^5\frac{\sqrt{x^2-16}}{x}\,dx = 2\pi(\sqrt{x^2-16} - 4\text{ Arcsec }\frac{x}{4})\Big|_4^5 = 2\pi[3 - 4\text{ Arcsec }\frac{5}{4}$

$\qquad\qquad - (0 - 4\text{ Arcsec }1)]$

$\qquad = 2\pi(3 - 4\text{ Arcsec }\frac{5}{4}) = 2.68$

23. $\frac{dy}{dx} = \frac{\sqrt{4+x^2}}{x^4}$; $dy = \frac{\sqrt{4+x^2}}{x^4}\,dx$; $\int dy = \int\frac{\sqrt{4+x^2}}{x^4}\,dx$; Let $x = 2\tan\theta$, $dx = 2\sec^2\theta\,d\theta$

$\int\frac{\sqrt{4+x^2}}{x^4}\,dx = \int\frac{\sqrt{4+4\tan^2\theta}}{16\tan^4\theta}(2\sec^2\theta\,d\theta) = \frac{1}{4}\int\frac{\sec^3\theta\,d\theta}{\tan^4\theta} = \frac{1}{4}\int\csc^3\theta\cot\theta\,d\theta$

$\qquad = -\frac{1}{4}\int\csc^2\theta(-\csc\theta\cot\theta\,d\theta) = -\frac{1}{12}\csc^3\theta + C = -\frac{1}{12}(\frac{\sqrt{4+x^2}}{x})^3 + C$

$y = -\frac{(4+x^2)^{3/2}}{12x^3} + C$; y=1 when x=2 ; $1 = -\frac{8^{3/2}}{12(8)} + C$; $C = 1 + \frac{\sqrt{8}}{12} = 1 + \frac{\sqrt{2}}{6}$

$y = -\frac{(4+x^2)^{3/2}}{12x^3} + \frac{6+\sqrt{2}}{6} = \frac{1}{12}(12 + 2\sqrt{2} - \frac{(4+x^2)^{3/2}}{x^3})$

Exercises 9-4 , p. 343

1. $\dfrac{x+3}{(x+1)(x+2)} = \dfrac{A}{x+1} + \dfrac{B}{x+2}$

$x+3 = A(x+2) + B(x+1)$

$x=-1:\ 2=A,\ A=2$

$x=-2:\ 1=-B,\ B=-1$

$\dfrac{x+3}{(x+1)(x+2)} = \dfrac{2}{x+1} - \dfrac{1}{x+2}$

$\displaystyle\int \dfrac{(x+3)dx}{(x+1)(x+2)} = \int\dfrac{2dx}{x+1} - \int\dfrac{dx}{x+2}$

$= 2\ln|x+1| - \ln|x+2| + C$

$= \ln\left|\dfrac{(x+1)^2}{x+2}\right| + C$

3. $x^2-4 = (x+2)(x-2)$

$\dfrac{1}{x^2-4} = \dfrac{A}{x+2} + \dfrac{B}{x-2}$

$1 = A(x-2) + B(x+2)$

$x=-2:\ 1=-4A,\ A=-1/4$

$x=2:\ \ 1=4B,\ B=1/4$

$\dfrac{1}{x^2-4} = \dfrac{-1/4}{x+2} + \dfrac{1/4}{x-2}$

$\displaystyle\int\dfrac{dx}{x^2-4} = -\dfrac{1}{4}\int\dfrac{dx}{x+2} + \dfrac{1}{4}\int\dfrac{dx}{x-2}$

$= -\dfrac{1}{4}\ln|x+2| + \dfrac{1}{4}\ln|x-2| + C$

$= \dfrac{1}{4}\ln\left|\dfrac{x-2}{x+2}\right| + C$

5. $\dfrac{x^2+3}{x^2+3x} = 1 + \dfrac{-3x+3}{x^2+3x}$

$x^2 + 3x = x(x+3)$

$\dfrac{-3x+3}{x^2+3x} = \dfrac{A}{x} + \dfrac{B}{x+3}$

$-3x+3 = A(x+3) + Bx$

$x=-3:\ 12=-3B,\ B=-4$

$x=0\ :\ \ 3=3A,\ A=1$

$\dfrac{-3x+3}{x^2+3x} = \dfrac{1}{x} - \dfrac{4}{x+3}$

$\displaystyle\int\dfrac{x^2+3}{x^2+3x}dx = \int dx + \int\dfrac{dx}{x} - \int\dfrac{4dx}{x+3}$

$= x + \ln|x| - 4\ln|x+3| + C$

$= x + \ln\left|\dfrac{x}{(x+3)^4}\right| + C$

7. $3x^2+5x+2 = (x+1)(3x+2)$

$\dfrac{2x+4}{3x^2+5x+2} = \dfrac{A}{x+1} + \dfrac{B}{3x+2}$

$2x+4 = A(3x+2) + B(x+1)$

$x=-1:\ 2=-A,\ A=-2$

$x=-\dfrac{2}{3}:\ \dfrac{8}{3}=\dfrac{1}{3}B,\ B=8$

$\dfrac{2x+4}{3x^2+5x+2} = -\dfrac{2}{x+1} + \dfrac{8}{3x+2}$

$\displaystyle\int\dfrac{2x+4}{3x^2+5x+2}dx = -\int\dfrac{2dx}{x+1} + \int\dfrac{8dx}{3x+2}$

$= -2\ln|x+1| + \dfrac{8}{3}\ln|3x+2| + C$

$\displaystyle\int_0^1\dfrac{2x+4}{3x^2+5x+2}dx = \ln\left|\dfrac{(3x+2)^{8/3}}{(x+1)^2}\right|\Big\|_0^1$

$= \ln\dfrac{5^{8/3}}{4} - \ln\dfrac{2^{8/3}}{1}$

$= \dfrac{8}{3}\ln 5 - \ln 4 - \dfrac{8}{3}\ln 2 = 1.057$

9. $\dfrac{4x^2-10}{x(x+1)(x-5)} = \dfrac{A}{x} + \dfrac{B}{x+1} + \dfrac{C}{x-5}$

$4x^2-10 = A(x+1)(x-5) + Bx(x-5) + Cx(x+1)$

$x=-1:\ -6=6B,\ B=-1$

$x=0\ :\ -10=-5A,\ A=2$

$x=5\ :\ 90=30C,\ C=3$

$\dfrac{4x^2-10}{x(x+1)(x-5)} = \dfrac{2}{x} - \dfrac{1}{x+1} + \dfrac{3}{x-5}$

$\displaystyle\int\dfrac{4x^2-10}{x(x+1)(x-5)}dx = \int\dfrac{2dx}{x} - \int\dfrac{dx}{x+1} + \int\dfrac{3dx}{x-5}$

$= 2\ln|x| - \ln|x+1| + 3\ln|x-5| + C_1$

$= \ln\left|\dfrac{x^2(x-5)^3}{x+1}\right| + C_1$

11. $4x^3 - x = x(2x+1)(2x-1)$

$\dfrac{6x^2-2x-1}{4x^3-x} = \dfrac{A}{x} + \dfrac{B}{2x+1} + \dfrac{C}{2x-1}$

$6x^2-2x-1 = A(2x+1)(2x-1) + Bx(2x-1) + Cx(2x+1)$

$x=-\dfrac{1}{2}:\ \dfrac{3}{2} = B$

$x=0:\ -1=-A,\ A=1$

$x=\dfrac{1}{2}:\ -\dfrac{1}{2} = C$

$\dfrac{6x^2-2x-1}{4x^3-x}dx = \dfrac{1}{x} + \dfrac{3/2}{2x+1} - \dfrac{1/2}{2x-1}$

$\displaystyle\int\dfrac{6x^2-2x-1}{4x^3-x}dx = \int\dfrac{dx}{x} + \dfrac{3}{2}\int\dfrac{dx}{2x+1} - \dfrac{1}{2}\int\dfrac{dx}{2x-1}$

$= \ln|x| + \dfrac{3}{4}\ln|2x+1| - \dfrac{1}{4}\ln|2x-1| + C_1$

$= \dfrac{1}{4}\ln\left|\dfrac{x^4(2x+1)^3}{2x-1}\right| + C_1$

13. $\dfrac{x^3+7x^2+9x+2}{x^3+3x^2+2x} = 1 + \dfrac{4x^2+7x+2}{x(x^2+3x+2)}$

$x(x^2+3x+2) = x(x+1)(x+2)$

$\dfrac{4x^2+7x+2}{x(x+1)(x+2)} = \dfrac{A}{x} + \dfrac{B}{x+1} + \dfrac{C}{x+2}$

$4x^2+7x+2 = A(x+1)(x+2) + Bx(x+2)$
$\qquad\qquad\qquad +Cx(x+1)$

$x=-2:\ 4=2C,\ C=2$
$x=-1:-1=-B,\ B=1$
$x=0\ :\ 2=2A,\ A=1$

$\dfrac{4x^2+7x+2}{x(x+1)(x+2)} = \dfrac{1}{x} + \dfrac{1}{x+1} + \dfrac{2}{x+2}$

$\displaystyle\int \dfrac{x^3+7x^2+9x+2}{x(x^2+3x+2)} dx = \int dx + \int\dfrac{dx}{x} + \int\dfrac{dx}{x+1} + \int\dfrac{2dx}{x+2}$

$\quad = x + \ln|x| + \ln|x+1| + 2\ln|x+2| + C_1$

$\quad = x + \ln|x(x+1)(x+2)^2| + C_1$

$\displaystyle\int_1^2 \dfrac{x^3+7x^2+9x+2}{x(x^2+3x+2)} dx = x + \ln\left|x(x+1)(x+2)^2\right|\Big|_1^2$

$\quad =2+\ln[2(3)(16)]-\big\{1+\ln[1(2)(9)]\big\}$

$\quad = 1 + \ln\dfrac{16}{3} = 2.674$

15. $(x^2-4)(x^2-9) = (x+2)(x-2)(x+3)(x-3)$

$\dfrac{1}{(x^2-4)(x^2-9)} = \dfrac{A}{x+2} + \dfrac{B}{x-2} + \dfrac{C}{x+3} + \dfrac{D}{x-3}$

$1 = A(x-2)(x+3)(x-3) + B(x+2)(x+3)(x-3)$
$\quad +C(x+2)(x-2)(x-3) + D(x+2)(x-2)(x+3)$

$x=-3:\ 1=-30C,\ C=-1/30$
$x=-2:\ 1=20A,\ A=1/20$
$x=2\ :\ 1=-20B,\ B=-1/20$
$x=3\ :\ 1=30D,\ D=1/30$

$\dfrac{1}{(x^2-4)(x^2-9)} = \dfrac{1/20}{x+2} - \dfrac{1/20}{x-2} - \dfrac{1/30}{x+3} + \dfrac{1/30}{x-3}$

$\displaystyle\int\dfrac{dx}{(x^2-4)(x^2-9)} = \dfrac{1}{20}\int\dfrac{dx}{x+2} - \dfrac{1}{20}\int\dfrac{dx}{x-2}$

$\qquad\qquad - \dfrac{1}{30}\int\dfrac{dx}{x+3} + \dfrac{1}{30}\int\dfrac{dx}{x-3}$

$\quad = \dfrac{1}{20}\ln|x+2| - \dfrac{1}{20}\ln|x-2| - \dfrac{1}{30}\ln|x+3|$

$\qquad + \dfrac{1}{30}\ln|x-3| + C_1$

$\quad = \dfrac{1}{60}\ln\left|\dfrac{(x+2)^3(x-3)^2}{(x-2)^3(x+3)^2}\right| + C_1$

17. $\dfrac{dy}{dx} = \dfrac{3x+5}{x^2+5x}$; $\displaystyle\int dy = \int\dfrac{3x+5}{x^2+5x}dx$

$x^2+5x = x(x+5);\ \dfrac{3x+5}{x^2+5x} = \dfrac{A}{x} + \dfrac{B}{x+5}$

$3x+5 = A(x+5) + Bx$

$x=-5:\ -10=-5B,\ B=2$
$x=0\ :\ \quad 5=5A,\ A=1$

$\dfrac{3x+5}{x^2+5x} = \dfrac{1}{x} + \dfrac{2}{x+5}$

$y = \displaystyle\int\dfrac{3x+5}{x^2+5x}dx = \int\dfrac{dx}{x} + \int\dfrac{2dx}{x+5}$

$\quad = \ln|x| + 2\ln|x+5| + C = \ln|x(x+5)^2| + C$

$y=0$ when $x=1$; $0 = \ln 36 + C$, $C = -\ln 36$

$y = \ln|x(x+5)^2| - \ln 36 = \ln\left|\dfrac{x(x+5)^2}{36}\right|$

19. $W = \displaystyle\int_0^{0.500} F dx = \int_0^{0.500}\dfrac{4x\,dx}{x^2+3x+2}$

$x^2+3x+2 = (x+1)(x+2)$

$\dfrac{4x}{x^2+3x+2} = \dfrac{A}{x+1} + \dfrac{B}{x+2}$

$4x = A(x+2) + B(x+1)$

$x=-2:\ -8=-B,\ B=8$; $x=-1:\ -4=A,\ A=-4$

$\dfrac{4x}{x^2+3x+2} = \dfrac{-4}{x+1} + \dfrac{8}{x+2}$

$\displaystyle\int\dfrac{4x\,dx}{x^2+3x+2} = -\int\dfrac{4}{x+1} + \int\dfrac{8}{x+2}$

$\quad = -4\ln|x+1| + 8\ln|x+2| + C$

$\displaystyle\int_0^{0.500}\dfrac{4x\,dx}{x^2+3x+2} = 4\ln\left|\dfrac{(x+2)^2}{x+1}\right|\Big|_0^{0.500}$

$\quad = 4\left(\ln\dfrac{6.25}{1.5} - \ln\dfrac{4}{1}\right) = 4\ln\dfrac{6.25}{6} = 0.1633\ \text{N·cm}$

21. $A = \displaystyle\int_1^3 y\,dx = \int_1^3\dfrac{dx}{x^3+3x^2+2x}$

$x^3+3x^2+2x = x(x+1)(x+2)$

$\dfrac{1}{x^3+3x^2+2x} = \dfrac{A}{x} + \dfrac{B}{x+1} + \dfrac{C}{x+2}$

$1 = A(x+1)(x+2) + Bx(x+2) + Cx(x+1)$

$x=-2:\ 1=2C,\ C=1/2$
$x=-1:\ 1=-B,\ B=-1$
$x=0\ :\ 1=2A,\ A=1/2$

$\dfrac{1}{x^3+3x^2+2x} = \dfrac{1/2}{x} - \dfrac{1}{x+1} + \dfrac{1/2}{x+2}$

$A = \displaystyle\int_1^3\dfrac{dx}{x^3+3x^2+2x} = \dfrac{1}{2}\int_1^3\dfrac{dx}{x} - \int_1^3\dfrac{dx}{x+1} + \dfrac{1}{2}\int_1^3\dfrac{dx}{x+2}$

$\quad = \dfrac{1}{2}\ln|x| - \ln|x+1| + \dfrac{1}{2}\ln|x+2|\Big|_1^3$

$\quad = \dfrac{1}{2}\ln\left|\dfrac{x(x+2)}{(x+1)^2}\right|\Big|_1^3 = \dfrac{1}{2}\left(\ln\dfrac{15}{16} - \ln\dfrac{3}{4}\right)$

$\quad = \dfrac{1}{2}\ln\dfrac{5}{4} = 0.1116$

Exercises 9-5, p. 349

1. $x^3 - 4x^2 + +4x = x(x-2)^2$

$$\frac{x-8}{x^3-4x^2+4x} = \frac{A}{x} + \frac{B}{x-2} + \frac{C}{(x-2)^2}$$

$x - 8 = A(x-2)^2 + Bx(x-2) + Cx$

$x = 0: \quad -8 = 4A, \quad A = -2$

$x = 2: \quad -6 = 2C, \quad C = -3$

Coeff. x^2: $0 = A + B, \quad B = 2$

$$\frac{x-8}{x^3-4x^2+4x} = -\frac{2}{x} + \frac{2}{x-2} - \frac{3}{(x-2)^2}$$

$$\int\frac{(x-8)dx}{x^3-4x^2+4x} = -\int\frac{2dx}{x} + \int\frac{2dx}{x-2} - \int\frac{3dx}{(x-2)^2}$$

$$= -2\ln|x| + 2\ln|x-2| + \frac{3}{x-2} + C_1$$

$$= 2\ln\left|\frac{x-2}{x}\right| + \frac{3}{x-2} + C_1$$

3. $x^2(x^2-1) = x^2(x+1)(x-1)$

$$\frac{2}{x^2(x^2-1)} = \frac{A}{x} + \frac{B}{x^2} + \frac{C}{x+1} + \frac{D}{x-1}$$

$2 = Ax(x+1)(x-1) + B(x+1)(x-1) + Cx^2(x-1)$
$\quad + Dx^2(x+1)$

$x = -1: 2 = -2C, \quad C = -1$; $x = 0, \quad 2 = -B, \quad B = -2$

$x = 1: \quad 2 = 2D, \quad D = 1$; Coeff. x^3: $0 = A + C + D, \quad A = 0$

$$\frac{2}{x^2(x^2-1)} = \frac{0}{x} - \frac{2}{x^2} - \frac{1}{x+1} + \frac{1}{x-1}$$

$$\int\frac{2dx}{x^2(x^2-1)} = -\int\frac{2dx}{x^2} - \int\frac{dx}{x+1} + \int\frac{dx}{x-1}$$

$$= \frac{2}{x} - \ln|x+1| + \ln|x-1| + C_1 = \frac{2}{x} + \ln\left|\frac{x-1}{x+1}\right| + C_1$$

5. Let $u = x-3, \quad x = u + 3, \quad dx = du$

$$\int\frac{2xdx}{(x-3)^3} = \int\frac{2(u+3)(du)}{u^3} = 2\int u^{-2}du + 6\int u^{-3}du = -\frac{2}{u} - \frac{3}{u^2} + C = -\frac{2u+3}{u^2} + C$$

$$= -\frac{2x-6+3}{(x-3)^2} + C = \frac{3-2x}{(x-3)^2} + C \quad ; \quad \int_1^2\frac{2xdx}{(x-3)^3} = \frac{3-2x}{(x-3)^2}\Big|_1^2 = \frac{-1}{(-1)^2} - \frac{1}{4} = -\frac{5}{4}$$

7. $\dfrac{x^3-2x^2-7x+28}{(x+1)^2(x-3)^2} = \dfrac{A}{x+1} + \dfrac{B}{(x+1)^2} + \dfrac{C}{x-3} + \dfrac{D}{(x-3)^2}$

$x^3 - 2x^2 - 7x + 28 = A(x+1)(x-3)^2 + B(x-3)^2 + C(x+1)^2(x-3) + D(x+1)^2$

$x = -1: 32 = 16B, \quad B = 2$ Coeff. x^3: $1 = A + C$ $A + C = 1 \Big\} A = 1$

$x = 3: \quad 16 = 16D, \quad D = 1$ Const: $28 = 9A + 9B - 3C + D$ $9A - 3C = 9 \quad C = 0$

$$\int\frac{x^3-2x^2-7x+28}{(x+1)^2(x-3)^2}dx = \int\frac{dx}{x+1} + \int\frac{2dx}{(x+1)^2} + \int\frac{dx}{(x-3)^2} = \ln|x+1| - \frac{2}{x+1} - \frac{1}{x-3} + C_1$$

9. $\dfrac{x^2+x+5}{(x+1)(x^2+4)} = \dfrac{A}{x+1} + \dfrac{Bx+C}{x^2+4}$

$x^2 + x + 5 = A(x^2+4) + Bx(x+1) + C(x+1)$

$x = -1: \quad 5 = 5A, \quad A = 1$

Coeff. x^2: $1 = A + B, \quad B = 0$

Const.: $5 = 4A + C, \quad C = 1$

$$\int\frac{x^2+x+5}{(x+1)(x^2+4)}dx = \int\frac{dx}{x+1} + \int\frac{dx}{x^2+4}$$

$$= \ln|x+1| + \frac{1}{2}\text{Arctan}\frac{x}{2} + C_1$$

$$\int_0^2\frac{(x^2+x+5)dx}{(x+1)(x^2+4)} = \ln|x+1| + \frac{1}{2}\text{Arctan}\frac{x}{2}\Big|_0^2$$

$$= \ln 3 + \frac{1}{2}\text{Arctan}\,1 - (\ln 1 + \frac{1}{2}\text{Arctan}\,0)$$

$$= \ln 3 + \frac{1}{2}\left(\frac{\pi}{4}\right) = \ln 3 + \frac{1}{8}\pi = 1.491$$

11. $\dfrac{5x^2+8x+16}{x^2(x^2+4x+8)} = \dfrac{A}{x} + \dfrac{B}{x^2} + \dfrac{Cx+D}{x^2+4x+8}$

$5x^2 + 8x + 16 = Ax(x^2+4x+8) + B(x^2+4x+8)$
$\quad + Cx^3 + Dx^2$

$x = 0: \quad 16 = 8B, \quad B = 2$

Coeff. x: $8 = 8A + 4B, \quad A = 0$

Coeff. x^2: $5 = 4A + B + D, \quad D = 3$

Coeff. x^3: $0 = A + C, \quad C = 0$

$$\int\frac{5x^2+8x+16}{x^2(x^2+4x+8)}dx = \int\frac{2dx}{x^2} + \int\frac{3dx}{(x^2+4x+4)+4}$$

$$= \int\frac{2dx}{x^2} + \int\frac{3dx}{(x+2)^2+4}$$

$$= -\frac{2}{x} + \frac{3}{2}\text{Arctan}\frac{x+2}{2} + C_1$$

13. $\dfrac{10x^3+40x^2+22x+7}{(4x^2+1)(x^2+6x+10)} = \dfrac{Ax+B}{4x^2+1} + \dfrac{Cx+D}{x^2+6x+10}$

$10x^3+40x^2+22x+7 = (Ax+B)(x^2+6x+10) + (Cx+D)(4x^2+1)$

Const: $7=10B+D$

Coeff.x: $22=10A+6B+C$

Coeff.x^2: $40=6A+B+4D$

Coeff.x^3: $10=A+4C$

$22=10A+6B+\dfrac{10-A}{4}$

$\underline{40=6A+B+4(7-10B)}$

$88=40A+24B+10-A$

$40=6A+B+28-40B$

$39A+24B=78$

$\underline{6A-39B=12}$

$13A+8B=26$

$\underline{2A-13B=4}$

$A=2,\ B=0$

$C=2,\ D=7$

$\displaystyle\int\dfrac{10x^3+40x^2+22x+7}{(4x^2+1)(x^2+6x+10)}dx = \int\dfrac{2xdx}{4x^2+1} + \int\dfrac{(2x+7)dx}{x^2+6x+10} = \int\dfrac{2xdx}{4x^2+1} + \int\dfrac{2x+6}{x^2+6x+10}dx + \int\dfrac{dx}{(x+3)^2+1}$

$= \dfrac{1}{4}\ln|4x^2+1| + \ln|x^2+6x+10| + \text{Arctan}(x+3) + C_1$

15. $\dfrac{2x^3}{(x^2+1)^2} = \dfrac{Ax+B}{x^2+1} + \dfrac{Cx+D}{(x^2+1)^2}$

$2x^3 = (Ax+B)(x^2+1) + Cx + D$

Coeff.x^3: $2=A,\ A=2$

Coeff.x^2: $0=B,\ B=0$

Coeff.x : $0=A+C,\ C=-2$

Const. : $0=B+D,\ D=0$

$\displaystyle\int\dfrac{2x^3dx}{(x^2+1)^2} = \int\dfrac{2xdx}{x^2+1} - \int\dfrac{2xdx}{(x^2+1)^2}$

$= \ln|x^2+1| + \dfrac{1}{x^2+1} + C_1$

17. $A = \displaystyle\int_1^3(-y)dx = \int_1^3\dfrac{-(x-3)dx}{x^3+x^2} = \int_1^3\dfrac{(3-x)dx}{x^3+x^2}$

$\dfrac{3-x}{x^3+x^2} = \dfrac{A}{x} + \dfrac{B}{x^2} + \dfrac{C}{x+1}$

$3-x = Ax(x+1) + B(x+1) + Cx^2$

$x=-1$: $4=C,\ C=4$

$x=0$: $3=B,\ B=3$

Coeff.x^2: $0=A+C,\ A=-4$

$\displaystyle\int\dfrac{(3-x)dx}{x^3+x^2} = -\int\dfrac{4dx}{x} + \int\dfrac{3dx}{x^2} + \int\dfrac{4dx}{x+1}$

$= -4\ln|x| - \dfrac{3}{x} + 4\ln|x+1| + C_1$

$\displaystyle\int_1^3\dfrac{(3-x)dx}{x^3+x^2} = -\dfrac{3}{x} + 4\ln\left|\dfrac{x+1}{x}\right|\Bigg|_1^3$

$= -1 + 4\ln\dfrac{4}{3} - (-3 + 4\ln 2) = 2 + 4\ln\dfrac{2}{3} = 0.3781$

19. $V = 2\pi\displaystyle\int_0^2 xy\,dx = 2\pi\int_0^2\dfrac{4xdx}{x^4+6x^2+5}$

$= \pi\displaystyle\int_0^2\dfrac{8xdx}{x^4+6x^2+5}$

$x^4 + 6x^2+5 = (x^2+5)(x^2+1)$

$\dfrac{8x}{x^4+6x^2+5} = \dfrac{Ax+B}{x^2+5} + \dfrac{Cx+D}{x^2+1}$

$8x = (Ax+B)(x^2+1) + (Cx+D)(x^2+5)$

Coeff.x^3: $0=A+C$ $A=-2$

Coeff.x : $8=A+5C$ $C=2$

Coeff.x^2: $0=B+D$ $B=0$

Const. : $0=B+5D$ $D=0$

$V=\pi\displaystyle\int_0^2\dfrac{8xdx}{x^4+6x^2+5} = -\pi\int_0^2\dfrac{2xdx}{x^2+5} + \pi\int_0^2\dfrac{2xdx}{x^2+1}$

$= -\pi\ln|x^2+5| + \pi\ln|x^2+1|\Big|_0^2$

$= \pi\ln\left|\dfrac{x^2+1}{x^2+5}\right|\Big|_0^2 = \pi\left(\ln\dfrac{5}{9} - \ln\dfrac{1}{5}\right)$

$= \pi\ln\dfrac{25}{9} = 3.210$

21. $v = \dfrac{ds}{dt} = \dfrac{t^2+14t+27}{(2t+1)(t+5)^2}$

$\displaystyle\int ds = \int\dfrac{t^2+14t+27}{(2t+1)(t+5)^2}dt$

$\dfrac{t^2+14t+27}{(2t+1)(t+5)^2} = \dfrac{A}{2t+1} + \dfrac{B}{t+5} + \dfrac{C}{(t+5)^2}$

$t^2+14t+27 = A(t+5)^2 + B(2t+1)(t+5) + C(2t+1)$

$t=-5$: $-18=-9C,\ C=2$

$t=-\dfrac{1}{2}$: $\dfrac{81}{4} = \dfrac{81}{4}A,\ A=1$

Coeff.t^2: $1=A+2B,\ B=0$

$s = \displaystyle\int_0^2\dfrac{t^2+14t+27}{(2t+1)(t+5)^2}dt = \int_0^2\dfrac{dt}{2t+1} + \int_0^2\dfrac{2dt}{(t+5)^2}$

$= \dfrac{1}{2}\ln|2t+1| - \dfrac{2}{t+5}\Big|_0^2$

$= \dfrac{1}{2}\ln 5 - \dfrac{2}{7} - \left(\dfrac{1}{2}\ln 1 - \dfrac{2}{5}\right)$

$= \dfrac{1}{2}\ln 5 - \dfrac{2}{7} + \dfrac{2}{5} = 0.9190\text{ m}$

23. $\bar{x} = \dfrac{\int_1^2 xy\,dx}{\int_1^2 y\,dx} = \dfrac{\int_1^2 \frac{4x\,dx}{x^3+x}}{\int_1^2 \frac{4\,dx}{x^3+x}} = \dfrac{\int_1^2 \frac{4\,dx}{x^2+1}}{\int_1^2 \frac{4\,dx}{x^3+x}}$

$x^3+x = x(x^2+1)$

$\dfrac{4}{x^3+x} = \dfrac{A}{x} + \dfrac{Bx+C}{x^2+1}$

$\dfrac{4}{x^3+x} = \dfrac{4}{x} - \dfrac{4x}{x^2+1}$

$4 = A(x^2+1) + Bx^2 + Cx$

$\int \dfrac{4\,dx}{x^3+x} = 4\ln|x| - 2\ln|x^2+1| + C_1 = 2\ln\dfrac{x^2}{x^2+1} + C_1$

$x=0:$ $4=A,\quad A=4$

Coeff. x^2: $\ 0=A+B,\ B=-4$

Coeff. x : $\ 0=C,\ C=0$

$\bar{x} = \dfrac{4\,\text{Arctan }x\big|_1^2}{2\ln\frac{x^2}{x^2+1}\big|_1^2} = \dfrac{4(\text{Arctan }2 - \text{Arctan }1)}{2(\ln\frac{4}{5} - \ln\frac{1}{2})} = 1.369$

Exercises 9-6 , p. 353

1. $\int \dfrac{3x\,dx}{2+5x} = 3\int \dfrac{x\,dx}{2+5x}$; Formula 1 ; $u=x$, $du=dx$, $a=2$, $b=5$

$3\int \dfrac{x\,dx}{2+5x} = \dfrac{3}{5^2}[(2+5x) - 2\ln(2+5x)] + C = \dfrac{3}{25}[2+5x - 2\ln(2+5x)] + C$

3. $\int 5x\sqrt{2+3x}\,dx = 5\int x\sqrt{2+3x}\,dx$; Formula 5 ; $u=x$, $du=dx$, $a=2$, $b=3$

$5\int x\sqrt{2+3x}\,dx = 5\left(\dfrac{-2(4-9x)(2+3x)^{3/2}}{15(3^2)}\right) + C = -\dfrac{2}{27}(4-9x)(2+3x)^{3/2} + C$

5. $\int \sqrt{4-x^2}\,dx$; Formula 15 ; $u=x$, $du=dx$, $a=2$

$\int \sqrt{4-x^2}\,dx = \dfrac{x}{2}\sqrt{4-x^2} + \dfrac{4}{2}\text{Arcsin }\dfrac{x}{2} + C = \dfrac{1}{2}x\sqrt{4-x^2} + 2\text{Arcsin }\dfrac{1}{2}x + C$

7. $\int \sin 2x \sin 3x\,dx$; Formula 39 ; $u=x$, $du=dx$, $a=2$, $b=3$

$\int \sin 2x \sin 3x\,dx = \dfrac{\sin(2-3)x}{2(2-3)} - \dfrac{\sin(2+3)x}{2(2+3)} + C = \dfrac{\sin(-x)}{-2} - \dfrac{\sin 5x}{10} + C$

$= \dfrac{1}{2}\sin x - \dfrac{1}{10}\sin 5x + C$

9. $\int \dfrac{\sqrt{4x^2-9}}{x}\,dx$; Formula 17 ; $u=2x$, $du=2dx$, $a=3$

$\int \dfrac{\sqrt{4x^2-9}}{x}\,dx = \int \dfrac{\sqrt{4x^2-9}}{2x}(2dx) = \sqrt{4x^2-9} - 3\text{Arcsec }\dfrac{2x}{3} + C$

11. $\int \cos^5 4x\,dx$; Formula 34 ; $u=4x$, $du=4dx$, $n=5$

$\int \cos^5 4x\,dx = \dfrac{1}{4}\int \cos^5 4x(4dx) = \dfrac{1}{4}[\dfrac{1}{5}\cos^4 4x \sin 4x + \dfrac{4}{5}\int \cos^3 4x(4dx)] + C$

Now use Formula 33; $u=4x$, $du=4dx$

$\int \cos^5 4x\,dx = \dfrac{1}{20}\cos^4 4x \sin 4x + \dfrac{1}{5}(\sin 4x - \dfrac{1}{3}\sin^3 4x) + C$

$= \dfrac{1}{20}\cos^4 4x \sin 4x + \dfrac{1}{5}\sin 4x - \dfrac{1}{15}\sin^3 4x + C$

13. $\int \text{Arctan }x^2(x\,dx)$; Formula 52 ; $u=x^2$, $du=2x\,dx$

$\int \text{Arctan }x^2(x\,dx) = \dfrac{1}{2}\int \text{Arctan }x^2(2x\,dx) = \dfrac{1}{2}[x^2\text{Arctan }x^2 - \dfrac{1}{2}\ln(1+x^4)] + C$

$= \dfrac{1}{2}x^2\text{Arctan }x^2 - \dfrac{1}{4}\ln(1+x^4) + C$

15. $\int (4 - x^2)^{3/2} dx$; Formula 20 ; $u = x$, $du = dx$, $a = 2$

$\int (4 - x^2)^{3/2} dx = \frac{x}{4}(4 - x^2)^{3/2} + \frac{3(4)(x)}{8}\sqrt{4 - x^2} + \frac{3(16)}{8} \text{Arcsin} \frac{x}{2} + C$

$\int_1^2 (4 - x^2)^{3/2} dx = \frac{1}{4} x (4 - x^2)^{3/2} + \frac{3}{2} x (4 - x^2)^{1/2} + 6 \text{Arcsin} \frac{1}{2} x \Big|_1^2$

$\qquad\qquad = \frac{1}{4}(2)(0) + \frac{3}{2}(2)(0) + 6 \text{Arcsin } 1 - [\frac{1}{4}(1)(3)^{3/2} + \frac{3}{2}(1)(3)^{1/2} + 6 \text{Arcsin} \frac{1}{2}]$

$\qquad\qquad = 6 \text{Arcsin } 1 - \frac{3}{4}\sqrt{3} - \frac{3}{2}\sqrt{3} - 6 \text{Arcsin} \frac{1}{2} = 6(\frac{\pi}{2}) - \frac{9}{4}\sqrt{3} - 6(\frac{\pi}{6})$

$\qquad\qquad = 2\pi - \frac{9}{4}\sqrt{3} = \frac{1}{4}(8\pi - 9\sqrt{3}) = 2.386$

17. $\int \frac{dx}{x\sqrt{4x^2 + 1}}$; Formula 11 ; $u = 2x$, $du = 2dx$, $a = 1$

$\int \frac{dx}{x\sqrt{4x^2 + 1}} = \int \frac{2dx}{2x\sqrt{4x^2 + 1}} = -\frac{1}{1} \ln(\frac{1 + \sqrt{4x^2 + 1}}{2x}) + C = -\ln(\frac{1 + \sqrt{4x^2 + 1}}{2x}) + C$

19. $\int \frac{dx}{x\sqrt{1 - 4x^2}}$; Formula 13 ; $u = 2x$, $du = 2dx$, $a = 1$

$\int \frac{dx}{x\sqrt{1 - 4x^2}} = \int \frac{2dx}{2x\sqrt{1 - 4x^2}} = -\frac{1}{1} \ln(\frac{1 + \sqrt{1 - 4x^2}}{2x}) + C = -\ln(\frac{1 + \sqrt{1 - 4x^2}}{2x}) + C$

21. $\int \sin x \cos 5x \, dx$; Formula 40 ; $u = x$, $du = dx$, $a = 1$, $b = 5$

$\int \sin x \cos 5x \, dx = -\frac{\cos(1-5)x}{2(1-5)} - \frac{\cos(1+5)x}{2(1+5)} + C = -\frac{\cos(-4x)}{-8} - \frac{\cos 6x}{12} + C$

$\qquad\qquad = \frac{1}{8} \cos 4x - \frac{1}{12} \cos 6x + C$

23. $\int x^5 \cos x^3 dx = \int x^3 \cos x^3 (x^2 dx)$; Formula 48 ; $u = x^3$, $du = 3x^2 dx$

$\int x^3 \cos x^3 (x^2 dx) = \frac{1}{3}\int x^3 \cos x^3 (3x^2 dx) = \frac{1}{3}(\cos x^3 + x^3 \sin x^3) + C$

25. $\int \frac{x dx}{(1 - x^4)^{3/2}}$; Formula 25 ; $u = x^2$, $du = 2x dx$, $a = 1$

$\int \frac{x dx}{(1 - x^4)^{3/2}} = \frac{1}{2}\int \frac{2x dx}{(1 - x^4)^{3/2}} = \frac{1}{2}(\frac{x^2}{1^2 \sqrt{1 - x^4}}) + C = \frac{x^2}{2\sqrt{1 - x^4}} + C$

27. $\int \frac{\sqrt{3 + 5x^2}}{x} dx$; Formula 16 ; $u = \sqrt{5}x$, $du = \sqrt{5}\, dx$, $a = \sqrt{3}$

$\int \frac{\sqrt{3 + 5x^2}}{x} dx = \int \frac{\sqrt{3 + 5x^2}}{\sqrt{5}x} \sqrt{5}\, dx = \sqrt{3 + 5x^2} - \sqrt{3} \ln(\frac{\sqrt{3} + \sqrt{3 + 5x^2}}{\sqrt{5}x}) + C$

$\int_1^3 \frac{\sqrt{3 + 5x^2}}{x} dx = \sqrt{3 + 5x^2} - \sqrt{3} \ln(\frac{\sqrt{3} + \sqrt{3 + 5x^2}}{\sqrt{5}x}) \Big|_1^3 = \sqrt{48} - \sqrt{3} \ln(\frac{\sqrt{3} + \sqrt{48}}{3\sqrt{5}}) - (\sqrt{8} - \sqrt{3} \ln \frac{\sqrt{3} + \sqrt{8}}{\sqrt{5}})$

$\qquad\qquad = 4\sqrt{3} - 2\sqrt{2} - \sqrt{3} \ln \frac{5\sqrt{3}}{3\sqrt{5}} + \sqrt{3} \ln \frac{\sqrt{3} + 2\sqrt{2}}{\sqrt{5}} = 4\sqrt{3} - 2\sqrt{2} + \sqrt{3} \ln \frac{3(\sqrt{3} + 2\sqrt{2})}{5\sqrt{3}} = 4.892$

29. $\int x^3 \ln x^2 dx = \int x^2 \ln x^2 (x dx)$; Formula 46 ; $u = x^2$, $du = 2x dx$, $n = 1$

$\int x^2 \ln x^2 (x dx) = \frac{1}{2}\int x^2 \ln x^2 (2x dx) = \frac{1}{2}(x^2)^2 (\frac{\ln x^2}{2} - \frac{1}{2^2}) + C = \frac{1}{4} x^4 (\ln x^2 - \frac{1}{2}) + C$

31. $\int \frac{x^2 dx}{(x^6 - 1)^{3/2}}$; Formula 24 ; $u = x^3$, $du = 3x^2 dx$, $a = 1$

$$\int \frac{x^2 dx}{(x^6 - 1)^{3/2}} = \frac{1}{3}\int \frac{3x^2 dx}{(x^6 - 1)^{3/2}} = -\frac{x^3}{3(1^2)\sqrt{x^6 - 1}} + C = -\frac{x^3}{3\sqrt{x^6 - 1}} + C$$

33. $y = x^2$; $\frac{dy}{dx} = 2x$; $s = \int_0^1 \sqrt{1 + (2x)^2}\,dx = \int_0^1 \sqrt{1 + 4x^2}\,dx$; Formula 14 ; $u = 2x$, $du = 2dx$, $a = 1$

$$s = \int_0^1 \sqrt{1 + 4x^2}\,dx = \frac{1}{2}\int_0^1 \sqrt{1 + 4x^2}\,(2dx) = \frac{1}{2}[\frac{2x}{2}\sqrt{1 + 4x^2} + \frac{1}{2}\ln(2x + \sqrt{1 + 4x^2})]\Big|_0^1$$

$$= \frac{1}{2}[(1)\sqrt{5} + \frac{1}{2}\ln(2 + \sqrt{5})] - \frac{1}{2}[0\sqrt{1} + \frac{1}{2}\ln(0 + 1)] = \frac{1}{4}[2\sqrt{5} + \ln(2 + \sqrt{5})] = 1.479$$

35. $F = w\int_0^3 \ell h\,dh = w\int_0^3 x(3 - y)\,dy = w\int_0^3 \frac{3 - y}{\sqrt{1 + y}}\,dy$ [Formula 6]

$$\int \frac{3 - y}{\sqrt{1 + y}}\,dy = 3\int \frac{dy}{\sqrt{1 + y}} - \int \frac{y\,dy}{\sqrt{1 + y}} = 3\frac{(1 + y)^{1/2}}{1/2} - [\frac{-2(2 - y)\sqrt{1 + y}}{3(1)^2}] + C$$

$$F = w\int_0^3 \frac{3 - y}{\sqrt{1 + y}}\,dy = w[6(1 + y)^{1/2} + \frac{2}{3}(2 - y)(1 + y)^{1/2}]\Big|_0^3 = w[6(2) + \frac{2}{3}(-1)(2) - 6(1) - \frac{2}{3}(2)(1)]$$

$$= w(12 - \frac{4}{3} - 6 - \frac{4}{3}) = \frac{10w}{3} = \frac{10(62.4)}{3} = 208 \text{ lb}$$

37. $A = \int_0^2 y\,dx = \int_0^2 \text{Arctan } 2x\,dx = \frac{1}{2}\int_0^2 \text{Arctan } 2x\,(2dx)$

$$= \frac{1}{2}[2x \text{ Arctan } 2x - \frac{1}{2}\ln(1 + 4x^2)]\Big|_0^2 \qquad \text{[Formula 52]}$$

$$= 2 \text{ Arctan } 4 - \frac{1}{4}\ln 17 - [\frac{1}{2}(0) - \frac{1}{4}\ln 1] = 2 \text{ Arctan } 4 - \frac{1}{4}\ln 17 = 1.943$$

39. $V = \pi\int_0^\pi y^2\,dx = \pi\int_0^\pi (e^x \sin x)^2\,dx = \pi\int_0^\pi e^{2x}\sin^2 x\,dx = \frac{\pi}{2}\int_0^\pi e^{2x}(1 - \cos 2x)\,dx$ [Formula 50]

$$= \frac{\pi}{4}\int_0^\pi e^{2x}(2dx) - \frac{\pi}{2}\int_0^\pi e^{2x}\cos 2x\,dx = \frac{\pi}{4}e^{2x} - \frac{\pi}{2}[\frac{e^{2x}(2\cos 2x + 2\sin 2x)}{4 + 4}]\Big|_0^\pi$$

$$= \frac{\pi}{4}e^{2x}[1 - \frac{1}{2}(\cos 2x + \sin 2x)]\Big|_0^\pi = \frac{\pi}{4}[e^{2\pi}[1 - \frac{1}{2}(\cos 2\pi + \sin 2\pi)]$$

$$- \frac{\pi}{4}e^0[1 - \frac{1}{2}(\cos 0 + \sin 0)] = \frac{\pi}{4}e^{2\pi}(\frac{1}{2}) - \frac{\pi}{4}(\frac{1}{2}) = \frac{\pi}{8}(e^{2\pi} - 1) = 209.9$$

Exercises 9-7, p. 358

1. $\int_1^{+\infty} \frac{dx}{(x + 2)^2} = \lim_{b \to +\infty}\int_1^b \frac{dx}{(x + 2)^2} = \lim_{b \to +\infty}\int_1^b (x + 2)^{-2}dx = \lim_{b \to +\infty} -\frac{1}{x + 2}\Big|_1^b = \lim_{b \to +\infty} -\frac{1}{b + 2} - (-\frac{1}{3}) = \frac{1}{3}$

3. $\int_{-\infty}^0 \frac{dx}{\sqrt{1 - 3x}} = \lim_{a \to -\infty}\int_a^0 \frac{dx}{\sqrt{1 - 3x}} = \lim_{a \to -\infty} -\frac{1}{3}\int_a^0 (1 - 3x)^{-1/2}(-3dx) = \lim_{a \to -\infty} -\frac{1}{3}\frac{(1 - 3x)^{1/2}}{1/2}\Big|_a^0$

$$= \lim_{a \to -\infty} -\frac{2}{3}(1 - 3x)^{1/2}\Big|_a^0 = \lim_{a \to -\infty} -\frac{2}{3}(1 - 3a)^{1/2} - (-\frac{2}{3})[1 - 3(0)] = -\infty \text{ (Divergent)}$$

5. $\int_1^{+\infty} \frac{x\,dx}{1 + x^2} = \lim_{b \to +\infty}\int_1^b \frac{x\,dx}{1 + x^2} = \lim_{b \to +\infty}\frac{1}{2}\int_1^b \frac{2x\,dx}{1 + x^2} = \lim_{b \to +\infty}\ln|1 + x^2|\Big|_1^b$

$$= \lim_{b \to +\infty}(\ln|1 + b^2| - \ln|1 + 1|) = +\infty \text{ (Divergent)}$$

7. $\int_0^{+\infty} xe^{-x^2}dx = \lim_{b\to+\infty}\int_0^b xe^{-x^2}dx = \lim_{b\to+\infty} -\frac{1}{2}\int_0^b e^{-x^2}(-2xdx) = \lim_{b\to+\infty} -\frac{1}{2}e^{-x^2}\Big|_0^b$

$= \lim_{b\to+\infty} -\frac{1}{2}(e^{-b^2} - e^0) = -\frac{1}{2}(0-1) = \frac{1}{2}$

9. $\int_{-\infty}^0 \frac{2dx}{(1-x)^3} = \lim_{a\to-\infty}\int_a^0 \frac{2dx}{(1-x)^3} = \lim_{a\to-\infty} -2\int_a^0 (1-x)^{-3}(-dx) = \lim_{a\to-\infty} -2\frac{(1-x)^{-2}}{-2}\Big|_a^0$

$= \lim_{a\to-\infty}(1-x)^{-2}\Big|_a^0 = \lim_{a\to-\infty}[\frac{1}{1^2} - \frac{1}{(1-a)^2}] = 1 - 0 = 1$

11. $\int_{-\infty}^{+\infty} e^{-x}dx = \lim_{a\to-\infty}\int_a^0 e^{-x}dx + \lim_{b\to+\infty}\int_0^b e^{-x}dx = \lim_{a\to-\infty} -e^{-x}\Big|_a^0 + \lim_{b\to+\infty} -e^{-x}\Big|_0^b$

$= \lim_{a\to-\infty}[-e^0 - (-e^{-a})] + \lim_{b\to+\infty}[-e^{-b} - (-e^0)] = +\infty$ (since $\lim_{a\to-\infty} e^{-a} = +\infty$)
(Divergent)

13. $\int_0^9 \frac{dx}{\sqrt{9-x}} = \lim_{h\to0}\int_0^{9-h} \frac{dx}{\sqrt{9-x}} = \lim_{h\to0} -\int_0^{9-h}(9-x)^{-1/2}(-dx) = \lim_{h\to0} -\frac{(9-x)^{1/2}}{1/2}\Big|_0^{9-h}$

$= \lim_{h\to0} -2(9-x)^{1/2}\Big|_0^{9-h} = \lim_{h\to0}[-2(h)^{1/2} - (-2)(9)^{1/2}] = 0 + 6 = 6$

15. $\int_{\pi/4}^{\pi/3} \tan 2x\, dx = \lim_{h\to0}\int_{\pi/4+h}^{\pi/3}\tan 2x\, dx = \lim_{h\to0}\frac{1}{2}\int_{\pi/4+h}^{\pi/3}\tan 2x(2dx) = \lim_{h\to0} -\frac{1}{2}\ln|\cos 2x|\Big|_{\pi/4+h}^{\pi/3}$

$= \lim_{h\to0} -\frac{1}{2}[\ln\cos\frac{2\pi}{3} - \ln\cos(\frac{\pi}{2}+2h)]$; $\ln\cos\frac{2\pi}{3}$ undefined since $\cos\frac{2\pi}{3} < 0$, same for $\ln\cos(\frac{\pi}{2}+2h)$; (Divergent)

17. $\int_0^3 \frac{dx}{(x-2)^2} = \lim_{h\to0}\int_0^{2-h} \frac{dx}{(x-2)^2} + \lim_{h'\to0}\int_{2+h'}^3 \frac{dx}{(x-2)^2} = \lim_{h\to0}[-\frac{1}{x-2}]\Big|_0^{2-h} + \lim_{h'\to0}[-\frac{1}{x-2}]\Big|_{2+h'}^3$

$= \lim_{h\to0}(\frac{1}{h} - \frac{1}{2}) + \lim_{h'\to0}(-1 + \frac{1}{h'}) = +\infty$ (Neither limit exists) (Divergent)

19. $\int_0^1 \frac{2dx}{1-x^2} = \lim_{h\to0}\int_0^{1-h} \frac{2dx}{1-x^2} = \lim_{h\to0}\int_0^{1-h}(\frac{1}{1+x} + \frac{1}{1-x})dx = \lim_{h\to0}\ln|1+x| - \ln|1-x|\Big|_0^{1-h}$

$= \lim_{h\to0}\ln\left|\frac{1+x}{1-x}\right|\Big|_0^{1-h} = \lim_{h\to0}[\ln\frac{2-h}{h} - \ln 1] = +\infty$ (Divergent)

21. $\int_1^2 \frac{dx}{\sqrt{x^2-1}} = \lim_{h\to0}\int_{1+h}^2 \frac{dx}{\sqrt{x^2-1}} = \lim_{h\to0}(\ln(x+\sqrt{x^2-1})\Big|_{1+h}^2$ (Trig. sub. or Formula 10)

$= \lim_{h\to0}[\ln(2+\sqrt{3}) - \ln(1+h+\sqrt{2h+h^2})] = \ln(2+\sqrt{3})$

23. $\int_1^3 \frac{dx}{(x-2)^{1/3}} = \lim_{h\to0}\int_1^{2-h} \frac{dx}{(x-2)^{1/3}} + \lim_{h'\to0}\int_{2+h'}^3 \frac{dx}{(x-2)^{1/3}} = \lim_{h\to0}\frac{3}{2}(x-2)^{2/3}\Big|_1^{2-h} + \lim_{h\to0}\frac{3}{2}(x-2)^{2/3}\Big|_{2+h'}^3$

$= \lim_{h\to0}[\frac{3}{2}(-h)^{2/3} - \frac{3}{2}(-1)^{2/3}] + \lim_{h'\to0}[\frac{3}{2}(1)^{2/3} - \frac{3}{2}(h)^{2/3}] = -\frac{3}{2} + \frac{3}{2} = 0$

25.(a) $A = \int_1^\infty y\,dx = \int_1^\infty \frac{dx}{x} = \lim_{b\to+\infty}\int_1^b \frac{dx}{x} = \lim_{b\to+\infty}\ln x\Big|_1^b$

$= \lim_{b\to+\infty}(\ln b - \ln 1) = +\infty$ (Divergent)

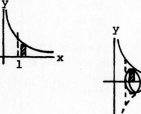

(b) $V = \pi\int_1^\infty y^2 dx = \pi\int_1^\infty \frac{dx}{x^2} = \lim_{b\to+\infty}\pi\int_1^b \frac{dx}{x^2} = \lim_{b\to+\infty} -\frac{\pi}{x}\Big|_1^b$

$= \lim_{b\to+\infty}(-\frac{\pi}{b} + \frac{\pi}{1}) = \pi$

27. $A = \int_1^2 y\,dx = \int_1^2 \frac{x\,dx}{\sqrt{x^2-1}} = \lim_{h\to 0}\int_{1+h}^2 \frac{x\,dx}{\sqrt{x^2-1}} = \lim_{h\to 0}(x^2-1)^{1/2}\Big|_{1+h}^2$

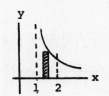

$= \lim_{h\to 0}(\sqrt{3}-\sqrt{2h+h^2}) = \sqrt{3}$

29. $W = \int_0^\infty F\,dx = \int_0^\infty \frac{10^{11}\,dx}{(x+3960)^2} = \lim_{b\to +\infty}\int_0^b \frac{10^{11}\,dx}{(x+3960)^2} = \lim_{b\to +\infty} -\frac{10^{11}}{x+3960}\Big|_0^b$

$= \lim_{b\to +\infty}[-\frac{10^{11}}{b+3960} - (-\frac{10^{11}}{3960})] = \frac{10^{11}}{3960} = 2.53 \times 10^7 \text{ mi·lb}$

31. $\int_0^{+\infty}\cos x\,dx = \lim_{b\to +\infty}\int_0^b \cos x\,dx = \lim_{b\to +\infty}\sin x\Big|_0^b = \lim_{b\to +\infty}(\sin b - \sin 0) = \lim_{b\to +\infty}\sin b$

(As $b\to +\infty$, $\sin b$ takes on all values from -1 to $+1$. Therefore, since the limit does not exist, the integral is divergent.)

Review Exercises for Chapter 9, p. 359

1. $\int x\csc^2 2x\,dx$; $u=x$, $du=dx$, $dv=\csc^2 2x\,dx$, $v=\frac{1}{2}\int\csc^2 2x(2dx) = -\frac{1}{2}\cot 2x$

$\int x\csc^2 2x\,dx = x(-\frac{1}{2}\cot 2x) - \int(-\frac{1}{2}\cot 2x)dx = -\frac{1}{2}x\cot 2x + \frac{1}{4}\int\cot 2x(2dx)$

$= -\frac{1}{2}x\cot 2x + \frac{1}{4}\ln|\sin 2x| + C$

3. $\int x\sqrt{x-4}\,dx$; $u=\sqrt{x-4}$, $x=u^2+4$, $dx=2u\,du$

$\int x\sqrt{x-4}\,dx = \int(u^2+4)(u)(2u\,du) = 2\int(u^4+4u^2)du = \frac{2}{5}u^5 + \frac{8}{3}u^3 + C$

$= \frac{2}{5}(x-4)^{5/2} + \frac{8}{3}(x-4)^{3/2} + C = 2(x-4)^{3/2}[\frac{1}{5}(x-4)+\frac{4}{3}]+C = \frac{2}{15}(3x+8)(x-4)^{3/2}+C$

5. $\int\frac{dx}{\sqrt{4x^2-9}}$; Let $x=\frac{3}{2}\sec\theta$, $dx=\frac{3}{2}\sec\theta\tan\theta\,d\theta$

$\int\frac{dx}{\sqrt{4x^2-9}} = \int\frac{(3/2)\sec\theta\tan\theta\,d\theta}{\sqrt{9\sec^2\theta-9}} = \frac{1}{2}\int\frac{\sec\theta\tan\theta\,d\theta}{\tan\theta} = \frac{1}{2}\int\sec\theta\,d\theta$

$= \frac{1}{2}\ln|\sec\theta+\tan\theta|+C_1 = \frac{1}{2}\ln\left|\frac{2x}{3}+\frac{\sqrt{4x^2-9}}{3}\right|+C_1 = \frac{1}{2}\ln|2x+\sqrt{4x^2-9}|+C$

$(C = C_1 - \frac{1}{2}\ln 3)$

7. $\int\frac{x+25}{x^2-25}\,dx$

$\frac{x+25}{x^2-25} = \frac{A}{x+5} + \frac{B}{x-5}$

$x+25 = A(x-5) + B(x+5)$

$x=-5$: $20=-10A$, $A=-2$

$x=5$: $30=10B$, $B=3$

$\int\frac{x+25}{x^2-25}dx = -\int\frac{2dx}{x+5} + \int\frac{3dx}{x-5}$

$= -2\ln|x+5| + 3\ln|x-5| + C$

$= \ln\left|\frac{(x-5)^3}{(x+5)^2}\right| + C$

9. $\int\frac{x^2-2x+3}{(x-1)^3}\,dx$; Let $u=x-1$, $x=u+1$, $dx=du$

$\int\frac{x^2-2x+3}{(x-1)^3}\,dx = \int\frac{(u+1)^2-2(u+1)+3}{u^3}\,du$

$= \int\frac{u^2+2}{u^3}\,du = \int\frac{du}{u} + 2\int\frac{du}{u^3} = \ln|u| - \frac{1}{u^2} + C$

$= \ln|x-1| - \frac{1}{(x-1)^2} + C$

11. $\int_1^{+\infty}\frac{dx}{2x+7} = \lim_{b\to +\infty}\int_1^b\frac{dx}{2x+7} = \lim_{b\to +\infty}\frac{1}{2}\ln|2x+7|\Big|_1^b$

$= \lim_{b\to +\infty}\frac{1}{2}\ln|2b+7| - \frac{1}{2}\ln 9 = +\infty$ (Divergent)

13. $\int \dfrac{xdx}{(x+3)^{2/3}}$; $u = (x+3)^{1/3}$, $x = u^3 - 3$, $dx = 3u^2 du$

$$\int \dfrac{xdx}{(x+3)^{2/3}} = \int \dfrac{(u^3-3)(3u^2 du)}{u^2} = 3\int(u^3-3)du = \frac{3}{4}u^4 - 9u + C = \frac{3}{4}(x+3)^{4/3} - 9(x+3)^{1/3} + C$$

$$= 3(x+3)^{1/3}[\frac{1}{4}(x+3) - 3] + C = \frac{3}{4}(x+3)^{1/3}(x-9) + C$$

15. $\int \dfrac{2dx}{(4x^2+1)^{3/2}}$; Let $x = \frac{1}{2}\tan\theta$, $dx = \frac{1}{2}\sec^2\theta\, d\theta$

$$\int \dfrac{2dx}{(4x^2+1)^{3/2}} = \int \dfrac{2(1/2)\sec^2\theta\, d\theta}{(\tan^2\theta+1)^{3/2}} = \int\dfrac{\sec^2\theta\, d\theta}{\sec^3\theta} = \int\cos\theta\, d\theta = \sin\theta + C = \dfrac{2x}{\sqrt{4x^2+1}} + C$$

17. $\int \dfrac{x^2+3}{x^4+3x^2+2}dx$; $x^4+3x^2+2 = (x^2+1)(x^2+2)$; $\dfrac{x^2+3}{x^4+3x^2+2} = \dfrac{Ax+B}{x^2+1} + \dfrac{Cx+D}{x^2+2}$

$$x^2+3 = (Ax+B)(x^2+2) + (Cx+D)(x^2+1)$$

Coeff. x^3: $0 = A+C$ $\Big\}$ $A=0$, $C=0$ $\int\dfrac{x^2+3}{x^4+3x^2+2}dx = \int\dfrac{2dx}{x^2+1} - \int\dfrac{dx}{x^2+2}$

Coeff. x : $0 = 2A+C$

Coeff. x^2: $1 = B+D$ $\Big\}$ $B=2$, $D=-1$ $= 2\operatorname{Arctan} x - \dfrac{1}{\sqrt{2}}\operatorname{Arctan}\dfrac{x}{\sqrt{2}} + C$

Const.   : $3 = 2B+D$

$$= 2\operatorname{Arctan} x - \frac{1}{2}\sqrt{2}\operatorname{Arctan}\frac{1}{2}x\sqrt{2} + C$$

19. $\int \ln(x+2)dx$; $u = \ln(x+2)$, $du = \dfrac{dx}{x+2}$, $dv = dx$, $v = \int dx = x$

$$\int \ln(x+2)dx = x\ln(x+2) - \int x(\dfrac{dx}{x+2}) = x\ln(x+2) - \int(1-\dfrac{2}{x+2})dx$$

$$= x\ln(x+2) - x + 2\ln(x+2) + C = (x+2)\ln(x+2) - x + C$$

$$\int_0^2 \ln(x+2)dx = (x+2)\ln(x+2) - x \Big|_0^2 = 4\ln 4 - 2 - (2\ln 2 - 0) = 8\ln 2 - 2\ln 2 - 2$$

$$= 6\ln 2 - 2 = 2.159$$

21. $\int \dfrac{6(2-x^2)}{(x^2-1)(x^2-4)}dx$

$$(x^2-1)(x^2-4) = (x+1)(x-1)(x+2)(x-2)$$

$$\dfrac{12-6x^2}{(x^2-1)(x^2-4)} = \dfrac{A}{x+1} + \dfrac{B}{x-1} + \dfrac{C}{x+2} + \dfrac{D}{x-2}$$

$$12-6x^2 = A(x-1)(x+2)(x-2) + B(x+1)(x+2)(x-2)$$
$$+ C(x+1)(x-1)(x-2) + D(x+1)(x-1)(x+2)$$

$x=-2$: $-12 = -12C$, $C=1$

$x=-1$: $6 = 6A$, $A=1$

$x=1$: $6 = -6B$, $B=-1$

$x=2$: $-12 = 12D$, $D=-1$

$$\int \dfrac{6(2-x^2)dx}{(x^2-1)(x^2-4)} = \int\dfrac{dx}{x+1} - \int\dfrac{dx}{x-1} + \int\dfrac{dx}{x+2} - \int\dfrac{dx}{x-2}$$

$$= \ln|x+1| - \ln|x-1| + \ln|x+2| - \ln|x-2| + C$$

$$= \ln\left|\dfrac{(x+1)(x+2)}{(x-1)(x-2)}\right| + C$$

25. $A = \int_2^6 ydx = \int_2^6 \dfrac{dx}{\sqrt{x-2}} = \lim\limits_{h\to 0}\int_{2+h}^6 \dfrac{dx}{\sqrt{x-2}}$

$$= \lim\limits_{h\to 0} 2(x-2)^{1/2}\Big|_{2+h}^6$$

$$= \lim\limits_{h\to 0} 2(2) - 2(h)^{1/2} = 4$$

23. $\int_0^8 \dfrac{xdx}{(8-x)^{1/3}} = \lim\limits_{h\to 0}\int_0^{8-h}\dfrac{xdx}{(8-x)^{1/3}}$

$$(8-x)^{1/3} = u, \quad x = 8-u^3, \quad dx = -3u^2 du$$

$$\int\dfrac{xdx}{(8-x)^{1/3}} = \int\dfrac{(8-u^3)(-3u^2 du)}{u}$$

$$= 3\int(u^4-8u)du = \frac{3}{5}u^5 - 12u^2 + C$$

$$= \frac{3}{5}(8-x)^{5/3} - 12(8-x)^{2/3} + C$$

$$= 3(8-x)^{2/3}[\frac{1}{5}(8-x) - 4] + C$$

$$= -\frac{3}{5}(8-x)^{2/3}(x+12) + C$$

$$\lim\limits_{h\to 0}\int_0^{8-h}\dfrac{xdx}{(8-x)^{1/3}}$$

$$= \lim\limits_{h\to 0} -\frac{3}{5}(8-x)^{2/3}(x+12)\Big|_0^{8-h}$$

$$= \lim\limits_{h\to 0} -\frac{3}{5}h^{2/3}(20-h) - [-\frac{3}{5}(8)^{2/3}(12)]$$

$$= 0 + \dfrac{144}{5} = \dfrac{144}{5}$$

27. $A = 2\int_3^5 y\,dx = 2\int_3^5 \sqrt{25 - x^2}\,dx$; Let $x = 5\sin\theta$, $dx = 5\cos\theta\,d\theta$

$\int (25-x^2)^{1/2}dx = \int(25-25\sin^2\theta)^{1/2}(5\cos\theta\,d\theta) = 25\int\cos^2\theta\,d\theta$

$= \dfrac{25}{2}\int(1+\cos 2\theta)d\theta = \dfrac{25}{2}\theta + \dfrac{25}{4}\sin 2\theta + C = \dfrac{25}{4}(2\theta + 2\sin\theta\cos\theta)+C$

$= \dfrac{25}{2}(\text{Arcsin}\,\dfrac{x}{5} + \dfrac{x}{5}\dfrac{\sqrt{25-x^2}}{5}) + C$

$A = 2[\dfrac{25}{2}(\text{Arcsin}\,\dfrac{x}{5} + \dfrac{x\sqrt{25-x^2}}{25})]\Big|_3^5 = 25\,\text{Arcsin}\,\dfrac{x}{5} + x\sqrt{25-x^2}\ \Big|_3^5$

$= 25\,\text{Arcsin}\,1 + 0 - [25\,\text{Arcsin}\,\dfrac{3}{5} + 3(4)] = 25(\dfrac{\pi}{2}) - 25(0.6435) - 12 = 11.18$

29. $V = 2\pi\int_0^2 xy\,dx = 2\pi\int_0^2 x^2 e^x dx$; $u = x^2$, $du = 2x\,dx$, $dv = e^x dx$, $v = e^x$

$\int x^2 e^x dx = x^2 e^x - \int e^x(2x\,dx) = x^2 e^x - 2\int xe^x dx$

$u=x$, $du=dx$, $dv=e^x dx$, $v=e^x$; $\int xe^x dx = xe^x - \int e^x dx = xe^x - e^x - \dfrac{C}{2}$

$\int x^2 e^x dx = x^2 e^x - 2(xe^x - e^x - \dfrac{C}{2}) = e^x(x^2 - 2x + 2) + C$

$V = 2\pi[e^x(x^2-2x+2)]\Big|_0^2 = 2\pi[e^2(4-4+2) - e^0(0-0+2)] = 4\pi(e^2 - 1) = 80.29$

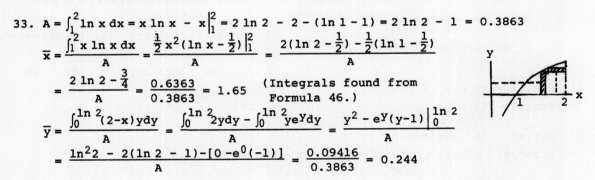

31. $W = \int_0^{0.500} F\,dx = \int_0^{0.500} \dfrac{1000x\,dx}{(2x+1)^2} = 1000\int_0^{0.500} \dfrac{x\,dx}{(2x+1)^2}$; $u = 2x+1$, $x = \dfrac{1}{2}(u-1)$, $dx = \dfrac{1}{2}du$

$\int \dfrac{x\,dx}{(2x+1)^2} = \int \dfrac{(1/2)(u-1)(1/2)(du)}{u^2} = \dfrac{1}{4}\int(\dfrac{1}{u} - \dfrac{1}{u^2})du = \dfrac{1}{4}(\ln|u| + \dfrac{1}{u}) + C$

When $x=0$, $u=1$, and when $x=0.500$, $u=2.00$.

$W = (1000)\dfrac{1}{4}(\ln u + \dfrac{1}{u})\Big|_1^{2.00} = 250[\ln 2.00 + \dfrac{1}{2.00} - (\ln 1 + 1)] = 48.3$ N·cm

33. $A = \int_1^2 \ln x\,dx = x\ln x - x\Big|_1^2 = 2\ln 2 - 2 - (\ln 1 - 1) = 2\ln 2 - 1 = 0.3863$

$\bar{x} = \dfrac{\int_1^2 x\ln x\,dx}{A} = \dfrac{\dfrac{1}{2}x^2(\ln x - \dfrac{1}{2})\Big|_1^2}{A} = \dfrac{2(\ln 2 - \dfrac{1}{2}) - \dfrac{1}{2}(\ln 1 - \dfrac{1}{2})}{A}$

$= \dfrac{2\ln 2 - \dfrac{3}{4}}{A} = \dfrac{0.6363}{0.3863} = 1.65$ (Integrals found from Formula 46.)

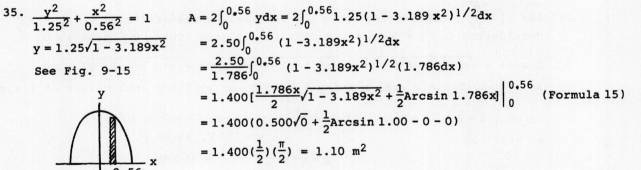

$\bar{y} = \dfrac{\int_0^{\ln 2}(2-x)y\,dy}{A} = \dfrac{\int_0^{\ln 2} 2y\,dy - \int_0^{\ln 2} ye^y dy}{A} = \dfrac{y^2 - e^y(y-1)\Big|_0^{\ln 2}}{A}$

$= \dfrac{\ln^2 2 - 2(\ln 2 - 1) - [0 - e^0(-1)]}{A} = \dfrac{0.09416}{0.3863} = 0.244$

35. $\dfrac{y^2}{1.25^2} + \dfrac{x^2}{0.56^2} = 1$

$y = 1.25\sqrt{1 - 3.189x^2}$

See Fig. 9-15

$A = 2\int_0^{0.56} y\,dx = 2\int_0^{0.56} 1.25(1 - 3.189x^2)^{1/2}dx$

$= 2.50\int_0^{0.56}(1 - 3.189x^2)^{1/2}dx$

$= \dfrac{2.50}{1.786}\int_0^{0.56}(1 - 3.189x^2)^{1/2}(1.786\,dx)$

$= 1.400[\dfrac{1.786x}{2}\sqrt{1 - 3.189x^2} + \dfrac{1}{2}\text{Arcsin}\,1.786x]\Big|_0^{0.56}$ (Formula 15)

$= 1.400(0.500\sqrt{0} + \dfrac{1}{2}\text{Arcsin}\,1.00 - 0 - 0)$

$= 1.400(\dfrac{1}{2})(\dfrac{\pi}{2}) = 1.10$ m^2

Exercises 10-1

1. From geometry: $V = \pi r^2 h$

3. From geometry: $A = \frac{1}{2} bh$

5. From geometry:

$A = 2\pi rh + 2\pi r^2$; $V = \pi r^2 h$, $h = \frac{V}{\pi r^2}$

$A = 2\pi r(\frac{V}{\pi r^2}) + 2\pi r^2 = \frac{2V}{r} + 2\pi r^2$

7. From Law of Cosines: See Fig. 10-4

$R^2 = F_1^2 + F_2^2 - 2F_1 F_2 \cos 150°$

$= F_1^2 + F_2^2 - 2F_1 F_2 (-0.8660)$

$R = \sqrt{F_1^2 + F_2^2 + 1.732 F_1 F_2}$

9. $f(x,y) = 2x - 6y$

$f(0,-4) = 2(0) - 6(-4) = 24$

11. $g(r,s) = r - 2rs - r^2 s$

$g(-2,1) = -2 - 2(-2)(1) - (-2)^2(1) = -2 + 4 - 4 = -2$

13. $Y(y,t) = \frac{2 - 3y}{t - 1} + 2y^2 t$

$Y(y,2) = \frac{2 - 3y}{2 - 1} + 2y^2(2)$

$= 2 - 3y + 4y^2$

15. $X(x,t) = -6xt + xt^2 - t^3$

$X(x,-t) = -6x(-t) + x(-t)^2 - (-t)^3$

$= 6xt + xt^2 + t^3$

17. $H(p,q) = p - \frac{p - 2q^2 - 5q}{p + q}$

$H(p,q+k) = p - \frac{p - 2(q+k)^2 - 5(q+k)}{p + q + k}$

$= \frac{p(p+q+k) - p + 2(q^2 + 2kq + k^2) + 5(q+k)}{p + q + k}$

$= \frac{p^2 + pq + pk - p + 2q^2 + 4kq + 2k^2 + 5q + 5k}{p + q + k}$

19. $f(x,y) = x^2 - 2xy - 4x$

$f(x+h, y+k) - f(x,y)$

$= (x+h)^2 - 2(x+h)(y+k) - 4(x+h)$
$\quad - (x^2 - 2xy - 4x)$

$= x^2 + 2hx + h^2 - 2xy - 2kx - 2hy - 2hk - 4x - 4h$
$\quad -x^2 + 2xy + 4x$

$= 2hx - 2kx - 2hy + h^2 - 2hk - 4h$

21. $f(x,y) = xy + x^2 - y^2$

$f(x,x) - f(x,0) = x(x) + x^2 - x^2 - [x(0) + x^2 - 0^2] = x^2 + x^2 - x^2 - x^2 = 0$

23. $g(y,z) = 3y^3 - y^2 z + 5z^2$

$g(3z^2, z) - g(z,z) = 3(3z^2)^3 - (3z^2)^2(z) + 5z^2 - [3z^3 - (z)^2 z + 5z^2]$

$= 81z^6 - 9z^5 + 5z^2 - 3z^3 + z^3 - 5z^2 = 81z^6 - 9z^5 - 2z^3$

25. $f(x,y) = \frac{\sqrt{y}}{2x}$; Considering $\sqrt{y}$, $y \geq 0$ for real values of $f(x,y)$;
Considering $2x$, $x \neq 0$ to avoid division by zero. Thus, $y \geq 0$ and $x \neq 0$.

27. $f(x,y) = \sqrt{x^2 + y^2 - x^2 y - y^3} = \sqrt{x^2(1-y) + y^2(1-y)} = \sqrt{(x^2 + y^2)(1 - y)}$

$x^2 + y^2 \geq 0$ for all real x and y ; $1-y \geq 0$, or $y \leq 1$ for real values of $f(x,y)$.

29. $v = iR$; $i = 3 A$, $R = 6 \Omega$

$v = 3(6) = 18 V$

31. $p = \frac{nRT}{V}$; $n = 3$ mol, $R = 8.31$ J/mol·K

$T = 300$ K, $V = 50$ m^3

$p = \frac{3(8.31)(300)}{50} = 150$ Pa

33. $\frac{1}{q} = \frac{1}{f} - \frac{1}{p}$; $q = \frac{pf}{p - f}$; $p = 20$ cm, $f = 5$ cm ; $q = \frac{20(5)}{20 - 5} = 6.67$ cm

35. $p = 2\ell + 2w$; $\ell = \dfrac{p - 2w}{2}$

$A = \ell w = \dfrac{p - 2w}{2} w = \dfrac{pw - 2w^2}{2}$

$p = 250$ cm, $w = 55$ cm

$A = \dfrac{250(55) - 2(55)^2}{2} = 3850$ cm^2

37. $L = \dfrac{kr^4}{\ell^2}$; $L = 20$ ton, $\ell = 20$ ft, $r = 0.5$ ft

$20 = \dfrac{k(0.5)^4}{20^2}$; $k = 1.28 \times 10^5$ ton/ft^2

$L = \dfrac{(1.28 \times 10^5)r^4}{\ell^2}$

Exercises 10-2 , p. 374

1. $z = x^2 + y^2$
   z=1: $x^2 + y^2 = 1$, circle, r=1
   z=4: $x^2 + y^2 = 4$, circle, r=2
   z=9: $x^2 + y^2 = 9$, circle, r=3

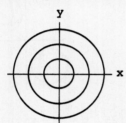

3. $z = y - x^2$
   z=0: $y = x^2$, parabola, V(0,0)
   z=2: $y - 2 = x^2$, parabola, V(0,2)
   z=4: $y - 4 = x^2$, parabola, V(0,4)

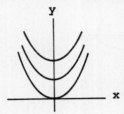

5. $x + y + 2z - 4 = 0$; plane
   <u>Intercepts</u>: (4,0,0),
   (0,4,0), (0,0,2)

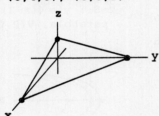

7. $4x - 2y + z - 8 = 0$; plane
   <u>Intercepts</u>: (2,0,0),
   (0,-4,0), (0,0,8)

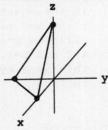

9. $z = y - 2x - 2$; plane
   <u>Intercepts</u>: (-1,0,0)
   (0,2,0), (0,0,-2)

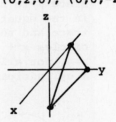

11. $x + 2y = 4$; plane
    <u>Intercepts</u>: (4,0,0),
    (0,2,0), no z-int.
    plane parallel to z-axis

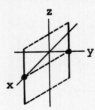

13. $x^2 + y^2 + z^2 = 4$
    <u>Intercepts</u>: (±2,0,0),
    (0,±2,0), (0,0,±2)
    <u>Traces</u>:
    yz-plane: $y^2 + z^2 = 4$,
      circle, r=2
    xz-plane: $x^2 + z^2 = 4$, circle, r=2
    xy-plane: $x^2 + y^2 = 4$, circle, r=2

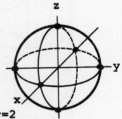

15. $z = 4 - x^2 - y^2$

Intercepts: $(\pm2,0,0)$, $(0,\pm2,0)$, $(0,0,4)$

Traces:

yz-plane: $z=4-y^2$
parabola, V(0,0,4)

xz-plane: $z=4-x^2$
parabola, V(0,0,4)

xy-plane: $x^2+y^2=4$
circle, r=2

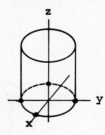

17. $z = 2x^2 + y^2 + 2$

Intercepts: No x-int., no y-int., $(0,0,2)$

Traces:

yz-plane: $z=y^2+2$
parabola, V(0,0,2)

xz-plane: $z=2x^2+2$
parabola, V(0,0,2)

xy-plane: No trace
$(2x^2+y^2+2\neq0)$

Section: For z=4
$2x^2+y^2=2$, ellipse

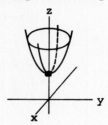

19. $x^2 - y^2 - z^2 = 9$

Intercepts: $(\pm3,0,0)$, no y-int., no z int.

Traces:

yz-plane: No trace
$(-y^2-z^2\neq9)$

xz-plane: $x^2-z^2=9$
hyperbola, a=3, b=3

xy-plane: $x^2-y^2=9$
hyperbola, a=3, b=3

Sections: For x=±5
$y^2+z^2=16$, circles, r=4

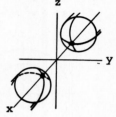

21. $x^2 + y^2 = 16$

Intercepts: $(\pm4,0,0)$ $(0,\pm4,0)$, no z-int.

Traces and Sections:
Since z is not present in the equation, the trace and sections are circles $x^2+ y^2=16$, with r=4, for all z. This is a cylindrical surface.

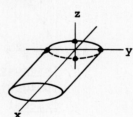

23. $y^2 + 9z^2 = 9$

Intercepts: No x-int., $(0,\pm3,0)$, $(0,0,\pm1)$

Traces and Sections:
Since x is not present in the equation, the trace and sections are ellipses $y^2+9z^2=9$, with a=3 and b=1, for all x. This is a cylindrical surface.

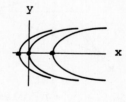

25. $t = 4x - y^2$

t=-4: $y^2= 4(x+1)$,
parabola, V(-1,0)

t=0: $y^2= 4x$,
parabola, V(0,0)

t=8: $y^2= 4(x-2)$,
parabola, V(2,0)

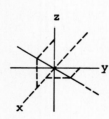

27. $x + 2y + 3z - 6 = 0$, plane

Intercepts: $(6,0,0)$, $(0,3,0)$, $(0,0,2)$

$2x + y + z - 4 = 0$, plane

Intercepts: $(2,0,0)$, $(0,4,0)$, $(0,0,4)$

The intersection of these two planes is the required line.

29. $2x^2 + 2y^2 + 3z^2 = 6$

Intercepts: $(\pm\sqrt{3},0,0)$,
$(0,\pm\sqrt{3},0)$, $(0,0,\pm\sqrt{2})$

Traces:

yz-plane: $2y^2+3z^2=6$
  ellipse, $a=\sqrt{3}$, $b=\sqrt{2}$

xz-plane: $2x^2+3z^2=6$
  ellipse, $a=\sqrt{3}$, $b=\sqrt{2}$

xy-plane: $2x^2+2y^2=6$
  circle, $x^2+y^2=3$, $r=\sqrt{3}$

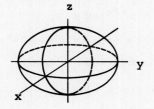

31. $z(2x^2+y^2+100) = 1500$

Intercepts: No x-int.,
no y-int., $(0,0,15)$

Traces:

yz-plane: $z = \dfrac{1500}{y^2 + 100}$

xz-plane: $z = \dfrac{1500}{2x^2 + 100}$

xy-plane: No trace $(z \neq 0)$

Sections and Contours:

z=3: $2x^2+y^2= 400$, ellipse
z=6: $2x^2+y^2= 150$, ellipse
z=9: $2x^2+y^2= 67$ ellipse
z=12: $2x^2+y^2= 25$, ellpise
z=15: $2x^2+y^2=0$, $(0,0,15)$

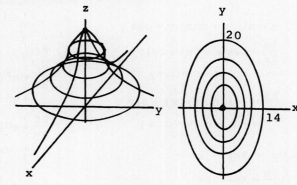

Exercises 10-3, p. 381

1. $z = 5x + 4x^2y$

$\dfrac{\partial z}{\partial x} = 5 + 4(2x)y = 5 + 8xy$

$\dfrac{\partial z}{\partial y} = 0 + 4x^2(1) = 4x^2$

3. $z = \dfrac{x^2}{y} - 2xy$

$\dfrac{\partial z}{\partial x} = \dfrac{2x}{y} - 2(1)y = \dfrac{2x}{y} - 2y$

$\dfrac{\partial z}{\partial y} = -\dfrac{x^2}{y^2} - 2x(1) = -\dfrac{x^2}{y^2} - 2x$

5. $f(x,y) = xe^{2y}$

$\dfrac{\partial f}{\partial x} = (1)e^{2y} = e^{2y}$

$\dfrac{\partial f}{\partial y} = xe^{2y}(2) = 2xe^{2y}$

7. $f(x,y) = \dfrac{2 + \cos x}{1 - \sec 3y}$

$\dfrac{\partial f}{\partial x} = \dfrac{-\sin x}{1 - \sec 3y}$

$\dfrac{\partial f}{\partial y} = -(2+\cos x)(1-\sec 3y)^{-2}(-\sec 3y \tan 3y)(3)$

$\qquad = \dfrac{3(2 + \cos x)\sec 3y \tan 3y}{(1 - \sec 3y)^2}$

9. $\phi = r\sqrt{1 + 2rs}$

$\dfrac{\partial \phi}{\partial r} = r(\tfrac{1}{2})(1+2rs)^{-1/2}(2s)+(1+2rs)^{1/2}$

$\qquad = \dfrac{rs}{(1+2rs)^{1/2}} + (1+2rs)^{1/2}$

$\qquad = \dfrac{1 + 3rs}{\sqrt{1 + 2rs}}$

$\dfrac{\partial \phi}{\partial s} = r(\tfrac{1}{2})(1+2rs)^{-1/2}(2r) = \dfrac{r^2}{\sqrt{1 + 2rs}}$

11. $z = (x^2 + xy^3)^4$

$\dfrac{\partial z}{\partial x} = 4(x^2 + xy^3)^3(2x + y^3)$
$\qquad = 4(2x + y^3)(x^2 + xy^3)^3$

$\dfrac{\partial z}{\partial y} = 4(x^2 + xy^3)^3[x(3y^2)]$
$\qquad = 12xy^2(x^2 + xy^3)^3$

13. $z = \sin xy$

$\dfrac{\partial z}{\partial x} = (\cos xy)(y)$
$\qquad = y \cos xy$

$\dfrac{\partial z}{\partial y} = (\cos xy)(x)$
$\qquad = x \cos xy$

15. $y = \ln(r^2 + s)$

$\dfrac{\partial y}{\partial r} = \dfrac{2r}{r^2+s}$

$\dfrac{\partial y}{\partial s} = \dfrac{1}{r^2 + s}$

17. $f(x,y) = \dfrac{2 \sin^3 2x}{1 - 3y}$

$\dfrac{\partial f}{\partial x} = \dfrac{2(3)(\sin^2 2x)(\cos 2x)(2)}{1 - 3y}$

$\quad = \dfrac{12 \sin^2 2x \cos 2x}{1 - 3y}$

$\dfrac{\partial f}{\partial y} = -(2 \sin^3 2x)(1-3y)^{-2}(-3)$

$\quad = \dfrac{6 \sin^3 2x}{(1-3y)^2}$

21. $z = \sin x + \cos xy - \cos y$

$\dfrac{\partial z}{\partial x} = \cos x - (\sin xy)(y)$

$\quad = \cos x - y \sin xy$

$\dfrac{\partial z}{\partial y} = -(\sin xy)(x) + \sin y$

$\quad = -x \sin xy + \sin y$

25. $z = 3xy - x^2 + y$

$\dfrac{\partial z}{\partial x} = 3y - 2x$

$\dfrac{\partial z}{\partial x}\bigg|_{(1,-2,-9)} = 3(-2) - 2(1)$

$\quad = -8$

29. $z = 2xy^3 - 3x^2y$

$\dfrac{\partial z}{\partial x} = 2y^3 - 6xy , \dfrac{\partial z}{\partial y} = 6xy^2 - 3x^2$

$\dfrac{\partial^2 z}{\partial x^2} = -6y , \dfrac{\partial^2 z}{\partial y^2} = 12xy$

$\dfrac{\partial^2 z}{\partial x \partial y} = \dfrac{\partial^2 z}{\partial y \partial x} = 6y^2 - 6x$

33. $z = 9 - x^2 - y^2$

$\dfrac{\partial z}{\partial y} = -2y$

$\dfrac{\partial z}{\partial y}\bigg|_{(1,2,4)} = -4$

$\dfrac{\partial z}{\partial y}\bigg|_{(2,2,1)} = -4$

35.

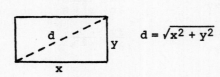

$d = \sqrt{x^2 + y^2}$

19. $z = \dfrac{\text{Arcsin } xy}{3 + x^2}$

$\dfrac{\partial z}{\partial x} = \dfrac{(3+x^2)\dfrac{1}{[1-(xy)^2]^{1/2}}(y) - (\text{Arcsin } xy)(2x)}{(3 + x^2)^2}$

$\quad = \dfrac{3y + x^2 y - 2x\sqrt{1 - x^2 y^2}\ \text{Arcsin } xy}{(3 + x^2)^2 \sqrt{1 - x^2 y^2}}$

$\dfrac{\partial z}{\partial y} = \dfrac{1}{3 + x^2}(\dfrac{1}{\sqrt{1 - x^2 y^2}})(x)$

$\quad = \dfrac{x}{(3 + x^2)\sqrt{1 - x^2 y^2}}$

23. $f(x,y) = e^x \cos xy + e^{-2x} \tan y$

$\dfrac{\partial f}{\partial x} = e^x[(-\sin xy)(y)] + (\cos xy)e^x + e^{-2x}\tan y(-2)$

$\quad = e^x(\cos xy - y \sin xy) - 2e^{-2x}\tan y$

$\dfrac{\partial f}{\partial y} = e^x[-(\sin xy)(x)] + e^{-2x}\sec^2 y$

$\quad = -xe^x \sin xy + e^{-2x}\sec^2 y$

27. $z = e^y \ln xy$

$\dfrac{\partial z}{\partial y} = e^y(\dfrac{1}{xy})(x) + e^y \ln xy = e^y(\dfrac{1}{y} + \ln xy)$

$\dfrac{\partial z}{\partial y}\bigg|_{(e,1,e)} = e(\dfrac{1}{1} + \ln e) = 2e$

31. $f(x,y) = y \ln(x + 2y)$

$\dfrac{\partial f}{\partial x} = \dfrac{y}{x + 2y} , \dfrac{\partial f}{\partial y} = \dfrac{2y}{x + 2y} + \ln(x + 2y)$

$\dfrac{\partial^2 f}{\partial x^2} = \dfrac{-y}{(x + 2y)^2}$

$\dfrac{\partial^2 f}{\partial y^2} = \dfrac{(x + 2y)(2) - 2y(2)}{(x + 2y)^2} + \dfrac{2}{x + 2y} = \dfrac{4x + 4y}{(x+2y)^2}$

$\dfrac{\partial^2 f}{\partial y \partial x} = \dfrac{(x + 2y) - y(2)}{(x + 2y)^2} = \dfrac{x}{(x + 2y)^2}$

$\dfrac{\partial^2 f}{\partial x \partial y} = \dfrac{-2y}{(x+2y)^2} + \dfrac{1}{x+2y} = \dfrac{-2y+x+2y}{(x+2y)^2} = \dfrac{x}{(x + 2y)^2}$

$\dfrac{\partial d}{\partial x} = \dfrac{1}{2}(x^2 + y^2)^{-1/2}(2x) = \dfrac{x}{\sqrt{x^2 + y^2}}$

For $x = 6.50$ ft, $y = 4.75$ ft

$\dfrac{\partial d}{\partial x} = \dfrac{6.50}{\sqrt{6.50^2 + 4.75^2}} = 0.807$

37. $\dfrac{1}{R_T} = \dfrac{1}{R_1} + \dfrac{1}{R_2}$

$R_T = \dfrac{R_1 R_2}{R_1 + R_2}$

$\dfrac{\partial R_T}{\partial R_1} = \dfrac{(R_1 + R_2)(R_2) - R_1 R_2(1)}{(R_1 + R_2)^2}$

$\quad = \dfrac{R_2^2}{(R_1 + R_2)^2}$

39. $a = \dfrac{M - m}{M + m}\, g$

$\dfrac{\partial a}{\partial M} = \dfrac{(M + m) - (M - m)}{(M + m)^2}\, g = \dfrac{2mg}{(M + m)^2}$

$\dfrac{\partial a}{\partial m} = \dfrac{(M + m)(-1) - (M - m)(1)}{(M + m)^2}\, g = \dfrac{-2Mg}{(M + m)^2}$

$M\left[\dfrac{2mg}{(M + m)^2}\right] + m\left[\dfrac{-2Mg}{(M + m)^2}\right] = 0 \; ; \; 0 = 0$

41. $i_b = 50(e_b + 5e_c)^{1.5}$

$\dfrac{\partial i_b}{\partial e_c} = 50(1.5)(e_b + 5e_c)^{0.5}(5)$

$\quad = 375(e_b + 5e_c)^{0.5}$

For $e_b = 200$ V and $e_c = -20$ v

$\dfrac{\partial i_b}{\partial e_c} = 375(200 - 100)^{0.5}$

$\quad = 3750 \ \mu A/V = 3.75 \ 10^{-3} \ 1/\Omega$

43. $L = L_0 + k_1 F + k_2 T + k_3 FT^2$

$\dfrac{\partial L}{\partial T} = k_2 + 2k_3 FT$

$= \dfrac{1}{L}\dfrac{\partial L}{\partial T} = \dfrac{k_2 + 2k_3 FT}{L_0 + k_1 F + k_2 T + k_3 FT^2}$

45. $pV = nRT, \quad p = \dfrac{nRT}{V}$

$\dfrac{\partial p}{\partial V} = -nRTV^{-2} = -\dfrac{nRT}{V^2}$

47. $u(x,t) = 5e^{-t}\sin 4x$

$\dfrac{\partial u}{\partial x} = 5e^{-t}(4\cos 4x) = 20e^{-t}\cos 4x$

$\dfrac{\partial^2 u}{\partial x^2} = 20e^{-t}(-4\sin 4x) = -80e^{-t}\sin 4x$

$\dfrac{\partial u}{\partial t} = -5e^{-t}\sin 4x$

$\dfrac{\partial u}{\partial t} = k\dfrac{\partial^2 u}{\partial x^2}, \quad k = \dfrac{1}{16}$

$-5e^{-t}\sin 4x = \dfrac{1}{16}(-80e^{-t}\sin 4x)$

$-5e^{-t}\sin 4x = -5e^{-t}\sin 4x$

Exercises 10-4, p. 390

1. $z = 2x^2 - y^2 + 3x$

$\dfrac{\partial z}{\partial x} = 4x + 3, \dfrac{\partial z}{\partial y} = -2y$

$dz = (4x+3)dx - 2y\,dy$

3. $z = xe^y - y^2$

$\dfrac{\partial z}{\partial x} = e^y, \dfrac{\partial z}{\partial y} = xe^y - 2y$

$dz = e^y dx + (xe^y - 2y)dy$

5. $z = x(y - 2x^2)^5$

$\dfrac{\partial z}{\partial x} = 5x(y-2x^2)^4(-4x) + (y-2x^2)^5$

$\quad = (y - 22x^2)(y - 2x^2)^4$

$\dfrac{\partial z}{\partial y} = 5x(y - 2x^2)^4$

$dz = (y-22x^2)(y-2x^2)^4 dx$
$\quad + 5x(y - 2x^2)^4 dy$

7. $z = \sin xy - y\cos x$

$\dfrac{\partial z}{\partial x} = y\cos xy + y\sin x, \dfrac{\partial z}{\partial y} = x\cos xy - \cos x$

$dz = (y\cos xy + y\sin x)dx + (x\cos xy - \cos x)dy$

9. $z = \dfrac{x - 3y^2}{1 - \sin y}$

$\dfrac{\partial z}{\partial x} = \dfrac{1}{1 - \sin y}, \dfrac{\partial z}{\partial y} = \dfrac{(1 - \sin y)(-6y) - (x - 3y^2)(-\cos y)}{(1 - \sin y)^2} = \dfrac{6y\sin y - 6y + x\cos y - 3y^2\cos y}{(1 - \sin y)^2}$

$dz = \dfrac{dx}{1 - \sin y} + \dfrac{(6y\sin y - 6y + x\cos y - 3y^2\cos y)dy}{(1 - \sin y)^2}$

11. $z = y \text{ Arctan } \dfrac{y}{x^2}$

$\dfrac{\partial z}{\partial x} = \dfrac{y}{1 + (y/x^2)^2}(\dfrac{-2y}{x^3}) = \dfrac{-2xy^2}{x^4 + y^2}$

$\dfrac{\partial z}{\partial y} = y[\dfrac{1}{1 + (y/x^2)^2}](\dfrac{1}{x^2}) + \text{Arctan } \dfrac{y}{x^2}$

$\qquad = \dfrac{yx^2}{x^4 + y^2} + \text{Arctan}\dfrac{y}{x^2}$

$dz = \dfrac{-2xy^2 dx}{x^4 + y^2} + (\dfrac{yx^2}{x^4 + y^2} + \text{Arctan}\dfrac{y}{x^2})dy$

13. $z = x^2 + y^2 - 2x - 6y + 10$

$\dfrac{\partial z}{\partial x} = 2x - 2, \ \dfrac{\partial z}{\partial y} = 2y - 6$

$2x-2=0, \ x=1 ; \ 2y-6=0, \ y=3$

$\dfrac{\partial^2 z}{\partial x^2} = 2, \ \dfrac{\partial^2 z}{\partial y^2} = 2, \ \dfrac{\partial^2 z}{\partial x \partial y} = 0$

$D = 2(2) - 0^2 = 4$

$D > 0, \dfrac{\partial^2 z}{\partial x^2} > 0, \ \text{Min. } (1,3,0)$

15. $z = x^2 + xy + y^2 - 3y + 2$

$\dfrac{\partial z}{\partial x} = 2x + y, \ \dfrac{\partial z}{\partial y} = x + 2y - 3$

$\begin{array}{l} 2x+y=0 \qquad 4x+2y=0 \\ \underline{x+2y=3} \qquad \underline{x+2y=3} \\ \qquad\qquad\quad\ \ 3x \quad\ =-3 \end{array} \quad x=-1, \ y=2$

$\dfrac{\partial^2 z}{\partial x^2} = 2, \ \dfrac{\partial^2 z}{\partial y^2} = 2, \ \dfrac{\partial^2 z}{\partial x \partial y} = 1$

$D = 2(2) - 1^2 = 3$

$D > 0, \dfrac{\partial^2 z}{\partial x^2} > 0, \ \text{Min. } (-1,2,-1)$

17. $z = 2y - xy + x^3$

$\dfrac{\partial z}{\partial x} = -y + 3x^2 , \ \dfrac{\partial z}{\partial y} = 2 - x$

$-y + 3x^2 = 0, \ y = 3x^2$

$2 - x = 0, \ x = 2, \ y = 12$

$\dfrac{\partial^2 z}{\partial x^2} = 6x , \ \dfrac{\partial^2 z}{\partial y^2} = 0, \ \dfrac{\partial^2 z}{\partial x \partial y} = -1$

$D = 12(0) - (-1)^2 = -1$

$D < 0, \ \text{no. max. or min.}$

19. $z = x^2 + 4 - 4x \cos y$

$\dfrac{\partial z}{\partial x} = 2x - 4 \cos y , \ \dfrac{\partial z}{\partial y} = 4x \sin y$

$2x - 4 \cos y = 0, \ \cos y = \dfrac{x}{2}$

$4x \sin y = 0, \ x=0, \ \sin y = 0$

$\sin y = 0, \ y = 0, \pi$

$x = 0, \ \cos y = 0, \ y = \dfrac{\pi}{2}, \dfrac{3\pi}{2}$

$\dfrac{\partial^2 z}{\partial x^2} = 2, \ \dfrac{\partial^2 z}{\partial y^2} = 4x \cos y , \ \dfrac{\partial^2 z}{\partial x \partial y} = 4 \sin y$

$D = 8x \cos y - 16 \sin^2 y$

For $y=0$, $x=2$: $D=16$, Min. $(2,0,0)$
For $y=\pi$, $x=-2$: $D=16$, Min. $(-2,\pi,0)$
For $y=\dfrac{\pi}{2}$, $x=0$: $D=-16$, No max. or min.
For $y=\dfrac{3\pi}{2}$, $x=0$: $D=-16$, No max. or min.

21. $a = r\omega^2$

$da = \omega^2 dr + 2r\omega \, d\omega$

$r = 10.0 \text{ cm}, \ dr = 0.3 \text{ cm}$

$\omega = 400 \text{ rad/min}, \ d\omega = 6 \text{ rad/min}$

$da = 400^2(0.3) + 2(10.0)(400)(6)$

$\qquad = 9.60 \times 10^4 \text{ rad/min}^2$

23. $n = \dfrac{\sin i}{\sin r}$

$dn = \dfrac{\cos i \, di}{\sin r} - \dfrac{\sin i \cos r \, dr}{\sin^2 r}$

$i = 45°, \ r = 30°, \ di = dr = -1° = -0.01745$

$dn = (\dfrac{\cos 45°}{\sin 30°} - \dfrac{\sin 45° \cos 30°}{\sin^2 30°})(-0.01745)$

$\qquad = 0.0181$

$n = \dfrac{\sin 45°}{\sin 30°} = 1.414 ; \ \dfrac{dn}{n} = 0.013 = 1.3 \%$

25.

$z^2 = x^2 + y^2$

$z = \sqrt{x^2 + y^2}$

$dz = \dfrac{x \, dx + y \, dy}{\sqrt{x^2 + y^2}}$

$x=2.30 \text{ cm}, \ y=2.10 \text{ cm}, \ dx=dy=0.10 \text{ cm}$

$dz = \dfrac{2.30(0.10) + 2.10(0.10)}{\sqrt{2.30^2 + 2.10^2}} = 0.141$

$z = \sqrt{2.30^2 + 2.10^2} = 3.11$

$\dfrac{dz}{z} = 0.045 = 4.5 \%$

27. Let x, y and z be the numbers.

$x + y + z = 120$, $x>0$, $y>0$, $z>0$

$P = xyz = xy(120 - x - y)$

$\quad = 120xy - x^2y - xy^2$

$\dfrac{\partial P}{\partial x} = 120y - 2xy - y^2$

$\dfrac{\partial P}{\partial y} = 120x - x^2 - 2xy$

$120y - 2xy - y^2 = 0$, $120 - 2x - y = 0$

$120x - x^2 - 2xy = 0$, $120 - x - 2y = 0$

$2x + y = 120$

$\underline{x + 2y = 120}$

$\qquad y = 40$, $x = 40$

$\dfrac{\partial^2 P}{\partial x^2} = -2y$, $\dfrac{\partial^2 P}{\partial y^2} = -2x$

$\dfrac{\partial^2 P}{\partial x \partial y} = 120 - 2x - 2y$

For x=40, y=40

$D = (-80)(-80) - (-40)^2 = 4800$

$\dfrac{\partial^2 P}{\partial x^2} = -80$; Max.(40,40,40)

33. $z = x^2 - xy + y^2$

$\dfrac{\partial z}{\partial x} = 2x - y$, $\dfrac{\partial z}{\partial y} = -x + 2y$

$x = 1 + t^2$, $y = 1 - t^2$

$\dfrac{dx}{dt} = 2t$, $\dfrac{dy}{dt} = -2t$

$\dfrac{dz}{dt} = (2x - y)(2t) + (-x + 2y)(-2t)$

$\quad = 4xt - 2yt + 2xt - 4yt$

$\quad = 6xt - 6yt$

29. $P = x^4 + y^2 - 4x + 20$

$\dfrac{\partial P}{\partial x} = 4x^3 - 4$, $\dfrac{\partial P}{\partial y} = 2y$

$4x^3 - 4 = 0$, $x=1$; $2y = 0$, $y = 0$

$\dfrac{\partial^2 P}{\partial x^2} = 12x^2$, $\dfrac{\partial^2 P}{\partial y^2} = 2$, $\dfrac{\partial^2 P}{\partial x \partial y} = 0$

For x = 1, y = 0

$D = 12(2) - 0^2 = 24$, $\dfrac{\partial^2 P}{\partial x^2} = 12$

Min. at (1,0) ; $P = 1^4 + 0^2 - 4(1) + 20$

$\qquad\qquad = 17$ Pa

31. $xyz = 500$ in.^{3}

$z = \dfrac{500}{xy}$

Least material
means surface
area is minimum.

No top (open)

$A = xy + 2yz + 2xz$

$\quad = xy + 2y(\dfrac{500}{xy}) + 2x(\dfrac{500}{xy}) = xy + \dfrac{1000}{x} + \dfrac{1000}{y}$

$\dfrac{\partial A}{\partial x} = y - \dfrac{1000}{x^2}$, $\dfrac{\partial A}{\partial y} = x - \dfrac{1000}{y^2}$

$y = \dfrac{1000}{x^2}$, $x = \dfrac{1000}{y^2} = \dfrac{1000}{(1000/x^2)^2} = \dfrac{x^4}{1000}$

$x^4 - 1000x = x(x^3 - 1000) = 0$; $x=0$, 10 $(x \neq 0)$

$\dfrac{\partial^2 A}{\partial x^2} = \dfrac{2000}{x^3}$, $\dfrac{\partial^2 A}{\partial y^2} = \dfrac{2000}{y^3}$, $\dfrac{\partial^2 A}{\partial x \partial y} = 1$

For x = 10, y = 10

$D = 2(2) - 1^2 = 3$, $\dfrac{\partial^2 A}{\partial x^2} = 2$

Minimum for x = 10 in., y = 10 in., z = 5 in.

Exercises 10-5, p. 396

1. $\int_2^4 \int_0^1 xy^2 \, dx \, dy = \int_2^4 y^2(\frac{1}{2}x^2)\Big|_0^1 dy = \int_2^4 y^2(\frac{1}{2} - 0)dy = \frac{1}{2}\int_2^4 y^2 dy = \frac{1}{6}y^3\Big|_2^4 = \frac{1}{6}(64 - 8) = \dfrac{28}{3}$

3. $\int_0^1 \int_0^x 2y \, dy \, dx = \int_0^1 y^2\Big|_0^x dx = \int_0^1 (x^2 - 0)dx = \int_0^1 x^2 dx = \frac{1}{3}x^3\Big|_0^1 = \frac{1}{3} - 0 = \dfrac{1}{3}$

5. $\int_1^2 \int_0^{y^2} xy^2 \, dx \, dy = \int_1^2 y^2(\frac{1}{2}x^2)\Big|_0^{y^2} dy = \int_1^2 y^2(\frac{1}{2}y^4)dy = \frac{1}{2}\int_1^2 y^6 dy = \frac{1}{14}y^7\Big|_1^2 = \dfrac{127}{14}$

7. $\int_0^1 \int_0^{\sqrt{1-x^2}} y \, dy \, dx = \int_0^1 \frac{1}{2}y^2\Big|_0^{\sqrt{1-x^2}} dx = \int_0^1 \frac{1}{2}(1 - x^2)dx = \frac{1}{2}x - \frac{1}{6}x^3\Big|_0^1 = \frac{1}{2} - \frac{1}{6} = \dfrac{1}{3}$

9. $\int_0^{\pi/6} \int_{\pi/3}^y \sin x \, dx \, dy = \int_0^{\pi/6} (-\cos x) \Big|_{\pi/3}^y \, dy = -\int_0^{\pi/6} (\cos y - \cos \frac{\pi}{3}) dy$

$\qquad = -\int_0^{\pi/6} (\cos y - \frac{1}{2}) dy = -\sin y + \frac{1}{2} y \Big|_0^{\pi/6} = -\sin \frac{\pi}{6} + \frac{\pi}{12} = \frac{\pi}{12} - \frac{1}{2} = \frac{\pi - 6}{12}$

11. $\int_1^e \int_1^y \frac{1}{x} dx \, dy = \int_1^e \ln x \Big|_1^y dy = \int_1^e \ln y \, dy = y(\ln y - 1) \Big|_1^e$ (Parts or Formula 46)

$\qquad = e(\ln e - 1) - (\ln 1 - 1) = e(0) + 1 = 1$

13. $\int_1^2 \int_0^x yx^3 e^{xy^2} dy \, dx = \frac{1}{2} \int_1^2 \int_0^x x^2 (2xye^{xy^2} dy) dx = \frac{1}{2} \int_1^2 x^2 (e^{xy^2}) \Big|_0^x dx = \frac{1}{2} \int_1^2 x^2 (e^{x^3} - 1) dx$

$\qquad = \frac{1}{6} \int_1^2 3x^2 e^{x^3} dx - \frac{1}{2} \int_1^2 x^2 dx = \frac{1}{6} e^{x^3} - \frac{1}{6} x^3 \Big|_1^2 = \frac{1}{6} [e^8 - 8 - (e - 1)] = 495.2$

15. $\int_0^{\ln 3} \int_0^x e^{2x+3y} dy \, dx = \frac{1}{3} \int_0^{\ln 3} e^{2x+3y} \Big|_0^x dx = \frac{1}{3} \int_0^{\ln 3} (e^{5x} - e^{2x}) dx = \frac{1}{15} e^{5x} - \frac{1}{6} e^{2x} \Big|_0^{\ln 3}$

$\qquad = \frac{1}{15} (e^{5 \ln 3} - 1) - \frac{1}{6} (e^{2 \ln 3} - 1) = \frac{1}{15} (3^5 - 1) - \frac{1}{6} (3^2 - 1)$

$\qquad = \frac{242}{15} - \frac{4}{3} = \frac{74}{5}$ $(e^{\ln a} = a)$

17. $V = \int_0^4 \int_0^{4-x} z \, dy \, dx = \int_0^4 \int_0^{4-x} (4 - x - y) dy \, dx = \int_0^4 (4y - xy - \frac{1}{2} y^2) \Big|_0^{4-x} dx$

$\qquad = \int_0^4 [4(4 - x) - x(4 - x) - \frac{1}{2}(4 - x)^2] dx = \int_0^4 (8 - 4x + \frac{1}{2} x^2) dx$

$\qquad = 8x - 2x^2 + \frac{1}{6} x^3 \Big|_0^4 = 32 - 32 + \frac{64}{6} = \frac{32}{3}$

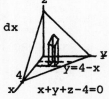

19. $V = 4 \int_0^2 \int_0^{\sqrt{4-y^2}} z \, dx \, dy = 4 \int_0^2 \int_0^{\sqrt{4-y^2}} (4 - x^2 - y^2) dx \, dy$

$\qquad = \int_0^2 (16x - \frac{4}{3} x^3 - 4y^2 x) \Big|_0^{\sqrt{4-y^2}} dy$

$\qquad = \int_0^2 (16\sqrt{4-y^2} - \frac{4}{3}(4-y^2)^{3/2} - 4y^2\sqrt{4-y^2}) dy$

$\qquad = \frac{1}{3} \int_0^2 (32 - 8y^2) \sqrt{4-y^2} \, dy = \frac{8}{3} \int_0^2 (4 - y^2)^{3/2} dy$

$\qquad = \frac{8}{3} [\frac{y}{4}(4-y^2)^{3/2} + \frac{3(4)y}{8}(4-y^2)^{1/2} + \frac{3(16)}{8} \text{Arcsin} \frac{y}{2}] \Big|_0^2$ (Formula 20)

$\qquad = \frac{8}{3} [0 + 0 + 6 \text{Arcsin } 1 - 0 - 0 - 0] = \frac{8}{3} (6)(\frac{\pi}{2}) = 8\pi$

Total volume is 4 times the volume in the first octant.

$x^2 + y^2 = 4$

$z = 4 - x^2 - y^2$

21. $V = \int_0^2 \int_x^2 z \, dy \, dx = \int_0^2 \int_x^2 (2 + x^2 + y^2) dy \, dx$

$\qquad = \int_0^2 (2y + x^2 y + \frac{1}{3} y^3) \Big|_x^2 dx = \int_0^2 (4 + 2x^2 + \frac{8}{3} - 2x - x^3 - \frac{1}{3} x^3) dx$

$\qquad = \int_0^2 (\frac{20}{3} - 2x + 2x^2 - \frac{4}{3} x^3) dx = \frac{20}{3} x - x^2 + \frac{2}{3} x^3 - \frac{1}{3} x^4 \Big|_0^2$

$\qquad = \frac{40}{3} - 4 + \frac{16}{3} - \frac{16}{3} = \frac{28}{3}$

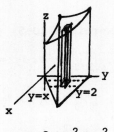

$z = 2 + x^2 + y^2$

23. $V = \int_0^3 \int_0^{\sqrt{9-x^2}} z\, dy\, dx = \int_0^3 \int_0^{\sqrt{9-x^2}} (x+y)dy\, dx$

$= \int_0^3 (xy + \frac{1}{2}y^2)\Big|_0^{\sqrt{9-x^2}} dx = \int_0^3 [x\sqrt{9-x^2} + \frac{1}{2}(9-x^2)]dx$

$= -\frac{1}{3}(9-x^2)^{3/2} + \frac{9}{2}x - \frac{1}{6}x^3\Big|_0^3 = 0 + \frac{27}{2} - \frac{27}{6} + \frac{1}{3}(27) - 0 + 0$

$= \frac{27}{2} - \frac{9}{2} + 9 = 18$

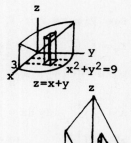

25. $V = \int_0^5 \int_0^{12} z\, dx\, dy = \int_0^5 \int_0^{12}(10-2y)dx\, dy = \int_0^5 (10-2y)x\Big|_0^{12} dy$

$= 12\int_0^5 (10-2y)dy = 12(10y - y^2)\Big|_0^5 = 12(50-25) = 300 \text{ cm}^3$

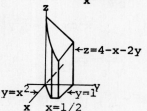

27. $z = 4 - x - 2y$ from $y = x^2$ to $y = 1$, from $x=0$ to $x = \frac{1}{2}$

Intercepts of plane: $(4,0,0)$, $(0,2,0)$, $(0,0,4)$

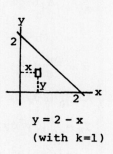

Exercises 10-6, p. 401

1. $A = \int_0^2 \int_0^{2-y} dx\, dy = \int_0^2 x\Big|_0^{2-y} dy = \int_0^2 (2-y)dy = 2y - \frac{1}{2}y^2\Big|_0^2 = 2$

$A\bar{x} = \int_0^2 \int_0^{2-y} x\, dx\, dy = \int_0^2 \frac{1}{2}x^2\Big|_0^{2-y} dy = \frac{1}{2}\int_0^2 (4-4y+y^2)dy$

$= \frac{1}{2}(4y - 2y^2 + \frac{1}{3}y^3)\Big|_0^2 = \frac{1}{2}(8-8+\frac{8}{3}) = \frac{8}{6} = \frac{4}{3}$; $\bar{x} = \frac{4}{3}(\frac{1}{2}) = \frac{2}{3}$

$A\bar{y} = \int_0^2 \int_0^{2-y} y\, dx\, dy = \int_0^2 x\Big|_0^{2-y} y\, dy = \int_0^2 (2y - y^2)dy$

$= y^2 - \frac{1}{3}y^3\Big|_0^2 = 4 - \frac{8}{3} = \frac{4}{3}$; $\bar{y} = \frac{4}{3}(\frac{1}{2}) = \frac{2}{3}$

$y = 2 - x$

(with k=1)

3. $A = \int_0^3 \int_0^{9-x^2} dy\, dx = \int_0^3 y\Big|_0^{9-x^2} dx = \int_0^3 (9-x^2)dx = 9x - \frac{1}{3}x^3\Big|_0^3 = 27 - 9 = 18$

$A\bar{x} = \int_0^3 \int_0^{9-x^2} x\, dy\, dx = \int_0^3 y\Big|_0^{9-x^2} x\, dx = \int_0^3 (9x - x^3)dx$

$= \frac{9}{2}x^2 - \frac{1}{4}x^4\Big|_0^3 = \frac{81}{2} - \frac{81}{4} = \frac{81}{4}$; $\bar{x} = \frac{81}{4}(\frac{1}{18}) = \frac{9}{8}$

$A\bar{y} = \int_0^3 \int_0^{9-x^2} y\, dy\, dx = \int_0^3 \frac{1}{2}y^2\Big|_0^{9-x^2} dx = \frac{1}{2}\int_0^3 (81 - 18x^2 + x^4)dx$

$= \frac{1}{2}(81x - 6x^3 + \frac{1}{5}x^5)\Big|_0^3 = \frac{1}{2}(243 - 162 + \frac{243}{5}) = \frac{324}{5}$; $\bar{y} = \frac{324}{5}(\frac{1}{18}) = \frac{18}{5}$

$y = 9 - x^2$

(with k=1)

5. See Figure for #1

$I_y = \int_0^2 \int_0^{2-x} x^2 dy\, dx = \int_0^2 y\Big|_0^{2-x} x^2 dx = \int_0^2 (2x^2 - x^3)dx = \frac{2}{3}x^3 - \frac{1}{4}x^4\Big|_0^2 = \frac{16}{3} - 4 = \frac{4}{3}$

$I_x = \int_0^2 \int_0^{2-y} y^2 dx\, dy = \int_0^2 x\Big|_0^{2-y} y^2 dy = \int_0^2 (2y^2 - y^3)dy = \frac{2}{3}y^3 - \frac{1}{4}y^4\Big|_0^2 = \frac{16}{3} - 4 = \frac{4}{3}$

7. See Figure for #3 ; $A = \int_0^3 \int_0^{9-x^2} dy\, dx = 18$ (see #3)

$I_y = \int_0^3 \int_0^{9-x^2} x^2 dy\, dx = \int_0^3 y \Big|_0^{9-x^2} x^2 dx = \int_0^3 (9x^2 - x^4) dx = 3x^3 - \frac{1}{5}x^5 \Big|_0^3 = 81 - \frac{243}{5} = \frac{162}{5}$

$R_y^2 = \frac{162}{5}(\frac{1}{18}) = \frac{9}{5}$, $R_y = \sqrt{\frac{9}{5}} = 1.34$

9. $A = \int_0^1 \int_{1-x^2}^1 dy\, dx = \int_0^1 y \Big|_{1-x^2}^1 dx = \int_0^1 (1 - 1 + x^2) dx = \int_0^1 x^2 dx = \frac{1}{3}x^3 \Big|_0^1 = \frac{1}{3}$

$A\overline{x} = \int_0^1 \int_{1-x^2}^1 x\, dy\, dx = \int_0^1 y \Big|_{1-x^2}^1 x\, dx = \int_0^1 x^3 dx = \frac{1}{4}x^4 \Big|_0^1 = \frac{1}{4}$; $\overline{x} = \frac{1}{4}(\frac{3}{1}) = \frac{3}{4}$

$A\overline{y} = \int_0^1 \int_{1-x^2}^1 y\, dy\, dx = \int_0^1 \frac{1}{2}y^2 \Big|_{1-x^2}^1 dx = \frac{1}{2}\int_0^1 [1-(1-x^2)^2] dx$

$= \frac{1}{2}\int_0^1 (2x^2 - x^4) dx = \frac{1}{3}x^3 - \frac{1}{10}x^5 \Big|_0^1 = \frac{1}{3} - \frac{1}{10} = \frac{7}{30}$; $\overline{y} = \frac{7}{30}(\frac{3}{1}) = \frac{7}{10}$

(with k=1)

$I_y = \int_0^1 \int_{1-x^2}^1 x^2 dy\, dx = \int_0^1 y \Big|_{1-x^2}^1 x^2 dx = \int_0^1 x^4 dx = \frac{1}{5}x^5 \Big|_0^1 = \frac{1}{5}$; $R_y^2 = \frac{1}{5}(\frac{3}{1}) = \frac{3}{5}$; $R_y = 0.775$

11. $A = \int_0^a \int_0^{b-\frac{b}{a}x} dy\, dx = \int_0^a y \Big|_0^{b-\frac{b}{a}x} dx = \int_0^a (b - \frac{b}{a}x) dx = bx - \frac{b}{2a}x^2 \Big|_0^a = \frac{1}{2}ab$

$A\overline{x} = \int_0^a \int_0^{b-\frac{b}{a}x} x\, dy\, dx = \int_0^a y \Big|_0^{b-\frac{b}{a}x} x\, dx = \int_0^a (bx - \frac{b}{a}x^2) dx = \frac{1}{2}bx^2 - \frac{b}{3a}x^3 \Big|_0^a = \frac{a^2 b}{6}$

$A\overline{y} = \int_0^a \int_0^{b-\frac{b}{a}x} y\, dy\, dx = \frac{1}{2}\int_0^a y^2 \Big|_0^{b-\frac{b}{a}x} dx = \frac{1}{2}\int_0^a (b^2 - \frac{2b^2}{a}x + \frac{b^2}{a^2}x^2) dx$

$= \frac{1}{2}(b^2 x - \frac{b^2}{a}x^2 + \frac{b^2}{3a^2}x^3) \Big|_0^a = \frac{b^2 a}{6}$ (with k=1)

$\overline{x} = \frac{ba^2}{6}(\frac{2}{ab}) = \frac{a}{3}$, $\overline{y} = \frac{ab^2}{6}(\frac{2}{ab}) = \frac{b}{3}$

13. $A = \int_0^a \int_0^b dy\, dx = \int_0^a y \Big|_0^b dx = \int_0^a b\, dx = bx \Big|_0^a = ab$; $m = A$ (with k=1)

$I_x = \int_0^a \int_0^b y^2 dy\, dx = \int_0^a \frac{1}{3}y^3 \Big|_0^b dx = \int_0^a \frac{1}{3}b^3 dx = \frac{1}{3}b^3 x \Big|_0^a = \frac{1}{3}ab^3$

$= \frac{1}{3}(ab)b^2 = \frac{1}{3}mb^2$

15. Solution for A and I_x same as in #13, with a=b=4.

$A = 16$; $I_x = \frac{1}{3}(4)(64) = \frac{256}{3}$; $R_x^2 = \frac{256}{3}(\frac{1}{16}) = \frac{16}{3}$; $R_x = \sqrt{\frac{16}{3}} = \frac{4}{3}\sqrt{3} = 2.31$

17. $A = \int_0^2 \int_y^{6-y} dx\, dy = \int_0^2 x \Big|_y^{6-y} dy = \int_0^2 (6 - 2y) dy = 6y - y^2 \Big|_0^2 = 8$

$A\overline{x} = \int_0^2 \int_y^{6-y} x\, dx\, dy = \int_0^2 \frac{1}{2}x^2 \Big|_y^{6-y} dy = \frac{1}{2}\int_0^2 (36 - 12y) dy = 18y - 3y^2 \Big|_0^2 = 36 - 12 = 24$

$A\overline{y} = \int_0^2 \int_y^{6-y} y\, dx\, dy = \int_0^2 x \Big|_y^{6-y} y\, dy = \int_0^2 (6y - 2y^2) dy = 3y^2 - \frac{2}{3}y^3 \Big|_0^2 = \frac{20}{3}$

$\overline{x} = \frac{24}{8} = 3$, $\overline{y} = \frac{20}{3}(\frac{1}{8}) = \frac{5}{6}$

(with k=1)

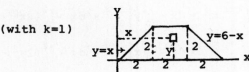

Review Exercises for Chapter 10, p. 402

1. $z = 5x^3y^2 - 2xy^4$

$\dfrac{\partial z}{\partial x} = 5y^2(3x^2) - 2y^4(1)$

$\quad = 15x^2y^2 - 2y^4$

$\dfrac{\partial z}{\partial y} = 5x^3(2y) - 2x(4y^3)$

$\quad = 10x^3y - 8xy^3$

3. $z = \sqrt{x^2 - 3y^2}$

$\dfrac{\partial z}{\partial x} = \dfrac{1}{2}(x^2 - 3y^2)^{-1/2}(2x)$

$\quad = \dfrac{x}{\sqrt{x^2 - 3y^2}}$

$\dfrac{\partial z}{\partial y} = \dfrac{1}{2}(x^2 - 3y^2)^{-1/2}(-6y)$

$\quad = \dfrac{-3y}{\sqrt{x^2 - 3y^2}}$

5. $z = \dfrac{2x - 3y}{x^2y + 1}$

$\dfrac{\partial z}{\partial x} = \dfrac{(x^2y+1)(2)-(2x-3y)(2xy)}{(x^2y + 1)^2}$

$\quad = \dfrac{2 - 2x^2y + 6xy^2}{(x^2y + 1)^2}$

$\dfrac{\partial z}{\partial y} = \dfrac{(x^2y+1)(-3)-(2x-3y)(x^2)}{(x^2y + 1)^2}$

$\quad = \dfrac{-(3 + 2x^3)}{(x^2y + 1)^2}$

7. $u = y \ln \sin(x^2 + 2y)$

$\dfrac{\partial u}{\partial x} = \dfrac{y \cos(x^2+2y)}{\sin(x^2+2y)}(2x) = 2xy \cot(x^2+2y)$

$\dfrac{\partial u}{\partial y} = y[\dfrac{\cos(x^2+2y)}{\sin(x^2+2y)}](2) + \ln \sin(x^2+2y)$

$\quad = 2y \cot(x^2+2y) + \ln \sin(x^2+2y)$

9. $z = \text{Arcsin}\sqrt{x+y}$

$\dfrac{\partial z}{\partial x} = \dfrac{1}{\sqrt{1 - (x+y)}}(\dfrac{1}{2})(x + y)^{-1/2}(1)$

$\quad = \dfrac{1}{2\sqrt{(x + y)(1 - x - y)}}$

$\dfrac{\partial z}{\partial y} = \dfrac{1}{\sqrt{1 - (x+y)}}(\dfrac{1}{2})(x + y)^{-1/2}(1)$

$\quad = \dfrac{1}{2\sqrt{(x + y)(1 - x - y)}}$

11. $z = 3x^2y - y^3 + 2xy$

$\dfrac{\partial z}{\partial x} = 6xy + 2y , \dfrac{\partial z}{\partial y} = 3x^2 - 3y^2 + 2x$

$\dfrac{\partial^2 z}{\partial x^2} = 6y , \dfrac{\partial^2 z}{\partial y^2} = -6y , \dfrac{\partial^2 z}{\partial y \partial x} = \dfrac{\partial^2 z}{\partial x \partial y} = 6x + 2$

13. $\int_0^2\int_1^2 (3y + 2xy)dx\,dy = \int_0^2 (3xy+x^2y)\Big|_1^2 dy = \int_0^2 (6y + 4y - 3y - y)dy = \int_0^2 6y\,dy = 3y^2\Big|_0^2 = 12$

15. $\int_0^3\int_1^x (x + 2y)dy\,dx = \int_0^3 (xy + y^2)\Big|_1^x dx = \int_0^3 (x^2 + x^2 - x - 1)dx = \int_0^3 (2x^2 - x - 1)dx$

$\quad = \dfrac{2}{3}x^3 - \dfrac{1}{2}x^2 - x\Big|_0^3 = \dfrac{2}{3}(27) - \dfrac{1}{2}(9) - 3 - 0 = \dfrac{21}{2}$

17. $\int_0^1\int_0^{2x} x^2 e^{xy}dy\,dx = \int_0^1\int_0^{2x} x(e^{xy}x\,dy)dx = \int_0^1 xe^{xy}\Big|_0^{2x} dx = \int_0^1 x(e^{2x^2} - 1)dx$

$\quad = \dfrac{1}{4}e^{2x^2} - \dfrac{1}{2}x^2\Big|_0^1 = \dfrac{1}{4}e^2 - \dfrac{1}{2} - \dfrac{1}{4} = \dfrac{1}{4}(e^2 - 3) = 1.097$

19. $\int_1^e\int_1^x \dfrac{\ln y}{xy}dy\,dx = \int_1^e\int_1^x \dfrac{1}{x}[\ln y(\dfrac{dy}{y})]dx = \int_1^e \dfrac{1}{x}(\dfrac{\ln^2 y}{2})\Big|_1^x dx = \int_1^e \dfrac{\ln^2 x}{2x}dx$

$\quad = \dfrac{1}{6}\ln^3 x\Big|_1^e = \dfrac{1}{6}\ln^3 e - 0 = \dfrac{1}{6}$

21. $z = \sqrt{x^2 + 4y^2}$

Intercepts: (0,0,0)

Traces:

$z \geq 0$ for all x and y

(defined by positive square root)

In yz-plane: $z = \pm 2y$

In xz-plane: $z = \pm x$

In xy-plane: (0,0,0)

Section: For $z=2$

$x^2 + 4y^2 = 4$ (ellipse)

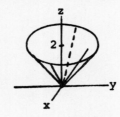

23. $z = \sqrt{x^2 + 4y^2}$

$dz = \frac{1}{2}(x^2 + 4y^2)^{-1/2}(2x)dx + \frac{1}{2}(x^2 + 4y^2)^{-1/2}(8y)dy = \frac{x\,dx + 4y\,dy}{\sqrt{x^2 + 4y^2}}$

For x=2.00, dx=0.04, y=1.00, dy=0.06

$dz = \frac{2.00(0.04) + 4(1.00)(0.06)}{\sqrt{2.00^2 + 4(1.00)^2}} = 0.113$

25. $z = e^{x+y}$

Parallel to xz-plane means y is constant.

$\frac{\partial z}{\partial x} = e^{x+y}$, $\frac{\partial z}{\partial x}\Big|_{(1,1,e^2)} = e^2$

$z - e^2 = e^2(x - 1)$

$z = e^2x$

27. $V = \int_0^1 \int_0^x z\,dy\,dx = \int_0^1 \int_0^x e^{x+y}dy\,dx$

$= \int_0^1 e^{x+y}\Big|_0^x dx = \int_0^1 (e^{2x} - e^x)dx$

$= \frac{1}{2}e^{2x} - e^x\Big|_0^1 = \frac{1}{2}e^2 - e - \frac{1}{2} + 1$

$= 1.48$

29. $v = \frac{rE}{r + R}$

$\frac{\partial v}{\partial r} = \frac{(r+R)E - rE(1)}{(r+R)^2} = \frac{ER}{(r+R)^2}$

$\frac{\partial v}{\partial R} = -rE(r+R)^{-2}(1) = \frac{-rE}{(r+R)^2}$

31. $i_c = i_e(1 - e^{-2v_c})$

$\alpha = \frac{\partial i_c}{\partial i_e} = 1 - e^{-2v_c}$

For $v_c = 2\,V$, $\alpha = 1 - e^{-4} = 0.982$

33. $T = 2\pi\sqrt{\frac{\ell}{g}}$

$\frac{\partial T}{\partial \ell} = 2\pi\sqrt{\frac{1}{g}}(\frac{1}{2}\ell^{-1/2}) = \frac{\pi}{\sqrt{g\ell}}$

$\frac{T}{2\ell} = \frac{2\pi\sqrt{\ell/g}}{2\ell} = \frac{\pi}{\sqrt{g\ell}}$

$\frac{\partial T}{\partial \ell} = \frac{T}{2\ell}$

35. $q = \frac{pf}{p - f}$

$dq = \frac{(p-f)f - pf(1)}{(p-f)^2}dp + \frac{(p-f)p - pf(-1)}{(p-f)^2}df$

$= \frac{-f^2 dp + p^2 df}{(p-f)^2}$

For f=20 cm, df=1 cm, p=100 cm, dp=5 cm

$dq = \frac{-(20^2)(5) + (100^2)(1)}{(100 - 20)^2} = 1.25$ cm

37. $T = 2xy - y^2 - 2x^2 + 3y + 2$

$\frac{\partial T}{\partial x} = 2y - 4x$, $\frac{\partial T}{\partial y} = 2x - 2y + 3$

$2y - 4x = 0$, $2x - 2y + 3 = 0$

$y = 2x$, $2x - 2(2x) + 3 = 0$

$x = \frac{3}{2}$, $y = 3$

$\frac{\partial^2 T}{\partial x^2} = -4$, $\frac{\partial^2 T}{\partial y^2} = -2$

$\frac{\partial^2 T}{\partial y \partial x} = 2$; $D = -4(-2) - 2^2 = 4$

$D > 0$, $\frac{\partial^2 T}{\partial x^2} < 0$, Max. at $(\frac{3}{2}, 3)$

39. $V = \int_0^2 \int_0^{4-x^2} z\,dy\,dx$

$= \int_0^2 \int_0^{4-x^2} (6-x-y)dy\,dx$

$= \int_0^2 (6y - xy - \frac{1}{2}y^2)\Big|_0^{4-x^2} dx$

$= \int_0^2 [6(4-x^2) - x(4-x^2) - \frac{1}{2}(4-x^2)^2]dx$

$= \int_0^2 (16 - 4x - 2x^2 + x^3 - \frac{1}{2}x^4)dx$

$= 16x - 2x^2 - \frac{2}{3}x^3 + \frac{1}{4}x^4 - \frac{1}{10}x^5\Big|_0^2$

$= 32 - 8 - \frac{16}{3} + 4 - \frac{32}{10} - 0$

$= 28 - \frac{16}{3} - \frac{16}{5} = \frac{292}{15}$

41. $V = \int_0^4 \int_0^{\sqrt{16-x^2}} z \, dy \, dx = \int_0^4 \int_0^{\sqrt{16-x^2}} (8-x) dy \, dx = \int_0^4 (8-x)y \Big|_0^{\sqrt{16-x^2}} dx$

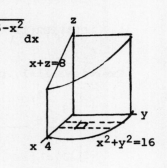

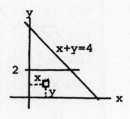

$= \int_0^4 (8-x)\sqrt{16-x^2} \, dx = 8\int_0^4 \sqrt{16-x^2} \, dx - \int_0^4 x\sqrt{16-x^2} \, dx$

$= 8[\frac{x}{2}\sqrt{16-x^2} + \frac{16}{2} \text{Arcsin} \frac{x}{4}] + \frac{1}{3}(16-x^2)^{3/2} \Big|_0^4$

$= 8[2(0) + 8 \text{Arcsin} 1] + \frac{1}{3}(0) - 8[0 + 8 \text{Arcsin} 0] - \frac{1}{3}(16)^{3/2}$

$= 64 \text{Arcsin} 1 - \frac{64}{3} = 64(\frac{\pi}{2}) - \frac{64}{3} = 32(\pi - \frac{2}{3}) = 79.20$

43. $A = \int_0^2 \int_0^{4-y} dx \, dy = \int_0^2 x \Big|_0^{4-y} dy = \int_0^2 (4-y) dy = 4y - \frac{1}{2}y^2 \Big|_0^2 = 6$

$A\bar{x} = \int_0^2 \int_0^{4-y} x \, dx \, dy = \frac{1}{2}\int_0^2 x^2 \Big|_0^{4-y} dy = \frac{1}{2}\int_0^2 (16 - 8y + y^2) dy$

$= 8y - 2y^2 + \frac{1}{6}y^3 \Big|_0^2 = 16 - 8 + \frac{8}{6} = \frac{28}{3} \; ; \; \bar{x} = \frac{28}{3}(\frac{1}{6}) = \frac{14}{9}$

$A\bar{y} = \int_0^2 \int_0^{4-y} y \, dx \, dy = \int_0^2 yx \Big|_0^{4-y} dy = \int_0^2 (4y - y^2) dy$

$= 2y^2 - \frac{1}{3}y^3 \Big|_0^2 = 8 - \frac{8}{3} = \frac{16}{3} \; ; \; \bar{y} = \frac{16}{3}(\frac{1}{6}) = \frac{8}{9}$

45. See Figure for #43

$I_y = \int_0^2 \int_0^{4-y} x^2 dx \, dy = \frac{1}{3}\int_0^2 x^3 \Big|_0^{4-y} dy = \frac{1}{3}\int_0^2 (4-y)^3 dy = -\frac{1}{12}(4-y)^4 \Big|_0^2$

$= -\frac{1}{12}(2^4 - 4^4) = 20 \; ; \; R_y^2 = \frac{20}{6} = \frac{10}{3} \; , \; R_y = 1.83$

47. $P = \frac{V^2}{R}$

$\frac{dP}{dt} = \frac{2V}{R}\frac{dV}{dt} - \frac{V^2}{R^2}\frac{dR}{dt}$

For $V=60$ V, $\frac{dV}{dt} = 1$ V/min, $R=20$ Ω , $\frac{dR}{dt} = 2$ Ω/min

$\frac{dP}{dt} = \frac{2(60)}{20}(1) - \frac{60^2}{20^2}(2) = -12$ W/min

Exercises 11-1, p. 409

1. 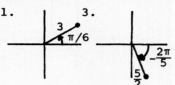 3. 5. 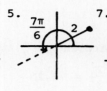 7. 9. 11.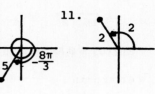

13. $(\sqrt{3}, 1)$; $x = \sqrt{3}$, $y = 1$

$r = \sqrt{x^2 + y^2} = \sqrt{(\sqrt{3})^2 + 1^2} = 2$

$\tan \theta = \frac{y}{x} = \frac{1}{\sqrt{3}}$, $\theta = \frac{\pi}{6}$ (x>0, y>0)

Polar coordinates: $(2, \frac{\pi}{6})$

15. $(-\frac{\sqrt{3}}{2}, -\frac{1}{2})$; $x = -\frac{\sqrt{3}}{2}$, $y = -\frac{1}{2}$

$r = \sqrt{(-\frac{1}{2}\sqrt{3})^2 + (-\frac{1}{2})^2} = 1$

$\tan \theta = -\frac{1}{2} \Big/ -\frac{\sqrt{3}}{2} = \frac{1}{\sqrt{3}}$, $\theta = \frac{7\pi}{6}$ (x<0, y<0)

Polar coordinates: $(1, \frac{7\pi}{6})$

17. $(8, \frac{4\pi}{3})$; $r = 8$, $\theta = \frac{4\pi}{3}$

$x = r \cos \theta = 8 \cos \frac{4\pi}{3} = 8(-\frac{1}{2}) = -4$

$y = r \sin \theta = 8 \sin \frac{4\pi}{3} = 8(-\sqrt{3}/2) = -4\sqrt{3}$

Rectangular coordinates: $(-4, -4\sqrt{3})$

19. $(3, -\frac{\pi}{8})$; $r = 3$, $\theta = -\frac{\pi}{8}$

$x = r \cos \theta = 3 \cos (-\frac{\pi}{8}) = 3(0.9239) = 2.77$

$y = r \sin \theta = 3 \sin (-\frac{\pi}{8}) = 3(-0.3827) = -1.15$

Rectangular coordinates: $(2.77, -1.15)$

21. $x = 3$

$r \cos \theta = 3$

$r = 3 \sec \theta$

23. $x^2 + y^2 = a^2$

$r^2 = a^2$

$r = a$

25. $y^2 = 4x$

$(r \sin \theta)^2 = 4(r \cos \theta)$

$r^2 = \frac{4r \cos \theta}{\sin^2 \theta}$; $r = 4 \cot \theta \csc \theta$

27. $x^2 + 4y^2 = 4$

$(r \cos \theta)^2 + 4(r \sin \theta)^2 = 4$

$r^2(\cos^2 \theta + \sin^2 \theta + 3 \sin^2 \theta) = 4$

$r^2 = \frac{4}{1 + 3 \sin^2 \theta}$

29. $r = \sin \theta$

$r^2 = r \sin \theta$

$x^2 + y^2 = y$

$x^2 + y^2 - y = 0$

31. $r \cos \theta = 4$

$x = 4$

33. $r = 2(1 + \cos \theta)$

$r = 2(1 + \frac{x}{r})$, $r^2 = 2r + 2x$, $x^2 + y^2 = 2\sqrt{x^2 + y^2} + 2x$

$(x^2 + y^2 - 2x)^2 = (2\sqrt{x^2 + y^2})^2$

$x^4 + y^4 - 4x^3 - 4xy^2 + 2x^2y^2 + 4x^2 = 4x^2 + 4y^2$

$x^4 + y^4 - 4x^3 + 2x^2y^2 - 4xy^2 - 4y^2 = 0$

35. $r^2 = \sin 2\theta$

$r^2 = 2 \sin \theta \cos \theta$

$r^4 = 2(r \sin \theta)(r \cos \theta)$

$(x^2 + y^2)^2 = 2xy$

37. $s = \theta r$, $A = \frac{1}{2} \theta r^2$

$s = (\arctan \frac{y}{x})(\sqrt{x^2 + y^2})$

$= \sqrt{x^2 + y^2} \arctan \frac{y}{x}$

$A = \frac{1}{2}(\arctan \frac{y}{x})(x^2 + y^2)$

$= \frac{1}{2}(x^2 + y^2)\arctan \frac{y}{x}$

39. See Exercise 29 of Exercises 1-6

$x^2 = 100y$

$(r \cos \theta)^2 = 100 \, r \sin \theta$

$r = \frac{100 \sin \theta}{\cos^2 \theta} = 100 \tan \theta \sec \theta$

Exercises 11-2, p. 412

1. $r = 4$

   $r = 4$ for all θ, circle

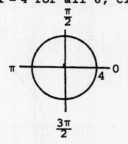

3. $\theta = \dfrac{3\pi}{4}$

   $\theta = \dfrac{3\pi}{4}$ for all r,

   straight line

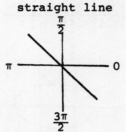

5. $r = 4 \sec \theta$

θ	0	$\frac{\pi}{3}$	$\frac{\pi}{2}$	$\frac{2\pi}{3}$	π
r	4	8	undef	-8	-4

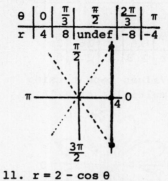

7. $r = 2 \sin \theta$

θ	0	$\frac{\pi}{4}$	$\frac{\pi}{2}$	$\frac{3\pi}{4}$	π	$\frac{-5\pi}{4}$
r	0	1.4	2	1.4	0	-1.4

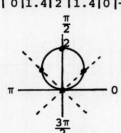

9. $r = 1 - \cos \theta$

θ	0	$\frac{\pi}{4}$	$\frac{\pi}{2}$	$\frac{3\pi}{4}$	π
r	0	0.3	1	1.7	2

θ	$\frac{5\pi}{4}$	$\frac{3\pi}{2}$	$\frac{7\pi}{4}$	2π
r	1.7	1	0.3	0

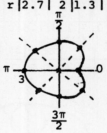

11. $r = 2 - \cos \theta$

θ	0	$\frac{\pi}{4}$	$\frac{\pi}{2}$	$\frac{3\pi}{4}$	π
r	1	1.3	2	2.7	3

θ	$\frac{5\pi}{4}$	$\frac{3\pi}{2}$	$\frac{7\pi}{4}$	2π
r	2.7	2	1.3	1

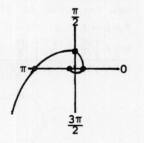

13. $r = 4 \sin 2\theta$

θ	0	$\frac{\pi}{8}$	$\frac{\pi}{4}$	$\frac{3\pi}{8}$	$\frac{\pi}{2}$
r	0	2.8	4	2.8	0

θ	$\frac{5\pi}{8}$	$\frac{3\pi}{4}$	$\frac{7\pi}{8}$	π
r	-2.8	-4	-2.8	0

θ	$\frac{9\pi}{8}$	$\frac{5\pi}{4}$	$\frac{11\pi}{8}$	$\frac{3\pi}{2}$
r	2.8	4	2.8	0

θ	$\frac{13\pi}{8}$	$\frac{7\pi}{4}$	$\frac{15\pi}{8}$	2π
r	-2.8	-4	-2.8	0

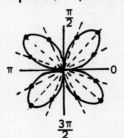

15. $r^2 = 4 \sin 2\theta$

θ	0	$\frac{\pi}{8}$	$\frac{\pi}{4}$	$\frac{3\pi}{8}$	$\frac{\pi}{2}$
r	0	1.7	2	1.7	0

θ	π	$\frac{9\pi}{8}$	$\frac{5\pi}{4}$	$\frac{11\pi}{8}$	$\frac{3\pi}{2}$
r	0	1.7	2	1.7	0

$\frac{\pi}{2} < \theta < \pi$, $\frac{3\pi}{2} < \theta < 2\pi$

r undefined

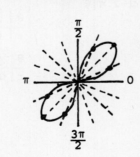

17. $r = 2^\theta$

θ	$-\pi$	0	$\frac{\pi}{2}$	π	2π
r	0.1	1	3.0	8.8	78

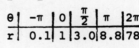

19. $r = -4 \sin 4\theta$

θ	0	$\frac{\pi}{16}$	$\frac{\pi}{8}$	$\frac{3\pi}{16}$	$\frac{\pi}{4}$
r	0	-2.8	-4	-2.8	0

θ	$\frac{5\pi}{16}$	$\frac{3\pi}{8}$	$\frac{7\pi}{16}$	$\frac{\pi}{2}$
r	2.8	4	2.8	0

Values repeat using multiples of π/16.

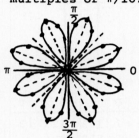

21. $r = \dfrac{1}{2 - \cos\theta}$

θ	0	$\frac{\pi}{4}$	$\frac{\pi}{2}$	$\frac{3\pi}{4}$	π
r	1	0.8	0.5	0.4	0.3

θ	$\frac{5\pi}{4}$	$\frac{3\pi}{2}$	$\frac{7\pi}{4}$	2π
r	0.4	0.5	0.8	1

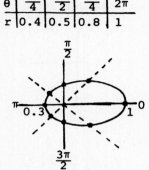

23. $r = \dfrac{6}{1 - 2\cos\theta}$

θ	0	$\frac{\pi}{4}$	$\frac{\pi}{3}$	$\frac{5\pi}{12}$	$\frac{\pi}{2}$	$\frac{3\pi}{4}$	π
r	-6	-14	undef	12	6	2.5	2

θ	$\frac{5\pi}{4}$	$\frac{3\pi}{2}$	$\frac{19\pi}{12}$	$\frac{5\pi}{3}$	$\frac{7\pi}{4}$	2π
r	2.5	6	12	undef	-14	-6

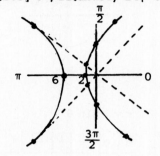

25. $r = 4\cos\frac{1}{2}\theta$

θ	0	$\frac{\pi}{4}$	$\frac{\pi}{2}$	$\frac{3\pi}{4}$	π	$\frac{5\pi}{4}$	$\frac{3\pi}{2}$	$\frac{7\pi}{4}$	2π
r	4	3.7	2.8	1.5	0	-1.5	-2.8	-3.7	-4

θ	$\frac{9\pi}{4}$	$\frac{5\pi}{2}$	$\frac{11\pi}{4}$	3π	$\frac{13\pi}{4}$	$\frac{7\pi}{2}$	$\frac{15\pi}{4}$	4π
r	-3.7	-2.8	-1.5	0	1.5	2.8	3.7	4

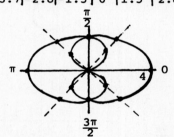

27. $r = 3 - \sin 3\theta$

θ	0	$\frac{\pi}{12}$	$\frac{\pi}{6}$	$\frac{\pi}{4}$	$\frac{\pi}{3}$	$\frac{5\pi}{12}$	$\frac{\pi}{2}$	$\frac{7\pi}{12}$	$\frac{2\pi}{3}$
r	3	2.3	2	2.3	3.7	4	3.7	3

Values repeat using multiples of π/12.

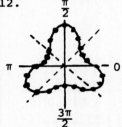

29. **See Exercise 41 of Exercises 1-5**

$a = R = 1.7\,m$, $r = 3.4\sin\theta$

θ	0	$\frac{\pi}{4}$	$\frac{\pi}{2}$	$\frac{3\pi}{4}$	π	$\frac{5\pi}{4}$	$\frac{3\pi}{2}$	$\frac{7\pi}{4}$	2π
r	0	2.4	3.4	2.4	0	-2.4	-3.4	-2.4	0

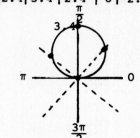

31. $r^2 = R^2\cos 2(\theta + \frac{\pi}{2})$

r is undefined for $0 \le \theta < \frac{\pi}{4}$,

$\frac{3\pi}{4} < \theta < \frac{5\pi}{4}$, $\frac{7\pi}{4} < \theta \le 2\pi$

θ	$\frac{\pi}{4}$	$\frac{3\pi}{8}$	$\frac{\pi}{2}$	$\frac{5\pi}{8}$	$\frac{3\pi}{4}$
r	0	0.8R	R	0.8R	0

θ	$\frac{5\pi}{4}$	$\frac{11\pi}{8}$	$\frac{3\pi}{2}$	$\frac{13\pi}{8}$	$\frac{7\pi}{4}$
r	0	0.8R	R	0.8R	0

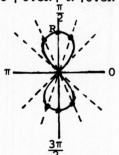

Exercises 11-3, p. 417

1. $r = 3t$, $\theta = t^3$, $t = 1$

$v_r = \dfrac{dr}{dt}$, $v_\theta = r\dfrac{d\theta}{dt}$

$v_r = 3$, $v_\theta = (3t)(3t^2) = 9t^3$

For $t = 1$, $v_r = 3$, $v_\theta = 9$

$v = \sqrt{3^2 + 9^2} = 9.49$

3. $r = e^{0.1t}$, $\theta = \cos 2t$, $t = 2$

$v_r = \dfrac{dr}{dt}$, $v_\theta = r\dfrac{d\theta}{dt}$

$v_r = 0.1e^{0.1t}$, $v_\theta = e^{0.1t}(-2\sin 2t)$

For $t = 2$: $v_r = 0.1e^{0.2} = 0.122$

$v_\theta = -2e^{0.2}\sin 4 = 1.849$

$v = \sqrt{0.122^2 + 1.849^2} = 1.85$

5. $r = 2$, $\theta = \dfrac{\pi}{3}$, $\omega = 2$ rad/s

$\dfrac{dr}{dt} = 0$, $\dfrac{d\theta}{dt} = 2$

$v_r = 0$, $v_\theta = 2(2) = 4$ ft/s

$v = \sqrt{0^2 + 4^2} = 4.0$ ft/s

7. $r = \sin\theta$, $\theta = \dfrac{\pi}{6}$, $\omega = 2$ rad/s

$\dfrac{dr}{dt} = \cos\theta\dfrac{d\theta}{dt}$

$v_r = (\cos\dfrac{\pi}{6})(2) = 2(\dfrac{\sqrt{3}}{2}) = \sqrt{3} = 1.7$ ft/s

$v_\theta = (\sin\dfrac{\pi}{6})(2) = 2(\dfrac{1}{2}) = 1.0$ ft/s

$v = \sqrt{1.7^2 + 1.0^2} = 2.0$ ft/s

9. $r = 4 - \cos\theta$, $\theta = \dfrac{\pi}{2}$, $\omega = 2$ rad/s

$\dfrac{dr}{dt} = \sin\theta\dfrac{d\theta}{dt}$

$v_r = (\sin\dfrac{\pi}{2})(2) = 2(1) = 2.00$ ft/s

$v_\theta = (4 - \cos\dfrac{\pi}{2})(2) = 8.00$ ft/s

$v = \sqrt{2.00^2 + 8.00^2} = 8.25$ ft/s

11. $r = \dfrac{1}{2 - \sin\theta}$, $\theta = \dfrac{\pi}{6}$, $\omega = 2$ rad/s

$\dfrac{dr}{dt} = \dfrac{\cos\theta}{(2 - \sin\theta)^2}\dfrac{d\theta}{dt}$

$v_r = \dfrac{\cos(\pi/6)}{[2 - \sin(\pi/6)]^2}(2) = \dfrac{2(\sqrt{3}/2)}{(2 - \frac{1}{2})^2} = 0.770$ ft/s

$v_\theta = \dfrac{1}{2 - \sin(\pi/6)}(2) = \dfrac{2}{2 - \frac{1}{2}} = 1.33$ ft/s

$v = \sqrt{0.770^2 + 1.33^2} = 1.54$ ft/s

13. $\theta = \dfrac{1}{6}\pi$, $\theta = \dfrac{2}{3}\pi$, $r = 2$

$A = \dfrac{1}{2}\int_{\pi/6}^{2\pi/3} 2^2 d\theta$

$= 2\theta\Big|_{\pi/6}^{2\pi/3}$

$= \dfrac{4\pi}{3} - \dfrac{\pi}{3} = \pi$

15. $\theta = 0$, $\theta = \dfrac{1}{4}\pi$, $r = \sec\theta$

$A = \dfrac{1}{2}\int_0^{\pi/4}\sec^2\theta\, d\theta = \dfrac{1}{2}\tan\theta\Big|_0^{\pi/4}$

$= \dfrac{1}{2}\tan\dfrac{\pi}{4} = \dfrac{1}{2}(1) = \dfrac{1}{2}$

17. $r = 2\cos\theta$

$A = 2[\dfrac{1}{2}\int_0^{\pi/2}(2\cos\theta)^2 d\theta]$

$= 4\int_0^{\pi/2}\cos^2\theta\, d\theta$

$= 2\int_0^{\pi/2}(1 + \cos 2\theta)d\theta$

$= 2\theta + \sin 2\theta\Big|_0^{\pi/2} = \pi + \sin\pi - 0 = \pi$

19. $\theta = 0$, $\theta = \dfrac{1}{2}\pi$, $r = e^{2\theta}$

$A = \dfrac{1}{2}\int_0^{\pi/2}(e^{2\theta})^2 d\theta$

$= \dfrac{1}{2}\int_0^{\pi/2}e^{4\theta}d\theta$

$= \dfrac{1}{8}e^{4\theta}\Big|_0^{\pi/2} = \dfrac{1}{8}(e^{2\pi} - 1) = 66.8$

21. $r = 2\sin 3\theta$

$A = 3[\dfrac{1}{2}\int_0^{\pi/3}(2\sin 3\theta)^2 d\theta = 6\int_0^{\pi/3}\sin^2 3\theta\, d\theta = 3\int_0^{\pi/3}(1 - \sin 6\theta)d\theta$

$= 3\theta + \dfrac{1}{2}\cos 6\theta\Big|_0^{\pi/3} = \pi + \dfrac{1}{2}\cos 2\pi - 0 - \dfrac{1}{2}\cos 0 = \pi$

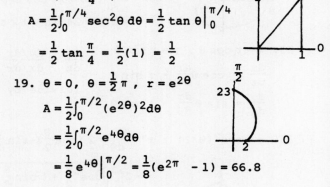

23. $r = 2 + \cos\theta$

$A = 2[\frac{1}{2}\int_0^\pi (2 + \cos\theta)^2 d\theta = \int_0^\pi (4 + 4\cos\theta + \cos^2\theta) d\theta$

$= \int_0^\pi [4 + 4\cos\theta + \frac{1}{2}(1 + \cos 2\theta)] d\theta = \frac{1}{2}\int_0^\pi (9 + 8\cos\theta + \cos 2\theta) d\theta$

$= \frac{1}{2}(9\theta + 8\sin\theta + \frac{1}{2}\sin 2\theta)\Big|_0^\pi = \frac{1}{2}(9\pi + 8\sin\pi + \frac{1}{2}\sin 2\pi - 0) = \frac{9}{2}\pi$

25. $r = 1 + \sin\theta, r = 1$

$A = \frac{1}{2}\int_0^\pi [(1 + \sin\theta)^2 - 1^2] d\theta = \frac{1}{2}\int_0^\pi (2\sin\theta + \sin^2\theta) d\theta$

$= \frac{1}{2}\int_0^\pi [2\sin\theta + \frac{1}{2}(1 - \cos 2\theta)] d\theta = \frac{1}{4}\int_0^\pi (1 + 4\sin\theta - \cos 2\theta) d\theta$

$= \frac{1}{4}(\theta - 4\cos\theta - \frac{1}{2}\sin 2\theta)\Big|_0^\pi = \frac{1}{4}(\pi - 4\cos\pi - \frac{1}{2}\sin 2\pi)$

$- \frac{1}{4}(0 - 4\cos 0 - \frac{1}{2}\sin 0) = \frac{1}{4}(\pi + 4 + 4) = \frac{1}{4}(8 + \pi) = 2.79$

27. $r = 2\sin\theta, r = 2\cos\theta$

$A = 2[\frac{1}{2}\int_0^{\pi/4} (2\sin\theta)^2 d\theta = 4\int_0^{\pi/4} \sin^2\theta \, d\theta$

$= 2\int_0^{\pi/4} (1 - \cos 2\theta) d\theta = 2\theta - \sin 2\theta\Big|_0^{\pi/4}$

$= \frac{\pi}{2} - \sin\frac{\pi}{2} - 0 = \frac{\pi}{2} - 1 = 0.571$

29. $r = 2\sin\theta$; $v_r = 2\cos\theta\frac{d\theta}{dt}$, $v_\theta = 2\sin\theta\frac{d\theta}{dt}$

For $\theta = \frac{\pi}{3}$: $v_r = (2\cos\frac{\pi}{3})(\frac{d\theta}{dt}) = 2(\frac{1}{2})\frac{d\theta}{dt} = \frac{d\theta}{dt}$; $v_\theta = (2\sin\frac{\pi}{3})(\frac{d\theta}{dt}) = 2(\frac{\sqrt{3}}{2})\frac{d\theta}{dt} = \sqrt{3}\frac{d\theta}{dt}$

$10 = \sqrt{(\frac{d\theta}{dt})^2 + (\sqrt{3}\frac{d\theta}{dt})^2} = 2\frac{d\theta}{dt}$; $\frac{d\theta}{dt} = 5$ rad/s

$v_r = (2\cos\frac{\pi}{3})(5) = 2(\frac{1}{2})(5) = 5$ ft/s, $v_\theta = (2\sin\frac{\pi}{3})(5) = 2(\frac{\sqrt{3}}{2})(5) = 5\sqrt{3}$ ft/s

31. $r = 6 - 2\cos\theta$

$A = 2[\frac{1}{2}\int_0^\pi (6 - 2\cos\theta)^2 d\theta = \int_0^\pi (36 - 24\cos\theta + 4\cos^2\theta) d\theta$

$= \int_0^\pi [36 - 24\cos\theta + 4(\frac{1}{2})(1 + \cos 2\theta)] d\theta$

$= \int_0^\pi (38 - 24\cos\theta + 4\cos 2\theta) d\theta = 38\theta - 24\sin\theta + 2\sin 2\theta\Big|_0^\pi$

$= 38\pi - 24\sin\pi + 2\sin 2\pi - 0 = 38\pi = 119$ cm^2

33. $x = r\cos\theta$, $y = r\sin\theta$

$\frac{dx}{dr} = \cos\theta\frac{dr}{d\theta} - r\sin\theta$

$\frac{dy}{dr} = \sin\theta\frac{dr}{d\theta} + r\cos\theta$

$\frac{dy}{dx} = \frac{dy/dr}{dx/dr} = \frac{r'\sin\theta + r\cos\theta}{r'\cos\theta - r\sin\theta} = \frac{r'\tan\theta + r}{r' - r\tan\theta}$

(divide by $\cos\theta$) , $r' = dr/d\theta$

35. $r = \cos^2(\theta/2)$; $r' = \frac{dr}{d\theta} = 2(\cos\frac{\theta}{2})(-\sin\frac{\theta}{2})(\frac{1}{2}) = -\sin\frac{\theta}{2}\cos\frac{\theta}{2}$

$s = 2\int_0^\pi \sqrt{\cos^4\frac{\theta}{2} + (-\sin\frac{\theta}{2}\cos\frac{\theta}{2})^2} \, d\theta = 2\int_0^\pi \sqrt{(\cos^2\frac{\theta}{2})(\cos^2\frac{\theta}{2} + \sin^2\frac{\theta}{2})} \, d\theta$

$= 2\int_0^\pi \cos\frac{\theta}{2} d\theta = 4\sin\frac{\theta}{2}\Big|_0^\pi = 4\sin\frac{\pi}{2} = 4$

Review Exercises for Chapter 11, p. 418

1. $y = 2x$
$r \sin \theta = 2r \cos \theta$
$\tan \theta = 2$
$\theta = \text{Arctan } 2 = 1.11$

3. $x^2 - y^2 = 16$
$r^2 \cos^2 \theta - r^2 \sin^2 \theta = 16$
$r^2 (\cos^2 \theta - \sin^2 \theta) = 16$
$r^2 \cos 2\theta = 16$

5. $r = 2 \sin 2\theta = 4 \sin \theta \cos \theta$
$r^3 = 4 (r \sin \theta)(r \cos \theta)$
$(x^2 + y^2)^{3/2} = 4yx$
$(x^2 + y^2)^3 = 16x^2 y^2$

7. $r = \dfrac{4}{2 - \cos \theta}$
$2r - r \cos \theta = 4$
$2\sqrt{x^2 + y^2} - x = 4$
$2\sqrt{x^2 + y^2} = x + 4$
$4(x^2 + y^2) = x^2 + 8x + 16$
$3x^2 + 4y^2 - 8x - 16 = 0$

9. $r = 4(1 + \sin \theta)$

θ	0	$\frac{\pi}{4}$	$\frac{\pi}{2}$	$\frac{3\pi}{4}$	π
r	4	6.8	8	6.8	4

θ	$\frac{5\pi}{4}$	$\frac{3\pi}{2}$	$\frac{7\pi}{4}$	2π
r	1.2	0	1.2	4

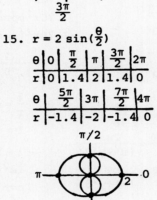

11. $r = 4 \cos 3\theta$

θ	0	$\frac{\pi}{12}$	$\frac{\pi}{6}$	$\frac{\pi}{4}$	$\frac{\pi}{3}$	$\frac{5\pi}{12}$	$\frac{\pi}{2}$	$\frac{7\pi}{12}$	$\frac{2\pi}{3}$
r	4	2.8	0	-2.8	-4	-2.8	0	2.8	4

Values repeat using
multiples of $\pi/12$.

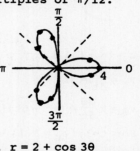

13. $r = \cot \theta$

θ	0	$\frac{\pi}{6}$	$\frac{\pi}{3}$	$\frac{\pi}{2}$
r	undef	1.7	0.6	0

θ	$\frac{2\pi}{3}$	$\frac{5\pi}{6}$	π	$\frac{7\pi}{6}$
r	-0.6	-1.7	undef	1.7

θ	$\frac{4\pi}{3}$	$\frac{3\pi}{2}$	$\frac{5\pi}{3}$	$\frac{11\pi}{6}$	2π
r	0.6	0	-1.7	-0.6	undef

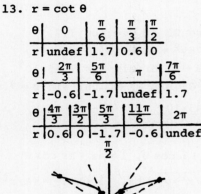

15. $r = 2 \sin\left(\frac{\theta}{2}\right)$

θ	0	$\frac{\pi}{2}$	π	$\frac{3\pi}{2}$	2π
r	0	1.4	2	1.4	0

θ	$\frac{5\pi}{2}$	3π	$\frac{7\pi}{2}$	4π
r	-1.4	-2	-1.4	0

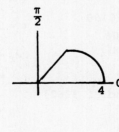

17. $r = 2 + \cos 3\theta$

θ	0	$\frac{\pi}{12}$	$\frac{\pi}{6}$	$\frac{\pi}{4}$	$\frac{\pi}{3}$	$\frac{5\pi}{12}$	$\frac{\pi}{2}$
r	3	2.7	2	1.3	1	1.3	2

θ	$\frac{7\pi}{12}$	$\frac{2\pi}{3}$
r	2.7	3

Values repeat
using multiples
of $\pi/12$.

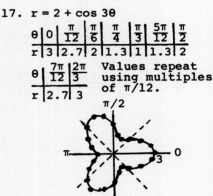

19. $r^2 = \theta$

θ	0	$\frac{\pi}{4}$	$\frac{\pi}{2}$	$\frac{3\pi}{4}$	π
r	0	± 0.9	± 1.3	± 1.5	± 1.8

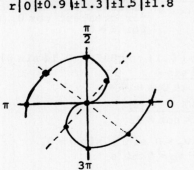

21. $r = 4 \cos \theta$, $\theta = 0$, $\theta = \frac{\pi}{4}$

$A = \dfrac{1}{2} \displaystyle\int_0^{\pi/4} (4 \cos \theta)^2 d\theta$

$= 8 \displaystyle\int_0^{\pi/4} \cos^2 \theta \, d\theta$

$= 4 \displaystyle\int_0^{\pi/4} (1 + \cos 2\theta) d\theta$

$= 4\theta + 2 \sin 2\theta \Big|_0^{\pi/4}$

$= \pi + 2 \sin \dfrac{\pi}{2} - 0 = \pi + 2$

23. $r = 3 - \sin\theta$ $A = \frac{1}{2}\int_0^{2\pi}(3 - \sin\theta)^2 d\theta = \frac{1}{2}\int_0^{2\pi}(9 - 6\sin\theta + \sin^2\theta)d\theta$

$= \frac{1}{2}\int_0^{2\pi}[9 - 6\sin\theta + \frac{1}{2}(1 - \cos2\theta)]d\theta = \frac{1}{4}\int_0^{2\pi}(19 - 12\sin\theta - \cos2\theta)d\theta$

$= \frac{1}{4}(19\theta + 12\cos\theta - \frac{1}{2}\sin2\theta)\Big|_0^{2\pi} = \frac{1}{4}(38\pi + 12\cos2\pi - \frac{1}{2}\sin\,4\pi)$

$-\frac{1}{4}(0 + 12\cos0 - \frac{1}{2}\sin0) = \frac{1}{4}(38\pi + 12 - 12) = \frac{19\pi}{2}$

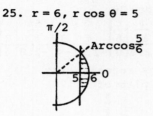

25. $r = 6$, $r\cos\theta = 5$ $A = 2[\frac{1}{2}\int_0^{Arccos\frac{5}{6}}[6^2 - (5\sec\theta)^2]d\theta$

$= \int_0^{Arccos\frac{5}{6}}(36 - 25\sec^2\theta)d\theta$

$= 36\theta - 25\tan\theta\Big|_0^{Arccos\frac{5}{6}}$

$= 36(Arccos\frac{5}{6}) - 25\tan(Arccos\frac{5}{6})$

$= 4.502$ $(Arccos\frac{5}{6} = 0.5856855)$

27. $r = \theta$, $\theta = \frac{\pi}{2}$, $\theta \geq 0$

$A = \frac{1}{2}\int_0^{\pi/2}\theta^2 d\theta$

$= \frac{1}{6}\theta^3\Big|_0^{\pi/2} = \frac{\pi^3}{48}$

29. $r = t^3 - 6t$, $\theta = \frac{t}{t+1}$, $t = 1$

$v_r = \frac{dr}{dt} = 3t^2 - 6$

$v_\theta = r\frac{d\theta}{dt} = (t^3 - 6t)[\frac{1}{(t+1)^2}]$

For t=1: $v_r = 3(1) - 6 = -3$

$v_\theta = (1 - 6)(\frac{1}{4}) = -\frac{5}{4}$

$v = \sqrt{(-3)^2 + (-\frac{5}{4})^2} = \frac{13}{4}$

31. See Exercise 29 of Section 1-7

$2.70x^2 + 2.76y^2 = 7.45\times10^7$

$2.70r^2\cos^2\theta + 2.76r^2\sin^2\theta = 7.45\times10^7$

$2.70r^2(\cos^2\theta + \sin^2\theta) + 0.06r^2\sin^2\theta = 7.45\times10^7$

$r^2 = \frac{7.45\times10^7}{2.70 + 0.06\sin^2\theta}$

33. $P = \frac{k\cos\theta}{r^2}$, $v_r = -\frac{\partial P}{\partial r}$, $v_\theta = -\frac{1}{r}\frac{\partial P}{\partial\theta}$

$v_r = -(\frac{-2k\cos\theta}{r^3}) = \frac{2kr\cos\theta}{r^4} = \frac{2kx}{(x^2+y^2)^2}$

$v_\theta = -\frac{1}{r}(\frac{-k\sin\theta}{r^2}) = \frac{kr\sin\theta}{r^4} = \frac{ky}{(x^2+y^2)^2}$

35. $\frac{1}{r} = a + b\cos\theta$, $1 = ar + br\cos\theta$

$1 = a\sqrt{x^2+y^2} + bx$, $(1 - bx)^2 = (a\sqrt{x^2+y^2})^2$

$1 - 2bx + b^2x^2 = a^2x^2 + a^2y^2$

$(a^2 - b^2)x^2 + a^2y^2 + 2bx - 1 = 0$

a=b, parabola (no x^2-term)

$a^2 > b^2$, ellipse (x^2 and y^2 terms positive, but different values)

$a^2 < b^2$, hyperbola (x^2 and y^2 terms of different signs)

37. See Figure 11-22, $r = 10\cos\theta$

$40° = 0.6981$, $90° = 1.5708$

$A = \frac{1}{2}\int_{0.6981}^{1.5708}(10\cos\theta)^2 d\theta$

$= 50\int_{0.6981}^{1.5708}\cos^2\theta\,d\theta$

$= 25\int_{0.6981}^{1.5708}(1 + \cos2\theta)d\theta$

$= 25\theta + \frac{25}{2}\sin2\theta\Big|_{0.6981}^{1.5708}$

$= 25(1.5708 - 0.6981)$

$+ 12.5(\sin3.1416 - \sin1.3962)$

$= 9.507\text{ mi}^2$

39. See Exercise 40 of Section 11-1

$r = \frac{4800}{1 + 0.14\cos\theta}$

$v = 5$ mi/s

Satellite closest for smallest r, which occurs for greatest $\cos\theta$, or for $\theta = 0$.

$v_r = 4800(1 + 0.14\cos\theta)^{-2}(-0.14\sin\theta)\frac{d\theta}{dt}$

$= \frac{-670\sin\theta}{(1 + 0.14\cos\theta)^2}\frac{d\theta}{dt}$

$v_\theta = \frac{4800}{1 + 0.14\cos\theta}\frac{d\theta}{dt}$

For $\theta = 0$, $v_r = 0$ since $\sin0 = 0$.

Therefore, $v_\theta = 5$ mi/s

Exercises 12-1, p. 427

1.

No.	2	3	4	5	6	7
Freq.	1	3	4	2	3	2

3.

No.	45	46	47	48	49	50	51	52	53	54	55	56	57
Freq.	1	1	1	2	2	0	1	0	1	0	2	0	1

5.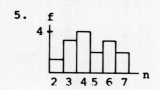

7.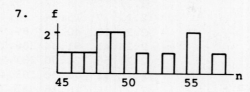

9. $\bar{x} = \dfrac{2(1) + 3(3) + 4(4) + 5(2) + 6(3) + 7(2)}{15} = \dfrac{69}{15} = 4.60$

11. $\bar{x} = \dfrac{45 + 46 + 47 + 48(2) + 49(2) + 51 + 53 + 55(2) + 57}{12} = \dfrac{603}{12} = 50.25$

13.

x	$x - \bar{x}$	$(x-\bar{x})^2$
3	-1.60	2.56
6	1.40	1.96
4	-0.60	0.36
2	-2.60	6.76
5	0.40	0.16
4	-0.60	0.36
7	2.40	5.76
6	1.40	1.96
3	-1.60	2.56
4	-0.60	0.36
6	1.40	1.96
4	-0.60	0.36
5	0.40	0.16
7	2.40	5.76
3	-1.60	2.56
69		33.60

$\bar{x} = \dfrac{69}{15} = 4.60$

$\overline{(x-\bar{x})^2} = \dfrac{33.60}{15}$
$= 2.24$

$s = \sqrt{2.24} = 1.50$

15.

x	$x-\bar{x}$	$(x-\bar{x})^2$
48	-2.25	5.0625
53	2.75	7.5625
49	-1.25	1.5625
45	-5.25	27.5625
55	4.75	22.5625
49	-1.25	1.5625
47	-3.25	10.5625
55	4.75	22.5625
48	-2.25	5.0625
57	6.75	45.5625
51	0.75	0.5625
46	-4.25	18.0625
603		168.2500

$\bar{x} = \dfrac{603}{12} = 50.25$

$\overline{(x-\bar{x})^2} = \dfrac{168.25}{12}$
$= 14.020833$

$s = \sqrt{14.020833} = 3.74$

17.

x	x^2
3	9
6	36
4	16
2	4
5	25
4	16
7	49
6	36
3	9
4	16
6	36
4	16
5	25
7	49
3	9
69	351

$\bar{x} = \dfrac{69}{15} = 4.60$

$\bar{x}^2 = 21.16$

$\overline{x^2} = \dfrac{351}{15} = 23.40$

$\overline{x^2} - \bar{x}^2 = 2.24$

$s = \sqrt{2.24} = 1.50$

19.

x	x^2
48	2304
53	2809
49	2401
45	2025
55	3025
49	2401
47	2209
55	3025
48	2304
57	3249
51	2601
46	2116
603	30469

$\bar{x} = \dfrac{603}{12} = 50.25$

$\bar{x}^2 = 2525.0625$

$\overline{x^2} = \dfrac{30469}{12} = 2539.0833$

$\overline{x^2} - \bar{x}^2 = 14.0208$

$s = \sqrt{14.0208} = 3.74$

21.

Vol. (mL)	746	747	748	749	750	751	752	753	754
No.	1	1	2	1	2	3	1	3	1

23. $\overline{V} = \dfrac{746 + 747 + 748(2) + 749 + 750(2) + 751(3) + 752 + 753(3) + 754}{15} = \dfrac{11256}{15} = 750.4 \text{ mL}$

25.

27.

t(s)	f	ft	ft^2
2.21	2	4.42	9.7682
2.22	7	15.54	34.4988
2.23	18	40.14	89.5122
2.24	41	91.84	205.7216
2.25	56	126.00	283.5000
2.26	32	72.32	163.4432
2.27	8	18.16	41.2232
2.28	3	6.84	15.5952
2.29	3	6.87	15.7323
	170	382.13	858.9947

$\overline{t} = \dfrac{\Sigma ft}{\Sigma f} = \dfrac{382.13}{170} = 2.2478235$

$\overline{t}^{\,2} = 5.0527106$

$\overline{t^2} = \dfrac{\Sigma ft^2}{\Sigma f} = \dfrac{858.9947}{170}$

$\quad = 5.05291$

$\overline{t^2} - \overline{t}^{\,2} = 0.0001994$

$s = \sqrt{0.0001994} = 0.014 \text{ s}$

29. See table for Exercise 28.

$\overline{m} = \dfrac{19.2(4) + 19.7(9) + 20.2(14) + 20.7(24) + 21.2(28) + 21.7(16) + 22.2(4) + 22.7}{100}$

$\quad = \dfrac{2086}{100} = 20.9 \text{ mi/gal}$

31. $\overline{S} = \dfrac{250 + 350 + 275 + 225 + 175 + 300 + 200 + 350 + 275 + 400 + 300 + 225 + 250 + 300}{14}$

$\quad = \dfrac{3875}{14} = \$277$

33. $\overline{P} = \dfrac{22(500) + 80(600) + 106(700) + 185(800) + 380(900) + 122(1000) + 90(1100) + 15(1200)}{1000}$

$\quad = \dfrac{862200}{1000} = 862 \text{ kW·h}$

35. $\overline{i} = \dfrac{\begin{array}{c}3.29 + 3.37 + 3.38 + 3.39 + 3.40 + 3.41(2) + 3.42 + 3.43 + 3.44(2) + 3.45 \\ + 3.46(2) + 3.47(2) + 3.48 + 3.50\end{array}}{18}$

$\quad = \dfrac{61.67}{18} = 3.43 \text{ μA}$

37. For set A: $\overline{x} = 4.60$ (Exer. 9)

$\qquad\qquad\quad s = 1.50$ (Exer. 13)

$\overline{x} - s = 3.10, \; \overline{x} + s = 6.10$

9 of 15 values, or 60 %, are in range 3.10 to 6.10.

39. For set C: $\overline{x} = 50.25$ (Exer. 11)

$\qquad\qquad\quad s = 3.74$ (Exer. 15)

$\overline{x} - s = 46.51, \; \overline{x} + s = 53.99$

7 of 12 values, or 58 %, are in range 46.51 to 53.99.

41. For times in Exercise 25:

$\overline{t} = 2.248$ s (Exer. 26)

$s = 0.014$ s (Exer. 27)

$\overline{t} - s = 2.234$ s, $\overline{t} + s = 2.262$ s

129 of 170 values, or 76 %, are in range 2.234 s to 2.262 s.

43. For salaries in Exercise 31:

$\overline{S} = \$277$ (Exer. 31)

$s = \$60$ (Exer. 32)

$\overline{S} - s = \$217, \; \overline{S} + s = \$337$

9 of 14 values, or 64 %, are in range $217 to $337.

Exercises 12-2, p. 433

1.

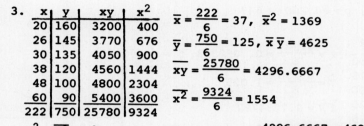
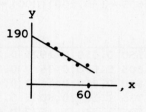

x	y	xy	x²
4	1	4	16
6	4	24	36
8	5	40	64
10	8	80	100
12	9	108	144
40	27	256	360

$\bar{x} = \dfrac{40}{5} = 8.0$, $\bar{x}^2 = 64$

$\bar{y} = \dfrac{27}{5} = 5.4$, $\bar{x}\,\bar{y} = 43.2$

$\overline{xy} = \dfrac{256}{5} = 51.2$

$\overline{x^2} = \dfrac{360}{5} = 72$

$s_x^2 = 72 - 64 = 8.0$

$m = \dfrac{51.2 - 43.2}{8.0} = 1.0$

$b = 5.4 - 1.0(8.0)$
$\quad = -2.6$

$y = 1.0x - 2.6$

3.

x	y	xy	x²
20	160	3200	400
26	145	3770	676
30	135	4050	900
38	120	4560	1444
48	100	4800	2304
60	90	5400	3600
222	750	25780	9324

$\bar{x} = \dfrac{222}{6} = 37$, $\bar{x}^2 = 1369$

$\bar{y} = \dfrac{750}{6} = 125$, $\bar{x}\,\bar{y} = 4625$

$\overline{xy} = \dfrac{25780}{6} = 4296.6667$

$\overline{x^2} = \dfrac{9324}{6} = 1554$

$s_x^2 = \overline{x^2} - \bar{x}^2 = 1554 - 1369 = 185$, $m = \dfrac{4296.6667 - 4625}{185} = -1.7747748$

$b = 125 - (-1.7747748)(37) = 190.66667$, $y = -1.77x + 191$

5.

i	V	iV	i²
15.0	3.00	45.00	225.0000
10.8	4.10	44.28	116.6400
9.30	5.60	52.08	86.4900
3.55	8.00	28.40	12.6025
4.60	10.50	48.30	21.1600
43.25	31.20	218.06	461.8925

$\bar{i} = \dfrac{43.25}{5} = 8.65$, $\bar{i}^2 = 74.8225$

$\bar{V} = \dfrac{31.20}{5} = 6.24$, $\bar{i}\,\bar{V} = 53.976$

$\overline{iV} = \dfrac{218.06}{5} = 43.612$

$\overline{i^2} = \dfrac{461.8925}{5} = 92.3785$

$s_i^2 = \overline{i^2} - \bar{i}^2 = 92.3785 - 74.8225 = 17.5560$, $m = \dfrac{43.612 - 53.976}{17.5560} = -0.5903395$

$b = 6.24 - (-0.5903395)(8.65) = 11.346437$, $V = -0.590i + 11.3$

7.

x	h	xh	x²
0	0	0	0
500	1130	565000	250000
1000	2250	2250000	1000000
1500	3360	5040000	2250000
2000	4500	9000000	4000000
2500	5600	14000000	6250000
7500	16840	30855000	13750000

$\bar{x} = \dfrac{7500}{6} = 1250$, $\bar{x}^2 = 1562500$

$\bar{h} = \dfrac{16840}{6} = 2806.6667$

$\bar{x}\,\bar{h} = 3508333.3$

$\overline{xh} = \dfrac{30855000}{6} = 5142500$

$\overline{x^2} = \dfrac{13750000}{6} = 2291666.7$

$s_x^2 = \overline{x^2} - \bar{x}^2 = 2291666.7 - 1562500 = 729166.7$, $m = \dfrac{5142500 - 3508333.3}{729166.7} = 2.2411429$

$b = 2806.6667 - (2.2411429)(1250) = 5.2380959$, $h = 2.24x + 5.24$

9.

x	p	xp	x^2
0	650	0	0
100	630	63000	10000
200	605	121000	40000
300	590	177000	90000
400	570	228000	160000
1000	3045	589000	300000

$\bar{x} = \dfrac{1000}{5} = 200$, $\bar{x}^2 = 40000$

$\bar{p} = \dfrac{3045}{5} = 609$, $\bar{x}\,\bar{p} = 121800$

$\overline{xp} = \dfrac{589000}{5} = 117800$

$\overline{x^2} = \dfrac{300000}{5} = 60000$

$s_x^2 = \overline{x^2} - \bar{x}^2 = 60000 - 40000 = 20000$, $m = \dfrac{117800 - 121800}{20000} = -0.200$

$b = 609 - (-0.200)(200) = 649$, $p = -0.200x + 649$

11.

f	V	fV	f^2
0.550	0.350	0.1925	0.302500
0.605	0.600	0.3630	0.366025
0.660	0.850	0.5610	0.435600
0.735	1.10	0.8085	0.540225
0.805	1.45	1.1672 5	0.648025
0.880	1.80	1.5840	0.774400
4.235	6.150	4.6762 5	3.066775

$\bar{f} = \dfrac{4.235}{6} = 0.7058333$, $\bar{f}^2 = 0.4982007$

$\bar{V} = \dfrac{6.150}{6} = 1.025$

$\bar{f}\,\bar{V} = 0.7234792$

$\overline{fV} = \dfrac{4.67625}{6} = 0.779375$

$\overline{f^2} = \dfrac{3.066775}{6} = 0.5111292$

$s_f^2 = \overline{f^2} - \bar{f}^2 = 0.5111292 - 0.4982007 = 0.0129285$, $m = \dfrac{0.779375 - 0.7234792}{0.0129285} = 4.3234678$

$b = 1.025 - (4.3234678)(0.7058333) = -2.0266477$, $V = 4.32 \times 10^{-15} f - 2.03$

For $V = 0$, $f_0 = 0.470$ PHz   (Factor 10^{-15} required if f is measured in hertz)

13. $r = m\dfrac{s_x}{s_y}$

For data of Exercise 1:

$m = 1.0$, $s_x = \sqrt{8.0} = 2.82843$

$s_y^2 = \overline{y^2} - \bar{y}^2 = 37.4 - 29.16 = 8.24$

$s_y = \sqrt{8.24} = 2.87054$

$r = 1.0\left(\dfrac{2.82843}{2.87054}\right) = 0.985$

15. $r = m\dfrac{s_x}{s_y}$

For data of Exercise 4:

$s_x = \sqrt{\overline{x^2} - \bar{x}^2} = 2.6551836$

$s_y = \sqrt{\overline{y^2} - \bar{y}^2} = 3.2619013$

$m = \dfrac{\overline{xy} - \bar{x}\,\bar{y}}{s_x^2} = -1.1063830$

$r = (-1.1063830)\left(\dfrac{2.6551836}{3.2619013}\right) = -0.901$

Exercises 12-3, p. 438

1.

x	y	$f(x)=x^2$	fy	f^2
2	12	4	48	16
4	38	16	608	256
6	72	36	2592	1296
8	135	64	8640	4096
10	200	100	20000	10000
	457	220	31888	15664

$\bar{f} = \dfrac{220}{5} = 44$, $\bar{f}^2 = 1936$

$\bar{y} = \dfrac{457}{5} = 91.4$, $\bar{f}\,\bar{y} = 4021.6$

$\overline{fy} = \dfrac{31888}{5} = 6377.6$, $\overline{f^2} = \dfrac{15664}{5}$

$= 3132.8$

$s_f^2 = 3132.8 - 1936 = 1196.8$, $m = \dfrac{6377.6 - 4021.6}{1196.8} = 1.9685829$

$b = 91.4 - (1.9685829)(44) = 4.7823530$, $y = 1.97x^2 + 4.8$

3.

x	y	f(x)=1/x	fy	f²
1.10	9.85	0.9090909	8.9545455	0.8264463
2.45	4.50	0.4081633	1.8367347	0.1665973
4.04	2.90	0.2475248	0.7178218	0.0612685
5.86	1.75	0.1706485	0.2986348	0.0291209
6.90	1.48	0.1449275	0.2144928	0.0210040
8.54	1.30	0.1170960	0.1522248	0.0137115
	21.78	1.9974510	12.1744544	1.1181485

$\bar{f} = \dfrac{1.9974510}{6} = 0.3329085$

$\bar{f}^2 = 0.1108281, \quad \bar{y} = \dfrac{21.78}{6} = 3.63$

$\bar{f}\,\bar{y} = 1.2084578$

$\overline{fy} = \dfrac{12.1744544}{6} = 2.0290757$

$\overline{f^2} = \dfrac{1.1181485}{6} = 0.1863581$

$s_f^2 = 0.1863581 - 0.1108281 = 0.0755300$

$m = \dfrac{2.0290757 - 1.2084578}{0.0755300} = 10.864794$

$b = 3.63 - (10.864794)(0.3329085) = 0.0130180, \quad y = \dfrac{10.9}{x}$

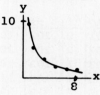

5.

t	y	f(t)=t²	fy	f²
1.0	6.0	1.0	6	1
2.0	23.0	4.0	92	16
3.0	55.0	9.0	495	81
4.0	98.0	16.0	1568	256
5.0	148.0	25.0	3700	625
	330.0	55.0	5861	979

$\bar{f} = \dfrac{55.0}{5} = 11.0, \quad \bar{f}^2 = 121$

$\bar{y} = \dfrac{330.0}{5} = 66.0, \quad \bar{f}\,\bar{y} = 726$

$\overline{fy} = \dfrac{5861}{5} = 1172.2$

$\overline{f^2} = \dfrac{979}{5} = 195.8$

$s_f^2 = 195.8 - 121 = 74.8, \quad m = \dfrac{1172.2 - 726}{74.8} = 5.9652406$

$b = 66.0 - (5.9652406)(11.0) = 0.3823530, \quad y = 5.97t^2 + 0.38$

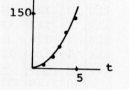

7.

x	y	f(x)=10^x/1000	fy	f²
50	1.00	1.1220185	1.1220185	1.2589254
100	4.40	1.2589254	5.5392718	1.5848932
150	9.40	1.4125375	13.277853	1.9952623
200	16.4	1.5848932	25.992248	2.5118864
250	24.0	1.7782794	42.678706	3.1622777
	55.2	7.1566540	88.610973	10.5132450

$\bar{f} = \dfrac{7.1566540}{5} = 1.4313308$

$\bar{f}^2 = 2.0487079, \quad \bar{y} = \dfrac{55.2}{5} = 11.04$

$\bar{f}\,\bar{y} = 15.801892$

$\overline{fy} = \dfrac{88.6100973}{5} = 17.722020$

$\overline{f^2} = \dfrac{10.5132450}{5} = 2.1026490$

$s_f^2 = 2.1026490 - 2.0487079 = 0.0539411$

$m = \dfrac{17.722020 - 15.801892}{0.0539411} = 35.596719$

$b = 11.04 - (35.596719)(1.4313308) = -39.910680$

$y = 35.6(10^z) - 39.9$

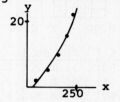

9.

L	f	F(L)=1/√L	Ff	F²
1.0	490	1.0000000	490.00000	1.0000000
2.0	360	0.7071068	254.55844	0.5000000
4.0	250	0.5000000	125.00000	0.2500000
6.0	200	0.4082483	81.649658	0.1666667
9.0	170	0.3333333	56.666667	0.1111111
	1470	2.9486884	1007.874765	2.0277778

$\bar{F} = \dfrac{2.9486884}{5} = 0.5897377$

$\bar{F}^2 = 0.3477905, \quad \bar{f} = \dfrac{1470}{5} = 294$

$\bar{F}\,\bar{f} = 173.38288$

$\overline{Ff} = \dfrac{1007.874765}{5} = 201.57495, \quad \overline{F^2} = \dfrac{2.0277778}{5} = 0.4055556$

$s_F^2 = 0.4055556 - 0.3477905 = 0.0577651, \quad m = 488.04750$

$b = 294 - (488.04750)(0.5897377) = 6.1800013$

$f = \dfrac{488}{\sqrt{L}} + 6$

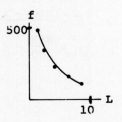

Review Exercises for Chapter 12, p. 439

1.

No.	2.3	2.4	2.5	2.6	2.7	2.8	2.9
Freq.	1	3	2	4	1	2	2

3. $\bar{x} = \dfrac{2.3 + 3(2.4) + 2(2.5) + 4(2.6) + 2.7 + 2(2.8) + 2(2.9)}{15} = \dfrac{39.0}{15} = 2.6$

5.

Int.	101–103	104–106	107–109	110–112	113–115
Freq.	5	4	3	3	5

7.

x	f	fx	fx^2
101	2	202	20402
102	2	204	20808
103	1	103	10609
105	1	105	11025
106	3	318	33708
107	1	107	11449
108	1	108	11664
109	1	109	11881
110	2	220	24200
112	1	112	12544
113	2	226	25538
114	1	114	12996
115	2	230	26450
	20	2158	233274

$\bar{x} = \dfrac{2158}{20} = 107.9$

$\bar{x}^2 = 11642.41$

$\overline{x^2} = \dfrac{233274}{20}$

$= 11663.7$

$s = \sqrt{11663.7 - 11642.41}$

$= 4.6$

9.

Visc.(Pa·s)	0.24	0.25	0.26
Freq.	1	2	4

Visc.(Pa·s)	0.27	0.28	0.29
Freq.	2	2	1

11. $\bar{v} = \dfrac{0.24 + 2(0.25) + 4(0.26) + 2(0.27) + 2(0.28) + 0.29}{12} = \dfrac{3.17}{12} = 0.264 \text{ Pa·s}$

13.

15.

t	f	ft	ft^2
0.90	3	2.70	2.4300
0.91	9	8.19	7.4529
0.92	31	28.52	26.2384
0.93	38	35.34	32.8662
0.94	12	11.28	10.6032
0.95	5	4.75	4.5125
0.96	2	1.92	1.8432
	100	92.70	85.9464

$\bar{t} = \dfrac{92.70}{100} = 0.927, \quad \bar{t}^2 = 0.859329$

$\overline{t^2} = \dfrac{85.9464}{100} = 0.859464$

$s_t = \sqrt{0.859464 - 0.859329} = 0.012 \text{ in.}$

17. $\bar{c} = \dfrac{3(0) + 10(1) + 25(2) + 45(3) + 29(4) + 39(5) + 26(6) + 11(7) + 7(8) + 2(9) + 3(10)}{200}$

$= \dfrac{843}{200} = 4.2 \text{ counts}$

19.

T	R	TR	T^2
0.0	25.0	0	0
20.0	26.8	536	400
40.0	28.9	1156	1600
60.0	31.2	1872	3600
80.0	32.8	2624	6400
100	34.7	3470	10000
300.0	179.4	9658	22000

$\overline{T} = \dfrac{300.0}{6} = 50.0$, $\overline{T}^2 = 2500$

$\overline{R} = \dfrac{179.4}{6} = 29.9$, $\overline{T}\,\overline{R} = 1495$

$\overline{TR} = \dfrac{9658}{6} = 1609.6667$

$\overline{T^2} = \dfrac{22000}{6} = 3666.6667$

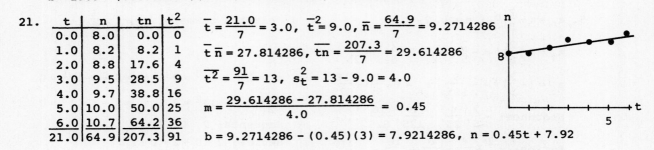

$s_T^2 = 3666.6667 - 2500 = 1166.6667$, $m = \dfrac{1609.6667 - 1495}{1166.6667} = 0.0982857$

$b = 29.9 - (0.0982857)(50.0) = 24.985714$, $R = 0.0983T + 25.0$

21.

t	n	tn	t^2
0.0	8.0	0.0	0
1.0	8.2	8.2	1
2.0	8.8	17.6	4
3.0	9.5	28.5	9
4.0	9.7	38.8	16
5.0	10.0	50.0	25
6.0	10.7	64.2	36
21.0	64.9	207.3	91

$\overline{t} = \dfrac{21.0}{7} = 3.0$, $\overline{t}^2 = 9.0$, $\overline{n} = \dfrac{64.9}{7} = 9.2714286$

$\overline{t}\,\overline{n} = 27.814286$, $\overline{tn} = \dfrac{207.3}{7} = 29.614286$

$\overline{t^2} = \dfrac{91}{7} = 13$, $s_t^2 = 13 - 9.0 = 4.0$

$m = \dfrac{29.614286 - 27.814286}{4.0} = 0.45$

$b = 9.2714286 - (0.45)(3) = 7.9214286$, $n = 0.45t + 7.92$

23.

x	y	f(x)=tan x	fy	f
10.0	0.16	0.1763270	0.0282123	0.0310912
20.0	0.34	0.3639702	0.1237499	0.1324743
30.0	0.55	0.5773503	0.3175426	0.3333333
40.0	0.85	0.8390996	0.7132347	0.7040882
50.0	1.24	1.1917536	1.4777745	1.4202766
60.0	1.82	1.7320508	3.1523325	3.0000000
70.0	2.80	2.7474774	7.6929368	7.5486322
	7.76	7.6280289	13.5057833	13.1698958

$\overline{f} = \dfrac{7.6280289}{7} = 1.0897184$

$\overline{f}\,^2 = 1.1874862$, $\overline{y} = \dfrac{7.76}{7} = 1.1085714$

$\overline{f}\,\overline{y} = 1.2080307$

$\overline{fy} = \dfrac{13.5057833}{7} = 1.9293976$

$\overline{f^2} = \dfrac{13.1698958}{7} = 1.8814137$

$s_f = 1.8814137 - 1.1874862 = 0.6939275$, $m = \dfrac{1.9293976 - 1.2080307}{0.6939275} = 1.0395422$

$b = 1.1085714 - (1.0395422)(1.0897184) = -0.0242369$, $y = 1.04 \tan x - 0.02$

25.

x	y	f(x)=√x	fy	f^2
1.00	1.10	1.0000000	1.1000000	1.00
3.00	1.90	1.7320508	3.2908965	3.00
5.00	2.50	2.2360680	5.5901699	5.00
7.00	2.90	2.6457513	7.6726788	7.00
9.00	3.30	3.0000000	9.9000000	9.00
	11.70	10.6138701	27.5537452	25.00

$\overline{f} = \dfrac{10.6138701}{5} = 2.1227740$

$\overline{f}\,^2 = 4.5061695$

$\overline{y} = \dfrac{11.70}{5} = 2.34$, $\overline{f}\,\overline{y} = 4.9672912$

$\overline{fy} = \dfrac{27.5537452}{5} = 5.5107491$

$\overline{f^2} = \dfrac{25.00}{5} = 5.00$, $s_f^2 = 5.00 - 4.5061695 = 0.4938305$

$m = \dfrac{5.5107491 - 4.9672912}{0.4938305} = 1.1004948$, $b = 2.34 - (1.1004948)(2.1227740)$

$\quad = 0.0038983$

$y = 1.10 \sqrt{x}$

Exercises 13-1 , p. 446

1. $a_n = n^2$, $n = 1, 2, 3, 4$
$a_1 = 1^2 = 1$; $a_2 = 2^2 = 4$
$a_3 = 3^2 = 9$; $a_4 = 4^2 = 16$
First 4 terms: 1, 4, 9, 16

3. $a_n = \dfrac{1}{n+2}$, $n = 0, 1, 2, 3$
$a_0 = \dfrac{1}{0+2} = \dfrac{1}{2}$; $a_1 = \dfrac{1}{1+2} = \dfrac{1}{3}$
$a_2 = \dfrac{1}{2+2} = \dfrac{1}{4}$; $a_3 = \dfrac{1}{3+2} = \dfrac{1}{5}$
First 4 terms: $\dfrac{1}{2}$, $\dfrac{1}{3}$, $\dfrac{1}{4}$, $\dfrac{1}{5}$

5. $a_n = (-\frac{2}{5})^n$; $n = 1, 2, 3, 4$
$a_1 = (-\frac{2}{5})^1 = -\frac{2}{5}$; $a_2 = (-\frac{2}{5})^2 = \frac{4}{25}$
$a_3 = (-\frac{2}{5})^3 = -\frac{8}{125}$; $a_4 = (-\frac{2}{5})^4 = \frac{16}{625}$
Sequence: $-\dfrac{2}{5}, \dfrac{4}{25}, -\dfrac{8}{125}, \dfrac{16}{625}, \cdots$
Series: $-\dfrac{2}{5} + \dfrac{4}{25} - \dfrac{8}{125} + \dfrac{16}{625} - \cdots$

7. $a_n = 1 + (-1)^n$; $n = 0, 1, 2, 3$
$a_0 = 1 + (-1)^0 = 2$; $a_1 = 1 + (-1)^1 = 0$
$a_2 = 1 + (-1)^2 = 2$; $a_3 = 1 + (-1)^3 = 0$
Sequence: 2, 0, 2, 0, ...
Series: 2 + 0 + 2 + 0 + ...

9. $\dfrac{1}{2} + \dfrac{1}{3} + \dfrac{1}{4} + \dfrac{1}{5} + \ldots$; $n = 1, 2, 3, 4$
Numerators are 1.
Denominators are n+1.
Therefore, $a_n = \dfrac{1}{n+1}$.

11. $\dfrac{1}{2\cdot3} + \dfrac{1}{3\cdot4} + \dfrac{1}{4\cdot5} + \dfrac{1}{5\cdot6} + \ldots$; $n = 1, 2, 3, 4$
Numerators are 1.
Denominators are (n+1)(n+2).
Therefore, $a_n = \dfrac{1}{(n+1)(n+2)}$.

13. $1 + \dfrac{1}{8} + \dfrac{1}{27} + \dfrac{1}{64} + \dfrac{1}{125} + \ldots$
$S_1 = 1$; $S_2 = 1 + \dfrac{1}{8} = 1.125$
$S_3 = 1 + \dfrac{1}{8} + \dfrac{1}{27} = 1.1620370$
$S_4 = 1 + \dfrac{1}{8} + \dfrac{1}{27} + \dfrac{1}{64} = 1.1776620$
$S_5 = 1 + \dfrac{1}{8} + \dfrac{1}{27} + \dfrac{1}{64} + \dfrac{1}{125} = 1.1856620$
Values appear to approach 1.2.
Convergent to approx. 1.2.

15. $1 + \dfrac{1}{2} + \dfrac{2}{3} + \dfrac{3}{4} + \dfrac{4}{5} + \ldots$
$S_1 = 1$; $S_2 = 1 + \dfrac{1}{2} = 1.5$
$S_3 = 1 + \dfrac{1}{2} + \dfrac{2}{3} = 2.1666667$
$S_4 = 1 + \dfrac{1}{2} + \dfrac{2}{3} + \dfrac{3}{4} = 2.9166667$
$S_5 = 1 + \dfrac{1}{2} + \dfrac{2}{3} + \dfrac{3}{4} + \dfrac{4}{5} = 3.7166667$
Values become increasingly larger.
Divergent.

17. $\sum\limits_{n=0}^{\infty} (-n) = -0 + (-1) + (-2) + (-3) + (-4) + \ldots = 0 - 1 - 2 - 3 - 4 - \ldots$
$S_1 = 0$; $S_2 = 0 - 1 = -1$; $S_3 = 0 - 1 - 2 = -3$; $S_4 = 0 - 1 - 2 - 3 = -6$
$S_5 = 0 - 1 - 2 - 3 - 4 = -10$; Values become increasingly more negative.
Divergent.

19. $\sum\limits_{n=1}^{\infty} \dfrac{2n+1}{n^2(n+1)^2} = \dfrac{3}{4} + \dfrac{5}{36} + \dfrac{7}{144} + \dfrac{9}{400} + \dfrac{11}{900} + \ldots$

$S_1 = \dfrac{3}{4} = 0.75$; $S_2 = \dfrac{3}{4} + \dfrac{5}{36} = 0.8888889$; $S_3 = \dfrac{3}{4} + \dfrac{5}{36} + \dfrac{7}{144} = 0.9375000$

$S_4 = \dfrac{3}{4} + \dfrac{5}{36} + \dfrac{7}{144} + \dfrac{9}{400} = 0.9600000$; $S_5 = \dfrac{3}{4} + \dfrac{5}{36} + \dfrac{7}{144} + \dfrac{9}{400} + \dfrac{11}{900} = 0.9722222$

Values appear to approach 1. Convergent to approximately 1.

21. $1 + 2 + 4 + \ldots + 2^n + \ldots$

Geometric series: $r = 2$

Since $r > 1$, divergent

23. $1 - \dfrac{1}{3} + \dfrac{1}{9} - \ldots + (-\dfrac{1}{3})^n + \ldots$

Geometric series: $r = -\dfrac{1}{3}$

$S = \dfrac{1}{1 - (-\dfrac{1}{3})} = \dfrac{1}{1 + \dfrac{1}{3}} = \dfrac{3}{4}$

25. $10 + 9 + 8.1 + 7.29 + 6.561 + \ldots$

Geometric series:

$r = 0.9$; $S = \dfrac{10}{1 - 0.9} = 100$

27. $512 - 64 + 8 - 1 + \dfrac{1}{8} - \ldots$

Geometric series: $r = -\dfrac{1}{8}$

$S = \dfrac{512}{1 - (-\dfrac{1}{8})} = \dfrac{512}{\dfrac{9}{8}} = \dfrac{4096}{9}$

29. $0.2222\ldots = 0.2 + 0.02 + 0.002 + 0.0002 + \ldots$

Geometric series: $r = 0.1$

$S = \dfrac{0.2}{1 - 0.1} = \dfrac{2}{9}$

31. $0.181818\ldots = 0.18 + 0.0018 + 0.000018 + \ldots$

Geometric series: $r = 0.01$; $S = \dfrac{0.18}{1 - 0.01} = \dfrac{18}{99} = \dfrac{2}{11}$

33. Distance $= 8.00 + 4.00 + 4.00 + 2.00 + 2.00 + 1.00 + 1.00 + \ldots$

$= 8.00 + 8.00 + 4.00 + 2.00 + \ldots = 8.00 + \dfrac{8.00}{1 - 0.5} = 8.00 + 16.00 = 24.0$ ft

35. Distance $= 90 + 60 + 40 + \ldots = \dfrac{90}{1 - \dfrac{2}{3}} = \dfrac{90}{\dfrac{1}{3}} = 270$ cm

37. $\sum\limits_{n=0}^{\infty} x^n = 1 + x + x^2 + x^3 + \ldots$; $|x| < 1$; Geometric series, $r = x$; $S = \dfrac{1}{1 - x}$

Therefore, $\sum\limits_{n=0}^{\infty} x^n = \dfrac{1}{1 - x}$, $|x| < 1$.

Exercises 13-2 , p. 452

1. $f(x) = e^x \qquad f(0) = 1$

$f'(x) = e^x \qquad f'(0) = 1$

$f''(x) = e^x \qquad f''(0) = 1$

$f(x) = 1 + (1)x + (1)\dfrac{x^2}{2!} + \ldots$

$e^x = 1 + x + \dfrac{1}{2}x^2 + \ldots$

3. $f(x) = \cos x \qquad f(0) = \cos 0 = 1$

$f'(x) = -\sin x \qquad f'(0) = -\sin 0 = 0$

$f''(x) = -\cos x \qquad f''(0) = -1$

$f'''(x) = \sin x \qquad f'''(0) = 0$

$f^{iv}(x) = \cos x \qquad f^{iv}(0) = 1$

$f(x) = 1 + (0)x + (-1)\dfrac{x^2}{2!} + (0)\dfrac{x^3}{3!} + (1)\dfrac{x^4}{4!} + \ldots$

$\cos x = 1 - \dfrac{1}{2}x^2 + \dfrac{1}{24}x^4 - \ldots$

5. $f(x) = \sqrt{1+x} = (1+x)^{1/2}, f(0) = 1$

$f'(x) = \dfrac{1}{2}(1+x)^{-1/2} \qquad f'(0) = \dfrac{1}{2}$

$f''(x) = -\dfrac{1}{4}(1+x)^{-3/2} \qquad f''(0) = -\dfrac{1}{4}$

$f(x) = 1 + (\dfrac{1}{2})x + (-\dfrac{1}{4})\dfrac{x^2}{2!} + \ldots$; $\sqrt{1+x} = 1 + \dfrac{1}{2}x - \dfrac{1}{8}x^2 + \ldots$

7. $f(x) = e^{-2x}$ $f(0) = 1$

   $f'(x) = -2e^{-2x}$ $f'(0) = -2$

   $f''(x) = 4e^{-2x}$ $f''(0) = 4$

   $f(x) = 1 + (-2)x + (4)\dfrac{x^2}{2!} + \ldots$

   $e^{-2x} = 1 - 2x + 2x^2 - \ldots$

11. $f(x) = \dfrac{1}{1-x} = (1-x)^{-1}$; $f(0)=1$

   $f'(x) = -(1-x)^{-2}(-1) = (1-x)^{-2}$

   $f'(0) = 1$

   $f''(x) = -2(1-x)^{-3}(-1) = 2(1-x)^{-3}$

   $f''(0) = 2$

   $f(x) = 1 + (1)x + (2)\dfrac{x^2}{2!} + \ldots$

   $\dfrac{1}{1-x} = 1 + x + x^2 + \ldots$

15. $f(x) = \cos \dfrac{1}{2}x$ $f(0) = 1$

   $f'(x) = -\dfrac{1}{2}\sin \dfrac{1}{2}x$ $f'(0) = 0$

   $f''(x) = -\dfrac{1}{4}\cos \dfrac{1}{2}x$ $f''(0) = -\dfrac{1}{4}$

   $f'''(x) = \dfrac{1}{8}\sin \dfrac{1}{2}x$ $f'''(0) = 0$

   $f^{iv}(x) = \dfrac{1}{16}\cos \dfrac{1}{2}x$ $f^{iv}(0) = \dfrac{1}{16}$

   $f(x) = 1 + (0)x + (-\dfrac{1}{4})\dfrac{x^2}{2!} + (0)\dfrac{x^3}{3!} + (\dfrac{1}{16})\dfrac{x^4}{4!} + \ldots$

   $\cos \dfrac{1}{2}x = 1 - \dfrac{1}{8}x^2 + \dfrac{1}{384}x^4 - \ldots$

19. $f(x) = \tan x$ $f(0) = 0$

   $f'(x) = \sec^2 x$ $f'(0) = 1$

   $f''(x) = 2\sec^2 x \tan x$ $f''(0) = 0$

   $f'''(x) = 4\sec^2 x \tan^2 x + 2\sec^4 x$

   $f'''(0) = 2$

   $f(x) = 0 + (1)x + (0)\dfrac{x^2}{2!} + (2)\dfrac{x^3}{3!} + \ldots$

   $\tan x = x + \dfrac{1}{3}x^3 + \ldots$

23. $f(x) = \sin^2 x$ $f(0) = 0$

   $f'(x) = 2\sin x \cos x = \sin 2x$ $f'(0)=0$

   $f''(x) = 2\cos 2x$ $f''(0) = 2$

   $f'''(x) = -4\sin 2x$ $f'''(0) = 0$

   $f^{iv}(x) = -8\cos 2x$ $f^{iv}(0) = -8$

   $f(x) = 0 + (0)x + (2)\dfrac{x^2}{2!} + (0)\dfrac{x^3}{3!} + (-8)\dfrac{x^4}{4!} + \ldots$

   $\sin^2 x = x^2 - \dfrac{1}{3}x^4 + \ldots$

9. $f(x) = \cos 4x$ $f(0) = 1$

   $f'(x) = -4\sin 4x$ $f'(0) = 0$

   $f''(x) = -16\cos 4x$ $f''(0) = -16$

   $f'''(x) = 64\sin 4x$ $f'''(0) = 0$

   $f^{iv}(x) = 256\cos 4x$ $f^{iv}(0) = 256$

   $f(x) = 1 + (0)x + (-16)\dfrac{x^2}{2!} + (0)\dfrac{x^3}{3!} + (256)\dfrac{x^4}{4!} + \ldots$

   $\cos 4x = 1 - 8x^2 + \dfrac{32}{3}x^4 - \ldots$

13. $f(x) = \ln(1 - 2x)$ $f(0) = \ln 1 = 0$

   $f'(x) = \dfrac{-2}{1 - 2x}$ $f'(0) = -2$

   $f''(x) = \dfrac{-4}{(1 - 2x)^2}$ $f''(0) = -4$

   $f'''(x) = \dfrac{-16}{(1 - 2x)^3}$ $f'''(0) = -16$

   $f(x) = 0 + (-2)x + (-4)\dfrac{x^2}{2!} + (-16)\dfrac{x^3}{3!} + \ldots$

   $\ln(1-2x) = -2x - 2x^2 - \dfrac{8}{3}x^3 + \ldots$

17. $f(x) = \text{Arctan } x$ $f(0) = 0$

   $f'(x) = \dfrac{1}{1 + x^2}$ $f'(0) = 1$

   $f''(x) = \dfrac{-2x}{(1 + x^2)^2}$ $f''(0) = 0$

   $f'''(x) = \dfrac{(1+x^2)^2(-2) - (-2x)(2)(1+x^2)(2x)}{(1+x^2)^4}$

   $= \dfrac{-2 + 6x^2}{(1+x^2)^3}$ $f'''(0) = -2$

   $f(x) = 0 + (1)x + (0)\dfrac{x^2}{2!} + (-2)\dfrac{x^3}{3!} + \ldots$

   $\text{Arctan } x = x - \dfrac{1}{3}x^3 + \ldots$

21. $f(x) = \ln \cos x$ $f(0) = \ln 1 = 0$

   $f'(x) = -\tan x$ $f'(0) = 0$

   $f''(x) = -\sec^2 x$ $f''(0) = -1$

   $f'''(x) = -2\sec^2 x \tan x$ $f'''(0) = 0$

   $f^{iv}(x) = -4\sec^2 x \tan^2 x - 2\sec^4 x$

   $f^{iv}(0) = -2$

   $f(x) = 0 + (0)x + (-1)\dfrac{x^2}{2!} + (0)\dfrac{x^3}{3!} + (-2)\dfrac{x^4}{4!} + \ldots$

   $\ln \cos x = -\dfrac{1}{2}x^2 - \dfrac{1}{12}x^4 - \ldots$

25.(a) $f(x) = \csc x$; $f(0)$ not defined

   (b) $f(x) = \sqrt{x}$; $f'(x) = \dfrac{1}{2\sqrt{x}}$;
       $f'(0)$ not defined

   (c) $f(x) = \ln x$; $f(0)$ not defined

27. $f(x) = x^3$; $f(0) = 0$ $f''(x) = 6x$; $f''(0) = 0$ $f^{iv}(x) = 0$; $f^{iv}(0) = 0$

$f'(x) = 3x^2$; $f'(0) = 0$ $f'''(x) = 6$; $f'''(0) = 6$ $f^{(n)}(x) = 0$, $n \geq 4$

$f(x) = 0 + (0)x + (0)\dfrac{x^2}{2!} + (6)\dfrac{x^3}{3!} + 0$; $x^3 = x^3$

Exercises 13-3, p. 458

1. $F(x) = e^x = 1 + x + \dfrac{1}{2}x^2 + \dfrac{1}{6}x^3 + \dots$

   $F(3x) = e^{3x} = 1 + 3x + \dfrac{1}{2}(3x)^2 + \dfrac{1}{6}(3x)^3 + \dots$

   $f(x) = e^{3x} = 1 + 3x + \dfrac{9}{2}x^2 + \dfrac{9}{2}x^3 + \dots$

3. $F(x) = \sin x = x - \dfrac{1}{3!}x^3 + \dfrac{1}{5!}x^5 - \dfrac{1}{7!}x^7 + \dots$

   $F(\tfrac{1}{2}x) = \sin \dfrac{1}{2}x = \dfrac{1}{2}x - \dfrac{1}{3!}(\tfrac{1}{2}x)^3 + \dfrac{1}{5!}(\tfrac{1}{2}x)^5$
   $\qquad\qquad - \dfrac{1}{7!}(\tfrac{1}{2}x)^7 + \dots$

   $f(x) = \sin \dfrac{1}{2}x = \dfrac{x}{2} - \dfrac{x^3}{3!2^3} + \dfrac{x^5}{5!2^5} - \dfrac{x^7}{7!2^7} + \dots$

5. $F(x) = \cos x = 1 - \dfrac{1}{2}x^2 + \dfrac{1}{24}x^4 - \dfrac{1}{720}x^6 + \dots$

   $F(4x) = \cos 4x = 1 - \dfrac{1}{2}(4x)^2 + \dfrac{1}{24}(4x)^4$
   $\qquad\qquad - \dfrac{1}{720}(4x)^6 + \dots$

   $f(x) = \cos 4x = 1 - 8x^2 + \dfrac{32}{3}x^4 - \dfrac{256}{45}x^6 + \dots$

7. $F(x) = \ln(1+x) = x - \dfrac{1}{2}x^2 + \dfrac{1}{3}x^3 - \dfrac{1}{4}x^4 + \dots$

   $F(x^2) = \ln(1+x^2) = x^2 - \dfrac{1}{2}(x^2)^2 - \dfrac{1}{3}(x^2)^3$
   $\qquad\qquad - \dfrac{1}{4}(x^2)^4 + \dots$

   $f(x) = \ln(1+x^2) = x^2 - \dfrac{1}{2}x^4 + \dfrac{1}{3}x^6 - \dfrac{1}{4}x^8 + \dots$

9. $f(x) = \sin x = x - \dfrac{1}{6}x^3 + \dfrac{1}{120}x^5 - \dots$

   $f(x^2) = \sin x^2 = x^2 - \dfrac{1}{6}x^6 + \dfrac{1}{120}x^{10} + \dots$

   $\displaystyle\int_0^1 \sin x^2\,dx = \int_0^1 (x^2 - \dfrac{1}{6}x^6 + \dfrac{1}{120}x^{10})\,dx$

   $\qquad = \dfrac{1}{3}x^3 - \dfrac{1}{42}x^7 + \dfrac{1}{1320}x^{11}\Big|_0^1$

   $\qquad = \dfrac{1}{3} - \dfrac{1}{42} + \dfrac{1}{1320} - 0 = 0.3103$

11. $f(x) = \cos x = 1 - \dfrac{1}{2}x^2 + \dfrac{1}{24}x^4 - \dots$

    $f(\sqrt{x}) = \cos \sqrt{x} = 1 - \dfrac{1}{2}x + \dfrac{1}{24}x^2 - \dots$

    $\displaystyle\int_0^{0.2} \cos \sqrt{x}\,dx = \int_0^{0.2} (1 - \dfrac{1}{2}x + \dfrac{1}{24}x^2)\,dx$

    $\qquad = x - \dfrac{1}{4}x^2 + \dfrac{1}{72}x^3\Big|_0^{0.2} = 0.2 - \dfrac{(0.2)^2}{4} + \dfrac{(0.2)^3}{72} - 0$

    $\qquad = 0.1901$

13. $3e^{0.5j} = 3(\cos 0.5 + j\sin 0.5) = 2.63 + 1.44j$ (0.5 is radian measure)

15. $6 + j$; $a = 6$, $b = 1$

    $r = \sqrt{6^2 + 1^2} = \sqrt{37} = 6.08$

    $\tan\theta = \dfrac{1}{6} = 0.1667$; $\theta = 9.5°$

    (θ in 1st qusd., $a>0, b>0$)

    $6 + j = 6.08(\cos 9.5° + j\sin 9.5°)$

17. $e^x = 1 + x + \dfrac{1}{2}x^2 + \dfrac{1}{6}x^3 + \dfrac{1}{24}x^4 + \dfrac{1}{120}x^5 + \dfrac{1}{720}x^6 + \dots$

    $e^{-x} = 1 - x + \dfrac{1}{2}x^2 - \dfrac{1}{6}x^3 + \dfrac{1}{24}x^4 - \dfrac{1}{120}x^5 + \dfrac{1}{720}x^6 - \dots$

    $e^x + e^{-x} = 2 + x^2 + \dfrac{1}{12}x^4 + \dfrac{1}{360}x^6 + \dots$

    $\dfrac{1}{2}(e^x + e^{-x}) = 1 + \dfrac{1}{2}x^2 + \dfrac{1}{24}x^4 + \dfrac{1}{720}x^6 + \dots$

19. $e^x \sin x = (1 + x + \dfrac{1}{2}x^2 + \dots)(x - \dfrac{1}{6}x^3 + \dots)$

    $\qquad = x - \dfrac{1}{6}x^3 + \dots + x^2 + \dots + \dfrac{1}{2}x^3 + \dots$

    $\qquad = x + x^2 + \dfrac{1}{3}x^3 + \dots$

21. $\sin x = x - \dfrac{1}{6}x^3 + \dfrac{1}{120}x^5 - \dots$

    $\dfrac{d\sin x}{dx} = 1 - \dfrac{1}{2}x^2 + \dfrac{1}{24}x^4 - \dots$

    $\qquad = \cos x$

23. $\cos x = 1 - \frac{1}{2}x^2 + \frac{1}{24}x^4 - \ldots$

$\int \cos x\, dx = \int (1 - \frac{1}{2}x^2 + \frac{1}{24}x^4 - \ldots)dx$

$= x - \frac{1}{6}x^3 + \frac{1}{120}x^5 - \ldots + C$

$= \sin x + C$

[C=0 if y=sin x passes thru (0,0)]

25.(a) $\int_0^1 e^x dx = e^x \Big|_0^1 = e - 1 = 1.7182818$

(b) $\int_0^1 (1 + x + \frac{1}{2}x^2 + \frac{1}{6}x^3)dx$

$= x + \frac{1}{2}x^2 + \frac{1}{6}x^3 + \frac{1}{24}x^4 \Big|_0^1$

$= 1 + \frac{1}{2} + \frac{1}{6} + \frac{1}{24} = 1.7083333$

27. $A = \int_0^{0.2} x^2 e^x dx = \int_0^{0.2} x^2(1 + x + \frac{1}{2}x^2)dx = \int_0^{0.2}(x^2 + x^3 + \frac{1}{2}x^4)dx$

$= \frac{1}{3}x^3 + \frac{1}{4}x^4 + \frac{1}{10}x^5 \Big|_0^{0.2} = \frac{1}{3}(0.2)^3 + \frac{1}{4}(0.2)^4 + \frac{1}{10}(0.2)^5 - 0$

$= 0.003099$

29. $V = 2\pi \int_0^{\pi/8} xy\, dx = 2\pi \int_0^{\pi/8} x \sin x\, dx = 2\pi \int_0^{\pi/8} x(x - \frac{1}{6}x^3)dx$

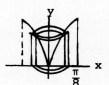

$= 2\pi \int_0^{\pi/8}(x^2 - \frac{1}{6}x^4)dx = 2\pi(\frac{1}{3}x^3 - \frac{1}{30}x^5) \Big|_0^{\pi/8}$

$= 2\pi[\frac{1}{3}(\frac{\pi}{8})^3 - \frac{1}{30}(\frac{\pi}{8})^5] - 0 = 0.1249$

31. $z = R + j(X_L - X_C) = |z|(\cos\theta + j\sin\theta)$

$R = |z|\cos\theta, \quad X_L - X_C = |z|\sin\theta$

$\frac{X_L - X_C}{R} = \frac{|z|\sin\theta}{|z|\cos\theta} = \tan\theta$

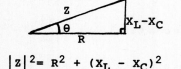

$|z|^2 = R^2 + (X_L - X_C)^2$

Exercises 13-4, p. 462

1. $e^x = 1 + x + \frac{1}{2}x^2 + \ldots$

$e^{0.2} = 1 + 0.2 + \frac{1}{2}(0.2)^2 = 1.22$

Calculator: $e^{0.2} = 1.2214028$

3. $\sin x = x - \frac{1}{6}x^3 + \ldots$

$\sin 0.1 = 0.1 - \frac{1}{6}(0.1)^3 = 0.0998333$

Calculator: $\sin 0.1 = 0.0998334$

5. $e^x = 1 + x + \frac{1}{2}x^2 + \frac{1}{6}x^3 + \frac{1}{24}x^4 + \frac{1}{120}x^5 + \frac{1}{720}x^6 + \ldots$

$e^1 = 1 + 1 + \frac{1}{2} + \frac{1}{6} + \frac{1}{24} + \frac{1}{120} + \frac{1}{720} = 2.7180556$

Calculator: $e = 2.7182818$

7. $\cos x = 1 - \frac{1}{2}x^2 + \ldots$

$\cos 3° = \cos\frac{\pi}{60} = 1 - \frac{1}{2}(\frac{\pi}{60})^2$

$= 0.9986292$

Calculator: $\cos 3° = 0.9986295$

9. $\ln(1+x) = x - \frac{1}{2}x^2 + \frac{1}{3}x^3 - \frac{1}{4}x^4 + \ldots$

$\ln 1.4 = 0.4 - \frac{1}{2}(0.4)^2 + \frac{1}{3}(0.4)^3$

$- \frac{1}{4}(0.4)^4 = 0.3349333$

Calculator: $\ln 1.4 = 0.3364722$

11. $\sin x = x - \frac{1}{6}x^3 + \frac{1}{120}x^5 - \ldots$

$\sin 0.3625 = 0.3625 - \frac{1}{6}(0.3625)^3 + \frac{1}{120}(0.3625)^5$

$= 0.3546130$

Calculator: $\sin 0.3625 = 0.3546129$

13. $\ln(1+x) = x - \frac{1}{2}x^2 + \frac{1}{3}x^3 - \ldots$

$\ln 0.9861 = \ln[1+(-0.0139)] = -0.0139 - \frac{1}{2}(-0.0139)^2 + \frac{1}{3}(-0.0139)^3$

$= -0.0139975$

Calculator: $\ln 0.9861 = -0.0139975$

15. $e^x = 1 + x + \frac{1}{2}x^2 + \frac{1}{6}x^3 + \ldots$

$e^{-0.3165} = 1 + (-0.3165) + \frac{1}{2}(-0.3165)^2$
$\qquad + \frac{1}{6}(-0.3165)^3 = 0.7283020$

Calculator: $e^{-0.3165} = 0.7286950$

17. $\sqrt{1+x} = (1+x)^{1/2} = 1 + \frac{1}{2}x - \frac{1}{8}x^2 + \ldots$

[Binomial expansion, $n = \frac{1}{2}$, see Eq.(13-10)]

$\sqrt{1.1076} = 1 + \frac{1}{2}(0.1076) - \frac{1}{8}(0.1076)^2$
$\qquad = 1.0523528$

19. $\sqrt[3]{1+x} = (1+x)^{1/3} = 1 + \frac{1}{3}x - \frac{1}{9}x^2 + \ldots$

[Binomial expansion, $n = \frac{1}{3}$, see Eq.(13-10)]

$\sqrt[3]{0.9628} = 1 + \frac{1}{3}(-0.0372) - \frac{1}{9}(-0.0372)^2$
$\qquad = 0.9874462$

21. First omitted term is $\frac{1}{120}x^5$

Maximum error $= \frac{1}{120}(0.1)^5 = 8.3 \times 10^{-8}$

23. First omitted term is $\frac{1}{24}x^4$

Maximum error $= \frac{1}{24}(\frac{\pi}{60})^4 = 3.1 \times 10^{-7}$

25. $i = \frac{E}{R}(1 - e^{-Rt/L})$

$e^{-Rt/L} = 1 + (-\frac{Rt}{L}) + \frac{1}{2}(-\frac{Rt}{L})^2$
$\qquad = 1 - \frac{Rt}{L} + \frac{R^2 t^2}{2L^2}$

$i = \frac{E}{R}[1 - (1 - \frac{Rt}{L} + \frac{R^2 t^2}{2L^2})]$
$\quad = \frac{E}{L}(t - \frac{Rt^2}{2L})$ Valid for small values of t.

27. In Example F, see derivation of

$x = 2000\,\theta^2$

$\theta = \tan\theta = \frac{10}{4000} = \frac{1}{400}$

$x = 2000(\frac{1}{400})^2$

$\quad = 0.0125\text{ mi} = 66\text{ ft}$

29. Arctan $x = x - \frac{1}{3}x^3 + \frac{1}{5}x^5 - \ldots$

$\frac{1}{4}\pi = $ Arctan $\frac{1}{2} + $ Arctan $\frac{1}{3} = [\frac{1}{2} - \frac{1}{3}(\frac{1}{2})^3 + \frac{1}{5}(\frac{1}{2})^5] + [\frac{1}{3} - \frac{1}{3}(\frac{1}{3})^3 + \frac{1}{5}(\frac{1}{3})^5] = 0.7863940$

$\pi = 3.146$

Exercises 13-5, p. 465

1. From Example A: $e^x = e[1 + (x-1) + \frac{(x-1)^2}{2} + \frac{(x-1)^3}{6} + \ldots]$; $e = 2.7183$, $x = 1.2$, $x - 1 = 0.2$

$e^{1.2} = 2.7183[1 + 0.2 + \frac{(0.2)^2}{2} + \frac{(0.2)^3}{6}] = 3.320$

3. From Example B: $\sqrt{x} = 2 + \frac{x-4}{4} - \frac{(x-4)^2}{64} + \frac{(x-4)^3}{512} - \ldots$; $x = 4.2$, $x - 4 = 0.2$

$\sqrt{4.2} = 2 + \frac{0.2}{4} - \frac{(0.2)^2}{64} + \frac{(0.2)^3}{512} = 2.049$

5. From Example D: $\sin x = \frac{1}{2} + \frac{1}{2}\sqrt{3}(x - \frac{\pi}{6}) - \frac{1}{4}(x - \frac{\pi}{6})^2 - \ldots$; $x = 31°$, $x - \frac{\pi}{6} = 1° = \frac{\pi}{180}$

$\sin 31° = \frac{1}{2} + \frac{1}{2}\sqrt{3}(\frac{\pi}{180}) - \frac{1}{4}(\frac{\pi}{180})^2 = 0.5150$

7. See solution for Exercise 5 for series: $x = 29.53°$, $x - \frac{\pi}{6} = -0.47° = -0.0082030$

$\sin 29.53° = \frac{1}{2} + \frac{1}{2}\sqrt{3}(-0.0082030) - \frac{1}{4}(-0.0082030)^2 = 0.49288$

9. $f(x) = e^{-x}$ $f(2) = e^{-2}$

   $f'(x) = -e^{-x}$ $f'(2) = -e^{-2}$

   $f''(x) = e^{-x}$ $f''(x) = e^{-2}$

   $f(x) = e^{-2} + (-e^{-2})(x-2) + (e^{-2})\dfrac{(x-2)^2}{2!} + \ldots$

   $e^{-x} = e^{-2}[1 + (x-2) + \dfrac{(x-2)^2}{2!} - \ldots]$

11. $f(x) = \sin x$ $f(\frac{\pi}{3}) = \frac{1}{2}\sqrt{3}$

    $f'(x) = \cos x$ $f'(\frac{\pi}{3}) = \frac{1}{2}$

    $f''(x) = -\sin x$ $f''(\frac{\pi}{3}) = -\frac{1}{2}\sqrt{3}$

    $f(x) = \frac{1}{2}\sqrt{3} + \frac{1}{2}(x - \frac{\pi}{3}) + (-\frac{1}{2}\sqrt{3})\dfrac{(x - \frac{\pi}{3})^2}{2!} - \ldots$

    $\sin x = \frac{1}{2}[\sqrt{3} + (x - \frac{\pi}{3}) - \frac{\sqrt{3}}{2!}(x - \frac{\pi}{3})^2 - \ldots]$

13. $f(x) = \sqrt[3]{x} = x^{1/3}$ $f(8) = 2$

   $f'(x) = \frac{1}{3}x^{-2/3}$ $f'(8) = \frac{1}{12}$

   $f''(x) = -\frac{2}{9}x^{-5/3}$ $f''(x) = -\dfrac{1}{144}$

   $f(x) = 2 + \frac{1}{12}(x-8) - \dfrac{1}{144}\dfrac{(x-8)^2}{2!} + \ldots$

   $\sqrt[3]{x} = 2 + \frac{1}{12}(x-8) - \frac{1}{288}(x-8)^2 + \ldots$

15. $f(x) = \tan x$ $f(\frac{\pi}{4}) = 1$

    $f'(x) = \sec^2 x$ $f'(\frac{\pi}{4}) = 2$

    $f''(x) = 2\sec^2 x \tan x$ $f''(\frac{\pi}{4}) = 4$

    $f(x) = 1 + 2(x - \frac{\pi}{4}) + 4\dfrac{(x - \frac{\pi}{4})^2}{2!} + \ldots$

    $\tan x = 1 + 2(x - \frac{\pi}{4}) + 2(x - \frac{\pi}{4})^2 + \ldots$

17. From the solution of Exercise 9: $e^{-x} = e^{-2}[1 - (x-2) + \dfrac{(x-2)^2}{2!} + \ldots]$

   $x = 2.2,\ x-2 = 0.2,\ e^{-2} = 0.1353$; $e^{-2.2} = e^{-2}[1 - 0.2 + \dfrac{(0.2)^2}{2!}] = 0.111$

19. $f(x) = \sqrt{x} = x^{1/2}$ $f(9) = 3$

   $f'(x) = \frac{1}{2}x^{-1/2}$ $f'(9) = \frac{1}{6}$ $\sqrt{x} = 3 + \frac{1}{6}(x-9) - \frac{1}{216}(x-9)^2 + \ldots$

   $f''(x) = -\frac{1}{4}x^{-3/2}$ $f''(9) = -\frac{1}{108}$ $\sqrt{9.3} = 3 + \frac{1}{6}(0.3) - \frac{1}{216}(0.3)^2 = 3.0496$

21. From the solution of Exercise 13: $\sqrt[3]{x} = 2 + \frac{1}{12}(x-8) - \frac{1}{288}(x-8)^2 + \ldots$

   $x = 8.3,\ x - 8 = 0.3$; $\sqrt[3]{8.3} = 2 + \frac{1}{12}(0.3) - \frac{1}{288}(0.3)^2 = 2.0247$

23. From the solution of Exercise 11: $\sin x = \frac{1}{2}[\sqrt{3} + (x - \frac{\pi}{3}) - \frac{\sqrt{3}}{2!}(x - \frac{\pi}{3})^2 - \ldots]$

   $x = 61°,\ x - \frac{\pi}{3} = 1° = \frac{\pi}{180}$; $\sin 61° = \frac{1}{2}[\sqrt{3} + \frac{\pi}{180} - \frac{\sqrt{3}}{2}(\frac{\pi}{180})^2] = 0.87462$

25. $f(x) = c_0 + c_1(x-a) + c_2(x-a)^2 + c_3(x-a)^3 + \ldots$ $f(a) = c_0$

   $f'(x) = c_1 + 2c_2(x-a) + 3c_3(x-a)^2 + \ldots$ $f'(a) = c_1$

   $f''(x) = 2c_2 + 6c_3(x-a) + \ldots$ $f''(a) = 2c_2,\ c_2 = \frac{1}{2!}f''(a)$

   $f(x) = f(a) + f'(a)(x-a) + \dfrac{f''(a)(x-a)^2}{2!} + \ldots$

27. $\sin x = x - \frac{1}{6}x^3 + \frac{1}{120}x^5 - \ldots$; $\sin 31° = \sin\frac{31\pi}{180} = \frac{31\pi}{180} - \frac{1}{6}(\frac{31\pi}{180})^3 + \frac{1}{120}(\frac{31\pi}{180})^5 = 0.5150408$

   $\sin x = \frac{1}{2} + \frac{\sqrt{3}}{2}(x - \frac{\pi}{6}) - \frac{1}{4}(x - \frac{\pi}{6})^2 - \ldots$; $\sin 31° = \frac{1}{2} + \frac{\sqrt{3}}{2}(\frac{\pi}{180}) - \frac{1}{4}(\frac{\pi}{180})^2 = 0.5150388$

   Calculator: $\sin 31° = 0.5150381$

Exercises 13-6, p. 474

1. $f(x) = \begin{cases} 1 & -\pi \le x < 0 \\ 0 & 0 \le x < \pi \end{cases}$ $a_o = \frac{1}{2\pi}\int_{-\pi}^{0}(1)dx + \frac{1}{2\pi}\int_{0}^{\pi}0\,dx = \frac{1}{2\pi}\int_{-\pi}^{0}dx = \left.\frac{x}{2\pi}\right|_{-\pi}^{0} = \frac{1}{2}$

$a_n = \frac{1}{\pi}\int_{-\pi}^{0}(1)\cos nx\,dx + \frac{1}{\pi}\int_{0}^{\pi}(0)\cos nx\,dx = \left.\frac{1}{n\pi}\sin nx\right|_{-\pi}^{0} = 0$ for all n

$b_n = \frac{1}{\pi}\int_{-\pi}^{0}(1)\sin nx\,dx + \frac{1}{\pi}\int_{0}^{\pi}(0)\sin nx\,dx = \left.-\frac{1}{n\pi}\cos nx\right|_{-\pi}^{0} = -\frac{1}{n\pi}[\cos 0 - \cos(-n\pi)]$

$= -\frac{1}{n\pi}(1 - \cos n\pi)$; $b_1 = -\frac{1}{\pi}(1 - \cos \pi) = -\frac{1}{\pi}(1+1) = -\frac{2}{\pi}$; $b_2 = -\frac{1}{2\pi}(1 - \cos 2\pi) = 0$;

$b_3 = -\frac{1}{3\pi}(1 - \cos 3\pi) = -\frac{2}{3\pi}$

$f(x) = \frac{1}{2} - \frac{2}{\pi}\sin x - \frac{2}{3\pi}\sin 3x - \ldots$

3. $f(x) = \begin{cases} 1 & -\pi \le x < 0 \\ 2 & 0 \le x < \pi \end{cases}$ $a_o = \frac{1}{2\pi}\int_{-\pi}^{0}(1)dx + \frac{1}{2\pi}\int_{0}^{\pi}2\,dx = \left.\frac{x}{2\pi}\right|_{-\pi}^{0} + \left.\frac{x}{\pi}\right|_{0}^{\pi} = \frac{1}{2} + 1 = \frac{3}{2}$

$a_n = \frac{1}{\pi}\int_{-\pi}^{0}(1)\cos nx\,dx + \frac{1}{\pi}\int_{0}^{\pi}2\cos nx\,dx = \left.\frac{1}{n\pi}\sin nx\right|_{-\pi}^{0} + \left.\frac{2}{n\pi}\sin nx\right|_{0}^{\pi} = 0$ for all n

$b_n = \frac{1}{\pi}\int_{-\pi}^{0}(1)\sin nx\,dx + \frac{1}{\pi}\int_{0}^{\pi}2\sin nx\,dx = \left.-\frac{1}{n\pi}\cos nx\right|_{-\pi}^{0} - \left.\frac{2}{n\pi}\cos nx\right|_{0}^{\pi}$

$= -\frac{1}{n\pi}[\cos 0 - \cos(-n\pi)] - \frac{2}{n\pi}[\cos n\pi - \cos 0] = -\frac{1}{n\pi}(1 - \cos n\pi + 2\cos n\pi - 2) = \frac{1}{n\pi}(1 - \cos n\pi)$

$b_1 = \frac{1}{\pi}(1 - \cos \pi) = \frac{2}{\pi}$, $b_2 = \frac{1}{2\pi}(1 - \cos 2\pi) = 0$, $b_3 = \frac{1}{3\pi}(1 - \cos 3\pi) = \frac{2}{3\pi}$

$f(x) = \frac{3}{2} + \frac{2}{\pi}\sin x + \frac{2}{3\pi}\sin 3x + \ldots$

5. $f(x) = \begin{cases} 0 & -\pi \le x < 0 \\ x & 0 \le x < \pi \end{cases}$ $a_o = \frac{1}{2\pi}\int_{-\pi}^{0}0\,dx + \frac{1}{2\pi}\int_{0}^{\pi}x\,dx = \left.\frac{x^2}{4\pi}\right|_{0}^{\pi} = \frac{\pi}{4}$

$a_n = \frac{1}{\pi}\int_{-\pi}^{0}0\cos nx\,dx + \frac{1}{\pi}\int_{0}^{\pi}x\cos nx\,dx = \frac{1}{n^2\pi}\int_{0}^{\pi}nx\cos nx(n\,dx) = \left.\frac{1}{n^2\pi}(\cos nx + nx\sin nx)\right|_{0}^{\pi}$

$= \frac{1}{n^2\pi}(\cos n\pi + n\pi\sin n\pi - \cos 0 - 0) = \frac{1}{n^2\pi}(\cos n\pi - 1)$

$a_1 = \frac{1}{\pi}(\cos \pi - 1) = -\frac{2}{\pi}$, $a_2 = \frac{1}{4\pi}(\cos 2\pi - 1) = 0$, $a_3 = \frac{1}{9\pi}(\cos 3\pi - 1) = -\frac{2}{9\pi}$

$b_n = \frac{1}{\pi}\int_{-\pi}^{0}0\sin nx + \frac{1}{\pi}\int_{0}^{\pi}x\sin nx\,dx = \frac{1}{n^2\pi}\int_{0}^{\pi}nx\sin nx\,(n\,dx) = \left.\frac{1}{n^2\pi}(\sin nx - nx\cos nx)\right|_{0}^{\pi}$

$= \frac{1}{n^2\pi}(\sin n\pi - n\pi\cos n\pi - \sin 0 - 0) = -\frac{1}{n}\cos n\pi$

$b_1 = -\cos \pi = 1$, $b_2 = -\frac{1}{2}\cos 2\pi = -\frac{1}{2}$

$f(x) = \frac{\pi}{4} - \frac{2}{\pi}(\cos x + \frac{1}{9}\cos 3x + \ldots) + (\sin x - \frac{1}{2}\sin 2x + \ldots)$

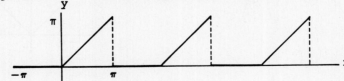

7. $f(x) = \begin{cases} -1 & -\pi \leq x < 0 \\ 0 & 0 \leq x < \pi/2 \\ 1 & \pi/2 \leq x < \pi \end{cases}$ $a_0 = \frac{1}{2\pi}\int_{-\pi}^{0} -dx + \frac{1}{2\pi}\int_{0}^{\pi/2} 0 dx + \frac{1}{2\pi}\int_{\pi/2}^{\pi} dx = \frac{1}{2\pi}\left[-x\Big|_{-\pi}^{0} + x\Big|_{\pi/2}^{\pi}\right]$

$= \frac{1}{2\pi}\left[0 - \pi + \pi - \frac{\pi}{2}\right] = -\frac{1}{4}$

$a_n = \frac{1}{\pi}\int_{-\pi}^{0} -\cos nx\, dx + \frac{1}{\pi}\int_{0}^{\pi/2} 0\cos nx + \frac{1}{\pi}\int_{\pi/2}^{\pi}\cos nx\, dx = \frac{1}{n\pi}(-\sin nx)\Big|_{-\pi}^{0} + \frac{1}{n\pi}\sin nx\Big|_{\pi/2}^{\pi}$

$= 0 + \frac{1}{n\pi}(\sin n\pi - \sin\frac{n\pi}{2}) = -\frac{1}{n\pi}\sin\frac{n\pi}{2}$; $a_1 = -\frac{1}{\pi}\sin\frac{\pi}{2} = -\frac{1}{\pi}$, $a_2 = -\frac{1}{2\pi}\sin\pi = 0$,

$a_3 = -\frac{1}{3\pi}\sin\frac{3\pi}{2} = \frac{1}{3\pi}$; $b_n = \frac{1}{\pi}\int_{-\pi}^{0} -\sin nx\, dx + \frac{1}{\pi}\int_{0}^{\pi/2} 0\sin nx + \frac{1}{\pi}\int_{\pi/2}^{\pi}\sin nx\, dx$

$b_n = \frac{1}{n\pi}\cos nx\Big|_{-\pi}^{0} - \frac{1}{n\pi}\cos nx\Big|_{\pi/2}^{\pi} = \frac{1}{n\pi}[\cos 0 - \cos(-n\pi)] - \frac{1}{n\pi}(\cos n\pi - \cos\frac{n\pi}{2})$

$= \frac{1}{n\pi}(1 - \cos n\pi - \cos n\pi + \cos\frac{n\pi}{2}) = \frac{1}{n\pi}(1 - 2\cos n\pi + \cos\frac{n\pi}{2})$

$b_1 = \frac{1}{\pi}(1 - 2\cos\pi + \cos\frac{\pi}{2}) = \frac{3}{\pi}$, $b_2 = \frac{1}{2\pi}(1 - 2\cos 2\pi + \cos\pi) = -\frac{1}{\pi}$,

$b_3 = \frac{1}{3\pi}(1 - 2\cos 3\pi + \cos\frac{3\pi}{2}) = \frac{1}{\pi}$

$f(x) = -\frac{1}{4} - \frac{1}{\pi}\cos x + \frac{1}{3\pi}\cos 3x - \ldots + \frac{3}{\pi}\sin x - \frac{1}{\pi}\sin 2x + \frac{1}{\pi}\sin 3x - \ldots$

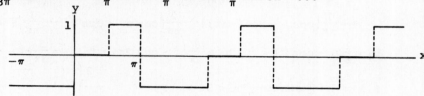

9. $f(x) = \begin{cases} -x & -\pi \leq x < 0 \\ x & 0 \leq x < \pi \end{cases}$ $a_0 = \frac{1}{2\pi}\int_{-\pi}^{0} -x\, dx + \frac{1}{2\pi}\int_{0}^{\pi} x\, dx = -\frac{x^2}{4\pi}\Big|_{-\pi}^{0} + \frac{x^2}{4\pi}\Big|_{0}^{\pi} = \frac{\pi}{4} + \frac{\pi}{4} = \frac{\pi}{2}$

$a_n = \frac{1}{\pi}\int_{-\pi}^{0} -x\cos nx\, dx + \frac{1}{\pi}\int_{0}^{\pi} x\cos nx\, dx = -\frac{1}{n^2\pi}[(\cos nx + nx\sin nx)\Big|_{-\pi}^{0} - (\cos nx + nx\sin nx)\Big|_{0}^{\pi}]$

$= -\frac{1}{n^2\pi}[\cos 0 + 0 - \cos(-n\pi) - n\pi\sin(-n\pi)] + \frac{1}{n^2\pi}(\cos n\pi + n\pi\sin n\pi - \cos 0 - 0)$

$= -\frac{1}{n^2\pi}(1 - \cos n\pi) + \frac{1}{n^2\pi}(\cos n\pi - 1) = \frac{2}{n^2\pi}(\cos n\pi - 1)$

$a_1 = \frac{2}{\pi}(\cos\pi - 1) = -\frac{4}{\pi}$, $a_2 = \frac{2}{4\pi}(\cos 2\pi - 1) = 0$, $a_3 = \frac{2}{9\pi}(\cos 3\pi - 1) = -\frac{4}{9\pi}$

$b_n = \frac{1}{\pi}\int_{-\pi}^{0} -x\sin nx\, dx + \frac{1}{\pi}\int_{0}^{\pi} x\sin nx\, dx = -\frac{1}{n^2\pi}[(\sin nx - nx\cos nx)\Big|_{-\pi}^{0} - (\sin nx - nx\cos nx)\Big|_{0}^{\pi}]$

$= -\frac{1}{n^2\pi}[\sin 0 - 0 - \sin(-n\pi) - n\pi\cos(-n\pi)] + \frac{1}{n^2\pi}(\sin n\pi - n\pi\cos n\pi - \sin 0 - 0)$

$= \frac{1}{n}\cos n\pi - \frac{1}{n}\cos n\pi = 0$ for all n

$f(x) = \frac{\pi}{2} - \frac{4}{\pi}\cos x - \frac{4}{9\pi}\cos 3x - \ldots$

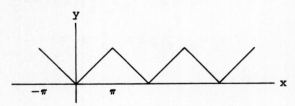

11. $f(x) = \begin{cases} 5 & -3 \le x < 0 \\ 0 & 0 \le x < 3 \end{cases}$ $\begin{array}{l} \text{period}=6 \\ L = 3 \end{array}$ $a_o = \frac{1}{6}\int_{-3}^0 5\,dx + \frac{1}{6}\int_0^3 0\,dx = \frac{5}{6}x\Big|_{-3}^0 = \frac{15}{6} = \frac{5}{2}$

$a_n = \frac{1}{3}\int_{-3}^0 5\cos\frac{n\pi x}{3}\,dx + \frac{1}{3}\int_0^3 0\cos\frac{n\pi x}{3}\,dx = \frac{5}{3}(\frac{3}{n\pi})\sin\frac{n\pi x}{3}\Big|_{-3}^0 = \frac{5}{n\pi}[\sin 0 - \sin(-n\pi)] = 0$ for all n

$b_n = \frac{1}{3}\int_{-3}^0 5\sin\frac{n\pi x}{3}\,dx + \frac{1}{3}\int_0^3 0\sin\frac{n\pi x}{3}\,dx = -\frac{5}{3}(\frac{3}{n\pi})\cos\frac{n\pi x}{3}\Big|_{-3}^0 = -\frac{5}{n\pi}[\cos 0 - \cos(-n\pi)]$

$= \frac{5}{n\pi}(\cos n\pi - 1)$; $b_1 = \frac{5}{\pi}(\cos\pi - 1) = -\frac{10}{\pi}$, $b_2 = \frac{5}{2\pi}(\cos 2\pi - 1) = 0$,

$b_3 = \frac{5}{3\pi}(\cos 3\pi - 1) = -\frac{10}{3\pi}$; $f(x) = \frac{5}{2} - \frac{10}{\pi}(\sin\frac{\pi x}{3} + \frac{1}{3}\sin\pi x - \ldots)$

13. $f(t) = \begin{cases} -\sin t & -\pi \le t < 0 \\ \sin t & 0 \le t < \pi \end{cases}$

$a_o = \frac{1}{2\pi}\int_{-\pi}^0 -\sin t\,dt + \frac{1}{2\pi}\int_0^\pi \sin t\,dt$

$= \frac{1}{2\pi}\cos t\Big|_{-\pi}^0 - \frac{1}{2\pi}\cos t\Big|_0^\pi = \frac{1}{2\pi}[\cos 0 - \cos(-\pi)] - \frac{1}{2\pi}(\cos\pi - \cos 0) = \frac{1}{2\pi}(1+1+1+1) = \frac{2}{\pi}$

$a_n = \frac{1}{\pi}\int_{-\pi}^0 -\sin t\cos nt\,dt + \frac{1}{\pi}\int_0^\pi \sin t\cos nt\,dt$

$= -\frac{1}{\pi}[-\frac{\cos(1-n)t}{2(1-n)} - \frac{\cos(1+n)t}{2(1+n)}]\Big|_{-\pi}^0 + \frac{1}{\pi}[-\frac{\cos(1-n)t}{2(1-n)} - \frac{\cos(1+n)t}{2(1+n)}]\Big|_0^\pi$

$= \frac{1}{\pi}[\frac{\cos 0}{2(1-n)} + \frac{\cos 0}{2(1+n)} - \frac{\cos(1-n)(-\pi)}{2(1-n)} - \frac{\cos(1+n)(-\pi)}{2(1+n)} - \frac{\cos(1-n)\pi}{2(1-n)} - \frac{\cos(1+n)\pi}{2(1+n)}$

$\qquad + \frac{\cos 0}{2(1-n)} + \frac{\cos 0}{2(1+n)}] = \frac{1}{\pi}[\frac{1}{1-n} + \frac{1}{1+n} - \frac{\cos(1-n)\pi}{1-n} - \frac{\cos(1+n)\pi}{1+n}]$ $(n \ne 1)$

$a_1 = \frac{1}{\pi}\int_{-\pi}^0 -\sin t\cos t\,dt + \frac{1}{\pi}\int_0^\pi \sin t\cos t\,dt = -\frac{1}{\pi}\sin^2 t\Big|_{-\pi}^0 + \frac{1}{\pi}\sin^2 t\Big|_0^\pi = 0$

$a_2 = \frac{1}{\pi}[\frac{1}{-1} + \frac{1}{3} - \frac{\cos(-\pi)}{-1} - \frac{\cos 3\pi}{3}] = \frac{1}{\pi}[-1 + \frac{1}{3} - 1 + \frac{1}{3}] = -\frac{4}{3\pi}$

$a_3 = \frac{1}{\pi}[\frac{1}{-2} + \frac{1}{4} - \frac{\cos(-2\pi)}{-2} - \frac{\cos 4\pi}{4}] = \frac{1}{\pi}(-\frac{1}{2} + \frac{1}{4} + \frac{1}{2} - \frac{1}{4}) = 0$

$a_4 = \frac{1}{\pi}[\frac{1}{-3} + \frac{1}{5} - \frac{\cos(-3\pi)}{-3} - \frac{\cos 5\pi}{5}] = \frac{1}{\pi}(-\frac{1}{3} + \frac{1}{5} - \frac{1}{3} + \frac{1}{5}) = -\frac{4}{15\pi}$

$b_n = \frac{1}{\pi}\int_{-\pi}^0 -\sin t\sin nt\,dt + \frac{1}{\pi}\int_0^\pi \sin t\sin nt\,dt$

$= -\frac{1}{\pi}[\frac{\sin(1-n)t}{2(1-n)} - \frac{\sin(1+n)t}{2(1+n)}]\Big|_{-\pi}^0 + \frac{1}{\pi}[\frac{\sin(1-n)t}{2(1-n)} - \frac{\sin(1+n)t}{2(1+n)}]\Big|_0^\pi$ $(n \ne 1)$

$= 0$ for all $n>1$ since $\sin n\pi = 0$ for all n

$b_1 = \frac{1}{\pi}\int_{-\pi}^0 -\sin t\sin t\,dt + \frac{1}{\pi}\int_0^\pi \sin t\sin t\,dt = -\frac{1}{\pi}\int_{-\pi}^0 \sin^2 t\,dt + \frac{1}{\pi}\int_0^\pi \sin^2 t\,dt$

$= -\frac{1}{\pi}(\frac{t}{2} - \frac{1}{2}\sin t\cos t)\Big|_{-\pi}^0 + \frac{1}{\pi}(\frac{t}{2} - \frac{1}{2}\sin t\cos t)\Big|_0^\pi = 0$

$f(t) = \frac{2}{\pi} - \frac{4}{3\pi}\cos 2t - \frac{4}{15\pi}\cos 4t - \ldots$ (See Fig. 13-9)

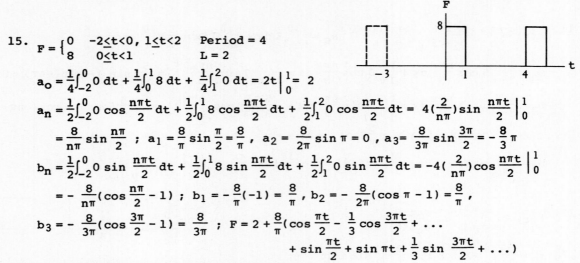

15. $F = \begin{cases} 0 & -2 \leq t < 0, \ 1 \leq t < 2 \\ 8 & 0 \leq t < 1 \end{cases}$ Period = 4   L = 2

$a_0 = \frac{1}{4}\int_{-2}^{0} 0 \, dt + \frac{1}{4}\int_{0}^{1} 8 \, dt + \frac{1}{4}\int_{1}^{2} 0 \, dt = 2t \Big|_{0}^{1} = 2$

$a_n = \frac{1}{2}\int_{-2}^{0} 0 \cos \frac{n\pi t}{2} \, dt + \frac{1}{2}\int_{0}^{1} 8 \cos \frac{n\pi t}{2} \, dt + \frac{1}{2}\int_{1}^{2} 0 \cos \frac{n\pi t}{2} \, dt = 4(\frac{2}{n\pi}) \sin \frac{n\pi t}{2} \Big|_{0}^{1}$

$= \frac{8}{n\pi} \sin \frac{n\pi}{2}$; $a_1 = \frac{8}{\pi} \sin \frac{\pi}{2} = \frac{8}{\pi}$, $a_2 = \frac{8}{2\pi} \sin \pi = 0$, $a_3 = \frac{8}{3\pi} \sin \frac{3\pi}{2} = -\frac{8}{3}\pi$

$b_n = \frac{1}{2}\int_{-2}^{0} 0 \sin \frac{n\pi t}{2} \, dt + \frac{1}{2}\int_{0}^{1} 8 \sin \frac{n\pi t}{2} \, dt + \frac{1}{2}\int_{1}^{2} 0 \sin \frac{n\pi t}{2} \, dt = -4(\frac{2}{n\pi}) \cos \frac{n\pi t}{2} \Big|_{0}^{1}$

$= -\frac{8}{n\pi}(\cos \frac{n\pi}{2} - 1)$; $b_1 = -\frac{8}{\pi}(-1) = \frac{8}{\pi}$, $b_2 = -\frac{8}{2\pi}(\cos \pi - 1) = \frac{8}{\pi}$,

$b_3 = -\frac{8}{3\pi}(\cos \frac{3\pi}{2} - 1) = \frac{8}{3\pi}$; $F = 2 + \frac{8}{\pi}(\cos \frac{\pi t}{2} - \frac{1}{3} \cos \frac{3\pi t}{2} + \dots$

$+ \sin \frac{\pi t}{2} + \sin \pi t + \frac{1}{3} \sin \frac{3\pi t}{2} + \dots)$

Review Exercises for Chapter 13, p. 475

1. $f(x) = \frac{1}{1 + e^x}$ $f(0) = \frac{1}{2}$

$f'(x) = \frac{-e^x}{(1 + e^x)^2}$ $f'(0) = -\frac{1}{4}$

$f''(x) = \frac{(1+e^x)^2(-e^x) - (-e^x)(2)(1+e^x)(e^x)}{(1 + e^x)^4}$

$= \frac{e^{2x} - e^x}{(1+e^x)^3}$ $f''(0) = 0$

$f'''(x) = \frac{(1+e^x)^3(2e^{2x}-e^x) - (e^{2x}-e^x)(3)(1+e^x)^2(e^x)}{(1 + e^x)^6}$

$f'''(0) = \frac{8}{2^6} = \frac{1}{8}$; $f(x) = \frac{1}{1+e^x} = \frac{1}{2} - \frac{1}{4} x + \frac{1}{48}x^3 - \dots$

3. $F(x) = \sin x = x - \frac{x^3}{3!} + \frac{x^5}{5!} - \dots$

$F(2x^2) = \sin 2x^2$

$= 2x^2 - \frac{(2x^2)^3}{3!} + \frac{(2x^2)^5}{5!} - \dots$

$f(x) = \sin 2x^2 = 2x^2 - \frac{4}{3} x^6 + \frac{4}{15}x^{10}$

$- \dots$

5. $f(x) = (x + 1)^{1/3}$, $f(0) = 1$

$f'(x) = \frac{1}{3}(x + 1)^{-2/3}, f'(0) = \frac{1}{3}$

$f''(x) = -\frac{2}{9}(x+1)^{-5/3}, f''(0) = -\frac{2}{9}$

$f(x) = 1 + \frac{1}{3} x - \frac{2}{9}(\frac{x^2}{2}) + \dots$

$(x + 1)^{1/3} = 1 + \frac{1}{3} x - \frac{1}{9} x^2 + \dots$

7. $f(x) = \text{Arcsin } x$ $f(0) = 0$

$f'(x) = \frac{1}{\sqrt{1 - x^2}}$ $f'(0) = 1$

$f''(x) = \frac{x}{(1 - x^2)^{3/2}}$ $f''(0) = 0$

$f'''(x) = \frac{(1-x^2)^{3/2}(1) - x(3/2)(1-x^2)^{1/2}(-2x)}{(1-x^2)^3}$

$= \frac{2x^2 + 1}{(1-x^2)^{5/2}}$ $f'''(0) = 1$

$f^{iv}(x) = \frac{(1-x^2)^{5/2}(4x) - (2x^2+1)(5/2)(1-x^2)^{3/2}(-2x)}{(1-x^2)^5} = \frac{6x^3 + 9x}{(1-x^2)^{7/2}}$ $f^{iv}(0) = 0$

$f^{v}(x) = \frac{(1-x^2)^{7/2}(18x^2+9) - (6x^3+9x)(7/2)(1-x^2)^{5/2}(-2x)}{(1-x^2)^7} = \frac{24x^4+72x^2+9}{(1-x^2)^{9/2}}$ $f^{v}(0) = 9$

$f(x) = x + \frac{x^3}{3!} + 9(\frac{x^5}{5!}) + \dots = x + \frac{1}{6} x^3 + \frac{3}{40} x^5 + \dots$

9. $e^x = 1 + x + \frac{1}{2} x^2 + \dots$

$e^{-0.2} = 1 + (-0.2) + \frac{1}{2}(-0.2)^2$

$= 0.82$

11. See Exercise 5.

$\sqrt[3]{1 + x} = 1 + \frac{1}{3} x - \frac{1}{9} x^2 + \dots$

$\sqrt[3]{1+0.3} = 1 + \frac{1}{3}(0.3) - \frac{1}{9}(0.3)^2$

$\sqrt[3]{1.3} = 1.09$

13. $f(x) = \sqrt{1 + x}$ $f(0) = 1$

$f'(x) = \frac{1}{2}(1 + x)^{-1/2}$ $f'(0) = \frac{1}{2}$

$f''(x) = -\frac{1}{4}(1 + x)^{-3/2}$ $f''(0) = -\frac{1}{4}$

$f(x) = 1 + \frac{1}{2}x - \frac{1}{4}(\frac{x^2}{2}) + \dots$

$\sqrt{1 + x} = 1 + \frac{1}{2}x - \frac{1}{8}x^2 + \dots$

$\sqrt{1 + 0.07} = 1 + \frac{1}{2}(0.07) - \frac{1}{8}(0.07)^2$

$\sqrt{1.07} = 1.0344$

19. $f(x) = \sqrt{x}$, $a = 144$

$f(x) = \sqrt{x}$, $f(144) = 12$

$f'(x) = \frac{1}{2}x^{-1/2}$, $f'(144) = \frac{1}{24}$

$f''(x) = -\frac{1}{4}x^{-3/2}$, $f''(144) = -\frac{1}{6912}$

$\sqrt{x} = 12 + \frac{1}{24}(x - 144) - \frac{1}{6912}\frac{(x - 144)^2}{2} + \dots$

$= 12 + \frac{x - 144}{24} - \frac{(x - 144)^2}{13824}$

$\sqrt{148} = 12 + \frac{4}{24} - \frac{4^2}{13824} = 12.1655$

23. $f(x) = \cos x$, $a = \frac{\pi}{3}$

$f(x) = \cos x$, $f(\frac{\pi}{3}) = \frac{1}{2}$

$f'(x) = -\sin x$, $f'(\frac{\pi}{3}) = \frac{1}{2}\sqrt{3}$

$f''(x) = -\cos x$, $f''(\frac{\pi}{3}) = -\frac{1}{2}$; $f(x) = \frac{1}{2} + \frac{1}{2}\sqrt{3}(x - \frac{\pi}{3}) - \frac{1}{4}(x - \frac{\pi}{3})^2 + \dots$

15. $\ln(1 + x) = x - \frac{x^2}{2} + \frac{x^3}{3} - \dots$

$\ln[1 + (-0.1828)] = -0.1828$

$\quad - \frac{(-0.1828)^2}{2} + \frac{(-0.1828)^3}{3} - \dots$

$\ln 0.8172 = -0.2015$

17. See Exercise 15 of Section 13-5.

$f(x) = 1 + 2(x - \frac{\pi}{4}) + 2(x - \frac{\pi}{4})^2 + \dots$

$x = 43.62° = 0.7613126$

$x - \frac{\pi}{4} = -0.0240855$

$\tan 43.62° = 1 + 2(-0.0240855)$

$\quad\quad + 2(-0.0240855)^2 = 0.95299$

21. $\int_{0.1}^{0.2} \frac{\cos x \, dx}{\sqrt{x}} = \int_{0.1}^{0.2} \frac{1}{\sqrt{x}}(1 - \frac{x^2}{2} + \frac{x^4}{24})dx$

$= \int_{0.1}^{0.2} (x^{-1/2} - \frac{x^{3/2}}{2} + \frac{x^{7/2}}{24})dx$

$= 2x^{1/2} - \frac{1}{5}x^{5/2} + \frac{1}{108}x^{9/2}\Big|_{0.1}^{0.2}$

$= 2(0.2)^{1/2} - \frac{1}{5}(0.2)^{5/2} + \frac{1}{108}(0.2)^{9/2}$

$\quad -2(0.1)^{1/2} + \frac{1}{5}(0.1)^{5/2} - \frac{1}{108}(0.1)^{9/2}$

$= 0.259$

25. $f(x) = \begin{cases} 0 & -\pi \leq x < -\pi/2, \ \pi/2 < x < \pi \\ 1 & -\pi/2 \leq x \leq \pi/2 \end{cases}$

$a_0 = \frac{1}{2\pi}\int_{-\pi}^{-\pi/2} 0 \, dx + \frac{1}{2\pi}\int_{-\pi/2}^{\pi/2} dx + \frac{1}{2\pi}\int_{\pi/2}^{\pi} 0 \, dx = \frac{x}{2\pi}\Big|_{-\pi/2}^{\pi/2} = \frac{1}{2\pi}(\frac{\pi}{2} + \frac{\pi}{2}) = \frac{1}{2}$

$a_n = \frac{1}{\pi}\int_{-\pi}^{-\pi/2} 0 \cos nx \, dx + \frac{1}{\pi}\int_{-\pi/2}^{\pi/2} \cos nx \, dx + \frac{1}{\pi}\int_{\pi/2}^{\pi} 0 \cos nx \, dx = \frac{1}{n\pi}\sin nx\Big|_{-\pi/2}^{\pi/2}$

$= \frac{1}{n\pi}[\sin \frac{n\pi}{2} - \sin \frac{(-n\pi)}{2}] = \frac{2}{n\pi}\sin \frac{n\pi}{2}$

$a_1 = \frac{2}{\pi}\sin \frac{\pi}{2} = \frac{2}{\pi}$, $a_2 = \frac{2}{2\pi}\sin \pi = 0$, $a_3 = \frac{2}{3\pi}\sin \frac{3\pi}{2} = -\frac{2}{3\pi}$

$b_n = \frac{1}{\pi}\int_{-\pi}^{-\pi/2} 0 \sin nx \, dx + \frac{1}{\pi}\int_{-\pi/2}^{\pi/2} \sin nx \, dx + \frac{1}{\pi}\int_{\pi/2}^{\pi} 0 \sin nx \, dx = -\frac{1}{n\pi}\cos nx\Big|_{-\pi/2}^{\pi/2}$

$= -\frac{1}{n\pi}[\cos \frac{n\pi}{2} - \cos \frac{(-n\pi)}{2}] = 0$

(for all n since $\cos \theta = \cos(-\theta)$)

$f(x) = \frac{1}{2} + \frac{2}{\pi}(\cos x - \frac{1}{3}\cos 3x + \dots)$

27. $f(x) = x$ $-2 \leq x < 2$, period = 4, $L = 2$; $a_0 = \frac{1}{4}\int_{-2}^{2} x\, dx = \frac{1}{8} x^2 \Big|_{-2}^{2} = \frac{1}{8}(4-4) = 0$

$a_n = \frac{1}{2}\int_{-2}^{2} x \cos \frac{n\pi x}{2}\, dx = \frac{1}{2}(\frac{2}{n\pi})^2 (\cos \frac{n\pi x}{2} + \frac{n\pi x}{2} \sin \frac{n\pi x}{2}) \Big|_{-2}^{2}$

$\qquad = \frac{2}{n^2\pi^2}[\cos n\pi + n\pi \sin n\pi - \cos(-n\pi) + n\pi \sin(-n\pi)] = 0$ for all n

$b_n = \frac{1}{2}\int_{-2}^{2} x \sin \frac{n\pi x}{2}\, dx = \frac{1}{2}(\frac{2}{n\pi})^2 (\sin \frac{n\pi x}{2} - \frac{n\pi x}{2} \cos \frac{n\pi x}{2}) \Big|_{-2}^{2}$

$\qquad = \frac{2}{n^2\pi^2}[\sin n\pi - n\pi \cos n\pi - \sin(-n\pi) - n\pi \cos(-n\pi)]$

$\qquad = \frac{2}{n^2\pi^2}(-n\pi \cos n\pi - n\pi \cos n\pi) = \frac{-4}{n\pi} \cos n\pi$; $b_1 = -\frac{4}{\pi} \cos \pi = \frac{4}{\pi}$,

$b_2 = -\frac{4}{2\pi} \cos 2\pi = -\frac{2}{\pi}$, $b_3 = \frac{-4}{3\pi} \cos 3\pi = \frac{4}{3\pi}$

$f(x) = \frac{4}{\pi}(\sin \frac{\pi x}{2} - \frac{1}{2} \sin \pi x + \frac{1}{3} \sin \frac{3\pi x}{2} - \ldots)$

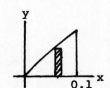

29. $\sin x = x - \frac{x^3}{3!} + \ldots$

$\sin(x+h) - \sin(x-h) = (x+h) - \frac{(x+h)^3}{3!} + \ldots -(x-h) + \frac{(x-h)^3}{3!} - \ldots$

$\qquad = x + h - \frac{x^3 + 3x^2 h + 3xh^2 + h^3}{3!} + \ldots -x + h + \frac{x^3 - 3x^2 h + 3xh^2 - h^3}{3!} - \ldots$

$\qquad = 2h - \frac{6x^2 h}{3!} - \frac{2h^3}{3!} + \ldots = 2h(1 - \frac{x^2}{2} + \ldots) - \frac{2h^3}{3!} + \ldots$

$\qquad = 2h \cos x$ for small h

31.

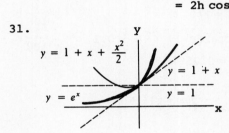

$y = 1 + x + \frac{x^2}{2}$

$y = 1 + x$

$y = e^x$

$y = 1$

33. $\sec x = \frac{1}{\cos x} = \frac{1}{1 - \frac{x^2}{2} + \frac{x^4}{24} - \ldots}$

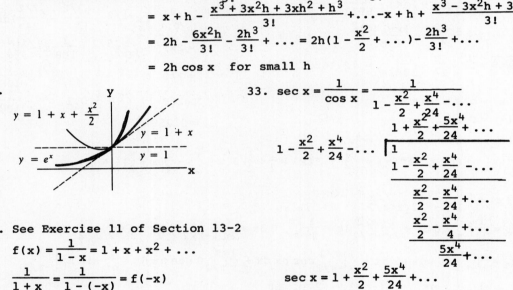

$\sec x = 1 + \frac{x^2}{2} + \frac{5x^4}{24} + \ldots$

35. See Exercise 11 of Section 13-2

$f(x) = \frac{1}{1-x} = 1 + x + x^2 + \ldots$

$\frac{1}{1+x} = \frac{1}{1-(-x)} = f(-x)$

$f(-x) = 1 + (-x) + (-x)^2 + \ldots$

$\qquad = 1 - x + x^2 - \ldots$

37. $y = \frac{x - \sin x}{x^2} = \frac{x - (x - x^3/6 + x^5/120 - \ldots)}{x^2} = \frac{x}{6} - \frac{x^3}{120} + \ldots$

$A = \int_0^{0.1} (\frac{x}{6} - \frac{x^3}{120})\, dx = \frac{x^2}{12} - \frac{x^4}{480} \Big|_0^{0.1} = \frac{0.1^2}{12} - \frac{0.1^4}{480}$

$\qquad = 0.0008331$

39. $f(t) = \begin{cases} 0 & -\pi \le t < 0, \ \pi/2 < t < \pi \\ \sin t & 0 < t < \pi/2 \end{cases}$

$a_o = \frac{1}{2\pi}\int_{-\pi}^{0} 0 \, dt + \frac{1}{2\pi}\int_{0}^{\pi/2}\sin t \, dt + \frac{1}{2\pi}\int_{\pi/2}^{\pi} 0 \, dt = -\frac{1}{2\pi}\cos t\Big|_{0}^{\pi/2} = -\frac{1}{2\pi}(\cos\frac{\pi}{2} - \cos 0) = \frac{1}{2\pi}$

$a_n = \frac{1}{\pi}\int_{-\pi}^{0} 0 \cos nt \, dt + \frac{1}{\pi}\int_{0}^{\pi/2}\sin t \cos nt \, dt + \frac{1}{\pi}\int_{\pi/2}^{\pi} 0 \cos nt \, dt$

$= \frac{1}{\pi}[-\frac{\cos(1-n)t}{2(1-n)} - \frac{\cos(1+n)t}{2(1+n)}]\Big|_{0}^{\pi/2} = \frac{1}{\pi}[-\frac{\cos(1-n)(\pi/2)}{2(1-n)} - \frac{\cos(1+n)(\pi/2)}{2(1+n)}$

$+ \frac{\cos 0}{2(1-n)} + \frac{\cos 0}{2(1+n)}] = \frac{1}{\pi}[-\frac{\cos(1-n)(\pi/2)}{2(1-n)} - \frac{\cos(1+n)(\pi/2)}{2(1+n)} + \frac{1}{1-n^2}]$ $(n \ne 1)$

$a_1 = \frac{1}{\pi}\int_{0}^{\pi/2}\sin t \cos t \, dt = \frac{1}{2\pi}\sin^2 t\Big|_{0}^{\pi/2} = \frac{1}{2\pi}\sin^2(\frac{\pi}{2}) = \frac{1}{2\pi}$

$a_2 = \frac{1}{\pi}[-\frac{\cos(-\pi/2)}{-2} - \frac{\cos(3\pi/2)}{6} + \frac{1}{1-4}] = -\frac{1}{3\pi}$

$b_n = \frac{1}{\pi}\int_{-\pi}^{0} 0 \sin nt \, dt + \frac{1}{\pi}\int_{0}^{\pi/2}\sin t \sin nt \, dt + \frac{1}{\pi}\int_{\pi/2}^{\pi} 0 \sin nt \, dt$

$= \frac{1}{\pi}[\frac{\sin(1-n)t}{2(1-n)} - \frac{\sin(1+n)t}{2(1+n)}]\Big|_{0}^{\pi/2} = \frac{1}{\pi}[\frac{\sin(1-n)(\pi/2)}{2(1-n)} - \frac{\sin(1+n)(\pi/2)}{2(1+n)}]$ $(n \ne 1)$

$b_1 = \frac{1}{\pi}\int_{0}^{\pi/2}\sin t \sin t \, dt = \frac{1}{\pi}\int_{0}^{\pi/2}\sin^2 t \, dt = \frac{1}{\pi}(\frac{t}{2} - \frac{1}{2}\sin t \cos t)\Big|_{0}^{\pi/2}$

$= \frac{1}{\pi}(\frac{\pi}{4} - 0 - 0 - 0) = \frac{1}{4}$, $b_2 = \frac{1}{\pi}[\frac{\sin(-\pi/2)}{-2} - \frac{\sin(3\pi/2)}{6}] = \frac{1}{\pi}(\frac{1}{2} + \frac{1}{6}) = \frac{2}{3\pi}$

$f(t) = \frac{1}{2\pi} + \frac{1}{\pi}(\frac{1}{2}\cos t - \frac{1}{3}\cos 2t + \dots) + \frac{1}{4}\sin t + \frac{2}{3\pi}\sin 2t + \dots$

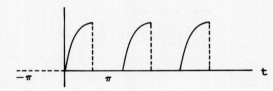

Exercises 14-1, p. 480

1. $y = x + 3$ $\dfrac{dy}{dx} = 1$ 3. $y = e^x - 1$ $\dfrac{dy}{dx} - y = 1$ 5. $y = cx^2$ $xy' = 2y$

$\dfrac{dy}{dx} = 1$ $1 = 1$ $\dfrac{dy}{dx} = e^x$ $e^x - (e^x - 1) = 1$ $y' = 2cx$ $x(2cx) = 2(cx^2)$

$1 = 1$ $2cx^2 = 2cx^2$

7. $y = ce^{-2x} + x - \dfrac{1}{2}$ $y' + 2y = 2x$

$y' = -2ce^{-2x} + 1$ $(-2ce^{-2x} + 1) + 2(ce^{-2x} + x - \dfrac{1}{2}) = 2x$

$-2ce^{-2x} + 1 + 2ce^{-2x} + 2x - 1 = 2x,$ $2x = 2x$

9. $y = 3 \cos 2x$ $\dfrac{d^2y}{dx^2} + 4y = 0$

$\dfrac{dy}{dx} = -6 \sin 2x$ $-12 \cos 2x + 4(3 \cos 2x) = 0$

$\dfrac{d^2y}{dx^2} = -12 \cos 2x$ $0 = 0$

11. $y = e^{2x}(c_1 + c_2 x + \dfrac{x^2}{2})$ $y'' - 4y' + 4y = e^{2x}$

$y' = e^{2x}(c_2 + x) + 2e^{2x}(c_1 + c_2 x + \dfrac{x^2}{2})$ $e^{2x}(1 + 4c_1 + 4c_2 + 4x + 4c_2 x + 2x^2)$

$\quad = e^{2x}(2c_1 + c_2 + x + 2c_2 x + x^2)$ $-4e^{2x}(2c_1 + c_2 + x + 2c_2 x + x^2)$

$y'' = e^{2x}(1 + 2c_2 + 2x) + 2e^{2x}(2c_1 + c_2 + x + 2c_2 x + x^2)$ $+4e^{2x}(c_1 + c_2 x + \dfrac{x^2}{2}) = e^{2x}$

$\quad = e^{2x}(1 + 4c_1 + 4c_2 + 4x + 4c_2 x + 2x^2)$ $e^{2x} = e^{2x}$

13. $xy = cx + cy$ $x^2 y' + y^2 = 0$

$xy' + y = c + cy'$ $x^2 [\dfrac{-c^2}{(x-c)^2}] + (\dfrac{cx}{x-c})^2 = 0$

$y = \dfrac{cx}{x-c}$ $0 = 0$

$y' = \dfrac{c - y}{x - c} = \dfrac{c - \dfrac{cx}{x-c}}{x-c} = \dfrac{-c^2}{(x-c)^2}$

15. $y = c_1 \ln x + c_2$ $x\dfrac{d^2y}{dx^2} + \dfrac{dy}{dx} = 0$ 17. $y = \sin x + \cos x - e^{-x}$

$\dfrac{dy}{dx} = \dfrac{c_1}{x}$ $x(-\dfrac{c_1}{x^2}) + \dfrac{c_1}{x} = 0$ $y' = \cos x - \sin x + e^{-x}$

$\dfrac{d^2y}{dx^2} = -\dfrac{c_1}{x^2}$ $0 = 0$ $\overline{\qquad\qquad\qquad\qquad}$

$y' + y = 2 \cos x$

$(\cos x - \sin x + e^{-x}) + (\sin x + \cos x - e^{-x})$

$= 2 \cos x$

$2 \cos x = 2 \cos x$

19. $y = e^{-x} - \dfrac{3}{5}\cos 2x - \dfrac{6}{5}\sin 2x$ $y'' + y' = 6 \sin 2x$

$y' = -e^{-x} + \dfrac{6}{5}\sin 2x - \dfrac{12}{5}\cos 2x$ $(e^{-x} + \dfrac{12}{5}\cos 2x + \dfrac{24}{5}\sin 2x) + (-e^{-x} + \dfrac{6}{5}\sin 2x - \dfrac{12}{5}\cos 2x)$

$y'' = e^{-x} + \dfrac{12}{5}\cos 2x + \dfrac{24}{5}\sin 2x$ $= 6 \sin 2x$

$6 \sin 2x = 6 \sin 2x$

21. $y = \dfrac{x+c}{\sec x + \tan x}$; $\dfrac{dy}{dx} = \dfrac{(\sec x + \tan x)(1) - (x+c)(\sec x \tan x + \sec^2 x)}{(\sec x + \tan x)^2}$

$\cos x \dfrac{dy}{dx} + \sin x = 1 - y$

$\cos x[\dfrac{\sec x + \tan x - (x+c)\sec x(\sec x + \tan x)}{(\sec x + \tan x)^2}] + \sin x = 1 - \dfrac{x+c}{\sec x + \tan x}$

$\dfrac{\cos x(1) - (x+c)\cos x \sec x (1)}{\sec x + \tan x} + \sin x = 1 - \dfrac{x+c}{\sec x + \tan x}$

$\dfrac{\cos x - (x+c) + \sin x \sec x + \sin x \tan x}{\sec x + \tan x} = 1 - \dfrac{x+c}{\sec x + \tan x}$

$\dfrac{\dfrac{\cos^2 x}{\cos x} + \tan x + \dfrac{\sin^2 x}{\cos x} - (x+c)}{\sec x + \tan x} = 1 - \dfrac{x+c}{\sec x + \tan x}$; $\dfrac{\dfrac{1}{\cos x} + \tan x - (x+c)}{\sec x + \tan x} = 1 - \dfrac{x+c}{\sec x + \tan x}$

$1 - \dfrac{x+c}{\sec x + \tan x} = 1 - \dfrac{x+c}{\sec x + \tan x}$

23. $y = cx + c^2$ $(y')^2 + xy' = y$

$y' = c$ $c^2 + c(x) = cx + c^2$; $cx + c^2 = cx + c^2$

Exercises 14-2 , p. 484

1. $2xdx + dy = 0$
$x^2 + y = c$
$y = c - x^2$

3. $y^2 dx + dy = 0$
$dx + \dfrac{dy}{y^2} = 0$
$x - \dfrac{1}{y} = c$

5. $x^2 + (x^3 + 5)y' = 0$
$\dfrac{x^2 dx}{x^3 + 5} + dy = 0$
$\dfrac{1}{3}\ln(x^3+5) + y = \dfrac{1}{3}c$
$\ln(x^3+5) + 3y = c$

7. $e^{x^2}dy = x\sqrt{1-y}\,dx$
$\dfrac{dy}{(1-y)^{1/2}} = xe^{-x^2}dx$
$-2(1-y)^{1/2} = -\dfrac{1}{2}e^{-x^2} - \dfrac{1}{2}c$
$4\sqrt{1-y} = e^{-x^2} + c$

9. $e^{x+y}dx + dy = 0$
$e^x e^y dx + dy = 0$
$e^x dx + e^{-y}dy = 0$
$e^x - e^{-y} = c$

11. $y' - y = 4$
$dy = (y+4)dx$
$\dfrac{dy}{y+4} = dx$
$\ln(y+4) = x + c$

13. $x\dfrac{dy}{dx} = y^2 + y^2 \ln x$
$xdy = y^2(1 + \ln x)dx$
$\dfrac{dy}{y^2} = \dfrac{(1 + \ln x)dx}{x}$
$-\dfrac{1}{y} = \dfrac{1}{2}(1 + \ln x)^2 + \dfrac{1}{2}c$
$-2 = y(1 + \ln x)^2 + cy$
$y(1 + \ln x)^2 + cy + 2 = 0$

15. $y \tan x\, dx + \cos^2 x\, dy = 0$
$\dfrac{\tan x\, dx}{\cos^2 x} + \dfrac{dy}{y} = 0$
$\tan x \sec^2 x\, dx + \dfrac{dy}{y} = 0$
$\dfrac{1}{2}\tan^2 x + \ln y = \dfrac{c}{2}$
$\tan^2 x + 2 \ln y = c$

17. $yx^2 dx = ydx - x^2 dy$
$yx^2 dx - ydx = -x^2 dy$
$y(x^2 - 1)dx = -x^2 dy$
$\dfrac{(x^2 - 1)dx}{x^2} = -\dfrac{dy}{y}$
$dx - \dfrac{dx}{x^2} = -\dfrac{dy}{y}$; $x + \dfrac{1}{x} = -\ln y - c$
$x^2 + 1 + x \ln y + cx = 0$

19. $y\sqrt{1 - x^2}\, dy + 2dx = 0$
$ydy + \dfrac{2dx}{\sqrt{1 - x^2}} = 0$
$\dfrac{1}{2}y^2 + 2\,\mathrm{Arcsin}\,x = \dfrac{1}{2}c$
$y^2 + 4\,\mathrm{Arcsin}\,x = c$

21. $2 \ln x \, dx + x \, dy = 0$

$2 \ln x \dfrac{dx}{x} + dy = 0$

$\ln^2 x + y = c$

$y = c - \ln^2 x$

23. $y^2 e^x + (e^x + 1)\dfrac{dy}{dx} = 0$

$\dfrac{e^x dx}{e^x + 1} + \dfrac{dy}{y^2} = 0$

$\ln(e^x + 1) - \dfrac{1}{y} = c$

25. $\dfrac{dy}{dx} + yx^2 = 0$

$\dfrac{dy}{y} + x^2 dx = 0$

$\ln y + \dfrac{1}{3} x^3 = \dfrac{1}{3} c$

$3 \ln y + x^3 = c$

$x = 0$ when $y = 1$

$3 \ln 1 + 0 = c, \quad c = 0$

$3 \ln y + x^3 = 0$

27. $(xy^2 + x)\dfrac{dy}{dx} = \ln x$

$x(y^2 + 1) dy = \ln x \, dx$

$(y^2 + 1) dy = \ln x \dfrac{dx}{x}$

$\dfrac{1}{3} y^3 + y = \dfrac{1}{2} \ln^2 x + c$

$x = 1$ when $y = 0$

$\dfrac{1}{3} 0^3 + 0 = \dfrac{1}{2} \ln^2 1 + c, \quad c = 0$

$\dfrac{1}{3} y^3 + y = \dfrac{1}{2} \ln^2 x$

29. $y' = (1 - y) \cos x$

$\dfrac{dy}{1 - y} = \cos x \, dx$

$-\ln(1 - y) = \sin x + c$

$x = \dfrac{\pi}{6}$ when $y = 0$

$-\ln(1 - 0) = \sin \dfrac{\pi}{6} + c \; ; \; 0 = \dfrac{1}{2} + c \; ; \; c = -\dfrac{1}{2}$

$-\ln(1 - y) = \sin x - \dfrac{1}{2}$

$2 \ln(1 - y) = 1 - 2 \sin x$

Exercises 14-3 , p. 487

1. $x \, dy + y \, dx + x \, dx = 0$

$(x \, dy + y \, dx) + x \, dx = 0$

$xy + \dfrac{1}{2} x^2 = \dfrac{c}{2}$

$2xy + x^2 = c$

3. $y \, dx - x \, dy + x^3 dx = 2dx$

$-\dfrac{x \, dy - y \, dx}{x^2} + x \, dx = \dfrac{2dx}{x^2}$

$-\dfrac{y}{x} + \dfrac{1}{2} x^2 = -\dfrac{2}{x} + \dfrac{c}{2}$

$x^3 - 2y = cx - 4$

5. $x^3 dy + x^2 y \, dx + y \, dx - x \, dy = 0$

$x^2(x \, dy + y \, dx) + y \, dx - x \, dy = 0$

$(x \, dy + y \, dx) - \dfrac{x \, dy - y \, dx}{x^2} = 0$

$xy - \dfrac{y}{x} = c \; ; \; x^2 y - y = cx$

7. $x^3 y^4 (x \, dy + y \, dx) = dy$

$(xy)^3 (x \, dy + y \, dx) = \dfrac{dy}{y}$

$\dfrac{1}{4}(xy)^4 = \ln y + \dfrac{1}{4} c$

$(xy)^4 = 4 \ln y + c$

9. $\sqrt{x^2 + y^2} \, dx - 2y \, dy = 2x \, dx$

$dx = \dfrac{2x \, dx + 2y \, dy}{\sqrt{x^2 + y^2}}$

$x = 2\sqrt{x^2 + y^2} - c$

$2\sqrt{x^2 + y^2} = x + c$

11. $\tan(x^2 + y^2) dy + x \, dx + y \, dy = 0$

$dy + \dfrac{x \, dx + y \, dy}{\tan(x^2 + y^2)} = 0$

$dy + \dfrac{1}{2} \cot(x^2 + y^2)(2x \, dx + 2y \, dy) = 0$

$y + \dfrac{1}{2} \ln \sin(x^2 + y^2) = c$

$y = c - \dfrac{1}{2} \ln \sin(x^2 + y^2)$

13. $y \, dy - x \, dx + (y^2 - x^2) dx = 0$

$\dfrac{1}{2} \dfrac{2y \, dy - 2x \, dx}{y^2 - x^2} + dx = 0$

$\dfrac{1}{2} \ln(y^2 - x^2) + x = \dfrac{c}{2}$

$\ln(y^2 - x^2) + 2x = c$

15. $10x \, dy + 5y \, dx + 3y \, dy = 0$

$5(2x \, dy + y \, dx) + 3y \, dy = 0$

$5(2yx \, dy + y^2 dx) + 3y^2 dy = 0$

$5xy^2 + y^3 = c$

17. $2(x \, dy + y \, dx) + 3x^2 dx = 0$

$2xy + x^3 = c$

$x = 1$ when $y = 2$

$2(1)(2) + 1^3 = c, \quad c = 5$

$2xy + x^3 = 5$

19. $y \, dx - x \, dy = y^3 dx + y^2 x \, dy$

$y \, dx - x \, dy = y^2(y \, dx + x \, dy)$

$\dfrac{y \, dx - x \, dy}{y^2} = y \, dx + x \, dy$

$\dfrac{x}{y} = xy + c$

$x = 2$ when $y = 4$

$\dfrac{2}{4} = 2(4) + c \; ; \; c = -\dfrac{15}{2}$

$\dfrac{x}{y} = xy - \dfrac{15}{2}$

$2x = 2xy^2 - 15y$

Exercises 14-4 , p. 490

1. $dy + ydx = e^{-x}dx$; $P = 1$, $Q = e^{-x}$, $e^{\int Pdx} = e^{\int dx} = e^x$

   $ye^x = \int e^{-x}e^x dx = \int dx = x + c$; $y = e^{-x}(x + c)$

3. $dy + 2ydx = e^{-4x}dx$; $P = 2$, $Q = e^{-4x}$, $e^{\int Pdx} = e^{\int 2dx} = e^{2x}$

   $ye^{2x} = \int e^{-4x}e^{2x}dx = \int e^{-2x}dx = -\frac{1}{2}e^{-2x} + c$; $y = -\frac{1}{2}e^{-4x} + ce^{-2x}$

5. $\frac{dy}{dx} - 2y = 4$; $dy - 2ydx = 4dx$; $P = -2$, $Q = 4$, $e^{\int Pdx} = e^{\int -2dx} = e^{-2x}$

   $ye^{-2x} = \int 4e^{-2x}dx = -2e^{-2x} + c$; $y = -2 + ce^{2x}$

7. $xdy - ydx = 3xdx$; $dy - \frac{1}{x}ydx = 3dx$; $P = -\frac{1}{x}$, $Q = 3$, $e^{\int Pdx} = e^{\int -dx/x} = e^{-\ln x} = e^{\ln x^{-1}} = \frac{1}{x}$

   $y(\frac{1}{x}) = \int 3(\frac{1}{x})dx = 3\ln x + c$; $y = x(3\ln x + c)$

9. $2xdy + ydx = 8x^3dx$; $dy + \frac{1}{2x}ydx = 4x^2dx$; $P = \frac{1}{2x}$, $Q = 4x^2$, $e^{\int Pdx} = e^{\int dx/2x} = e^{(\ln x)/2}$

   $yx^{1/2} = \int 4x^2(x^{1/2})dx = \int 4x^{5/2}dx = \frac{8}{7}x^{7/2} + c$; $y = \frac{8}{7}x^3 + \frac{c}{\sqrt{x}}$ $= e^{\ln x^{1/2}} = x^{1/2}$

11. $dy + y\cot x \, dx = dx$; $P = \cot x$, $Q = 1$, $e^{\int Pdx} = e^{\int \cot x \, dx} = e^{\ln \sin x} = \sin x$

    $y\sin x = \int \sin x \, dx = -\cos x + c$; $y = -\cot x + c\csc x$

13. $\sin x \frac{dy}{dx} = 1 - y\cos x$; $dy + \frac{\cos x}{\sin x}ydx = \frac{dx}{\sin x}$; $dy + y\cot x \, dx = \csc x \, dx$

    $P = \cot x$, $Q = \csc x$, $e^{\int Pdx} = e^{\int \cot x \, dx} = e^{\ln \sin x} = \sin x$

    $y\sin x = \int \csc x \sin x \, dx = \int dx = x + c$; $y = (x + c)\csc x$

15. $y' + y = 3$; $dy + ydx = 3dx$; $P = 1$, $Q = 3$, $e^{\int Pdx} = e^{\int dx} = e^x$

    $ye^x = \int 3e^x dx = 3e^x + c$; $y = 3 + ce^{-x}$

17. $\frac{dy}{dx} = xe^{4x} + 4y$; $dy - 4ydx = xe^{4x}dx$; $P = -4$, $Q = xe^{4x}$, $e^{\int Pdx} = e^{\int -4dx} = e^{-4x}$

    $ye^{-4x} = \int xe^{4x}e^{-4x}dx = \int xdx = \frac{1}{2}x^2 + \frac{c}{2}$; $2y = e^{4x}(x^2 + c)$

19. $y' = x^3(1 - 4y)$; $dy + 4x^3ydx = x^3dx$; $P = 4x^3$, $Q = x^3$, $e^{\int Pdx} = e^{\int 4x^3dx} = e^{x^4}$

    $ye^{x^4} = \int x^3 e^{x^4}dx = \frac{1}{4}\int e^{x^4}(4x^3dx) = \frac{1}{4}e^{x^4} + c$; $y = \frac{1}{4} + ce^{-x^4}$

21. $x\frac{dy}{dx} = y + (x^2-1)^2$; $dy - \frac{1}{x}ydx = \frac{1}{x}(x^2-1)^2dx$; $P = -\frac{1}{x}$, $Q = \frac{1}{x}(x^2-1)^2$; $e^{\int Pdx} = e^{\int -dx/x}$

    $= e^{-\ln x} = \frac{1}{x}$

    $y(\frac{1}{x}) = \int \frac{1}{x}[\frac{1}{x}(x^2-1)^2]dx = \int(x^2-2+\frac{1}{x^2})dx = \frac{1}{3}x^3 - 2x - \frac{1}{x} + \frac{c}{3}$

    $3y = x^4 - 6x^2 - 3 + cx$

23. $xdy + (1 - 3x)ydx = 3x^2e^{3x}dx$; $dy + (\frac{1}{x} - 3)ydx = 3xe^{3x}dx$

    $P = \frac{1}{x} - 3$, $Q = 3xe^{3x}$, $e^{\int Pdx} = e^{\int(1/x - 3)dx} = e^{\ln x - 3x} = e^{\ln x}e^{-3x} = xe^{-3x}$

    $y(xe^{-3x}) = \int(3xe^{3x})(xe^{-3x})dx = \int 3x^2dx = x^3 + c$; $xy = e^{3x}(x^3 + c)$

25. $\dfrac{dy}{dx} + 2y = e^{-x}$; $dy + 2ydx = e^{-x}dx$; $P = 2$, $Q = e^{-x}$, $e^{\int Pdx} = e^{\int 2dx} = e^{2x}$

$ye^{2x} = \int e^{-x}e^{2x}dx = \int e^x dx = e^x + c$; $y = e^{-x} + ce^{-2x}$

$x=0$ when $y=1$; $1 = e^0 + ce^0$, $c=0$; $y = e^{-x}$

27. $xdy - 2ydx = x^3\cos x\, dx$; $dy - \dfrac{2}{x}ydx = x^2\cos x\, dx$; $P = -\dfrac{2}{x}$, $Q = x^2\cos x\, dx$

$e^{\int Pdx} = e^{\int -2dx/x} = e^{-2\ln x} = e^{\ln x^{-2}} = \dfrac{1}{x^2}$

$y(\dfrac{1}{x^2}) = \int (x^2\cos x)(\dfrac{1}{x^2})dx = \int \cos x\, dx = \sin x + c$; $y = x^2(\sin x + c)$

$y = \pi^2$ when $x = \dfrac{\pi}{2}$, $\pi^2 = (\dfrac{\pi}{2})^2(\sin\dfrac{\pi}{2} + c)$; $4 = 1 + c$, $c = 3$; $y = x^2(3 + \sin x)$

29. $\sqrt{x}\, y' + \dfrac{1}{2}y = e^{\sqrt{x}}$; $dy + \dfrac{1}{2\sqrt{x}}ydx = \dfrac{1}{\sqrt{x}}e^{\sqrt{x}}$; $P = \dfrac{1}{2}x^{-1/2}$, $Q = \dfrac{1}{\sqrt{x}}e^{\sqrt{x}}$

$e^{\int Pdx} = e^{\int (1/2)x^{-1/2}dx} = e^{x^{1/2}} = e^{\sqrt{x}}$

$ye^{\sqrt{x}} = \int \dfrac{1}{\sqrt{x}}e^{\sqrt{x}}e^{\sqrt{x}}dx = \int \dfrac{1}{\sqrt{x}}e^{2\sqrt{x}}dx = e^{2\sqrt{x}} + c$; $y = e^{\sqrt{x}} + ce^{-\sqrt{x}}$

$x=1$ when $y=3$; $3 = e + ce^{-1}$, $c = 3e - e^2$; $y = e^{\sqrt{x}} + (3e - e^2)e^{-\sqrt{x}}$

Exercises 14-5, p. 496

1. $\dfrac{dy}{dx} = \dfrac{2x}{y}$ Curve passes through 2,3).

$ydy = 2xdx$

$\dfrac{1}{2}y^2 = x^2 + c$

$\dfrac{1}{2}(3)^2 = 2^2 + c$, $c = \dfrac{1}{2}$

$\dfrac{1}{2}y^2 = x^2 + \dfrac{1}{2}$

$y^2 = 2x^2 + 1$

3. $\dfrac{dy}{dx} = y + x$; $dy - ydx = xdx$; $P = -1$, $Q = x$, $e^{\int -dx} = e^{-x}$

$ye^{-x} = \int xe^{-x}dx = \dfrac{e^{-x}(-x-1)}{(-1)^2} + c = -(x+1)e^{-x} + c$

$y = ce^x - (x+1)$ (Formula 44)

Curve passes through $(0,1)$; $1 = ce^0 - (0+1)$;

$1 = c - 1$, $c = 2$; $y = 2e^x - x - 1$

5. See Example B.

$y = ce^x$, $c = ye^{-x}$

$y' = ce^x = ye^{-x}e^x$

$y' = y$

$y'|_{OT} = -\dfrac{1}{y}$

$ydy = -dx$

$\dfrac{1}{2}y^2 = -x + \dfrac{c'}{2}$

$y^2 = c' - 2x$

7. See Example B.

$y = c(\sec x + \tan x)$

$y' = c(\sec x\tan x + \sec^2 x)$

$= c\sec x(\sec x + \tan x)$

$= y\sec x$

$y'|_{OT} = -\dfrac{1}{y\sec x}$

$ydy = -\cos x\, dx$

$\dfrac{1}{2}y^2 = -\sin x + \dfrac{c'}{2}$

$y^2 = c' - 2\sin x$

9. See Example C.

$\dfrac{dN}{dt} = kN$; $N = N_0 e^{kt}$

$N = \dfrac{N_0}{2}$ when $t = 2$ h

$\dfrac{N_0}{2} = N_0 e^{2k}$; $0.5 = e^{2k}$

$e^k = (0.5)^{1/2}$; $N = N_0(0.5)^{t/2}$

For $t = 3$ h, $N = N_0(0.5)^{3/2}$

$= 0.354 N_0$

11. See Example C.

$\dfrac{dN}{dt} = kN$; $N = N_0 e^{kt}$

$N = 0.5N_0$ for $t = 5600$ yr

$0.5N_0 = N_0 e^{5600k}$

$0.5 = e^{5600k}$

$e^k = (0.5)^{1/5600}$

$N = N_0(0.5)^{t/5600}$

For $N = 0.6N_0$

$0.6N_0 = N_0(0.5)^{t/5600}$

$0.6 = 0.5^{t/5600}$

$\ln 0.6 = \ln 0.5^{t/5600}$

$= \dfrac{t}{5600}\ln 0.5$

$t = 5600\dfrac{\ln 0.6}{\ln 0.5} = 4130$ yr

13. Let T = temperature at time t.

$$\frac{dT}{dt} = k(T - 80)$$

$$\frac{dT}{T - 80} = kdt$$

$$\ln(T-80) = kt + \ln c$$

$$\ln\frac{T - 80}{c} = kt$$

$$T = 80 + ce^{kt}$$

T = 200°F when t = 0

$200 = 80 + c$, $c = 120$

$\underline{T = 80 + 120e^{kt}}$

T = 140°F when t = 5 min

$140 = 80 + 120e^{5k}$

$0.5 = e^{5k}$; $e^k = (0.5)^{1/5}$

$T = 80 + 120(0.5)^{t/5}$

For T = 100°F

$100 = 80 + 120(0.5)^{t/5}$

$\frac{1}{6} = (0.5)^{t/5}$

$\ln\frac{1}{6} = \ln(0.5)^{t/5} = \frac{t}{5}\ln 0.5$

$t = \frac{5\ln(1/6)}{\ln 0.5} = 12.9$ min

15. Let V = amount in account at time t. Since V earns 8% annual interest, V increases at an 8% rate, or k = 0.08.

$$\frac{dV}{dt} = kV$$

$$\frac{dV}{V} = kdt$$

$$\ln V = kt + c$$

V = $1000 when t=0

$\ln 1000 = c$

$\ln V = kt + \ln 1000$

$\ln\frac{V}{1000} = kt$; $V = 1000e^{kt}$

$V = 1000e^{0.08t}$

For t = 1 year

$V = 1000e^{0.08} = \$1083.29$

17. See Example D.

$$L\frac{di}{dt} + Ri = E$$

$$i = \frac{E}{R}(1 - e^{-Rt/L})$$

$\lim_{t\to\infty} i = \lim_{t\to\infty}\frac{E}{R}(1 - e^{-Rt/L})$

$= \frac{E}{R}(1 - 0) = \frac{E}{R}$

19. See Example D

$$L\frac{di}{dt} + Ri = E$$

$$i = \frac{E}{R}(1 - e^{-Rt/L})$$

R=10 Ω, L=0.1 H

E=100 V, t=0.001 s

$i = 10(1 - e^{-0.1})$

$= 0.952$ A

21. See Example D.

$$Ri + \frac{q}{C} = 0$$

$$R\frac{dq}{dt} + \frac{q}{C} = 0$$

$$\frac{dq}{q} + \frac{1}{RC}dt = 0$$

$\ln q + \frac{t}{RC} = c$

q=q_0 when t=0

$\ln q_0 = c$

$\ln q + \frac{t}{RC} = \ln q_0$

$\ln\frac{q}{q_0} = -\frac{t}{RC}$

$q = q_0 e^{-t/RC}$

23. See Example E. Let x = no. pounds of salt at time t. Each gal. contains $\frac{x}{100}$ lb.

$$-dx = 5\left(\frac{x}{100}\right)dt$$

$$\frac{dx}{x} = -\frac{dt}{20}$$

$$\ln x = -\frac{t}{20} + c$$

x=30 lb when t=0

$\ln 30 = c$

$\ln x = -\frac{t}{20} + \ln 30$

$\ln\frac{x}{30} = -\frac{t}{20}$

$x = 30 e^{-t/20}$

For t = 20 min

$x = 30e^{-1} = 11.04$ lb

25.

$$\frac{dv}{dt} = 32 - v$$

$$\frac{dv}{32 - v} = dt$$

$$-\ln(32-v) = t + c$$

v=0 when t=0

$-\ln 32 = c$

$-\ln(32-v) = t - \ln 32$

$\ln(32-v) - \ln 32 = -t$

$\ln\frac{32 - v}{32} = -t$

$\frac{32 - v}{32} = e^{-t}$

$v = 32(1 - e^{-t})$

$\lim_{t\to\infty} v = 32$

27. See Example F ; m = 10 slugs, F = 20 lb, k = 2 (velocity in ft/s)

$$dv + \frac{1}{5}vdt = 2dt$$

$$e^{\int dt/5} = e^{t/5}$$

$$ve^{t/5} = 2\int e^{t/5}dt = 10e^{t/5} + c$$

$v = 10 + ce^{-t/5}$

v=8 mi/h = 11.7 ft/s

when t=0

$11.7 = 10 + c$, $c = 1.7$

$v = 10 + 1.7e^{-t/5}$

When t = 3 min = 180 s

$v = 10 + 1.7e^{-180/5} = 10.0$ ft/s

$= 10.0$ ft/s

29. $\dfrac{dx}{dt} = 2t$ $x=1$ when $t=0$ $xy = 1$

$dx = 2tdt$ $1 = 0 + c, \ c = 1$ $y = \dfrac{1}{x} = \dfrac{1}{t^2 + 1}$

$x = t^2 + c$ $x = t^2 + 1$

31. Let p = pressure at height h

$\dfrac{dp}{dh} = kp$ $p = 15 \ lb/in^2$ for $h = 0$ $p = 10 \ lb/in^2$ for $h = 10^4$ ft

      $\ln 15 = c, \ \ln p = kh + \ln 15$ $10 = 15e^{10^4 k}, \ e^k = (\dfrac{2}{3})^{10^{-4}}$

$\dfrac{dp}{p} = kdh$ $\ln \dfrac{p}{15} = kh$

$\ln p = kh + c$ $p = 15e^{kh}$ $p = 15(0.667)^{10^{-4}h}$

33. Let V = value at time t.

$\dfrac{dV}{dt} = kV$ $V = \$8200$ for $t = 0$ $V = 8200e^{kt}$ $V = 8200(0.600)^{t/3}$

      $\ln 8200 = c$ $V = \$4920$ for $t = 3$ yr   For $t = 11$ years

$\dfrac{dV}{V} = kdt$ $\ln V = kt + \ln 8200$ $4920 = 8200e^{3k}$ $V = 8200(0.600)^{11/3}$

$\ln V = kt + c$ $\ln \dfrac{V}{8200} = kt$ $e^k = (\dfrac{4920}{8200})^{1/3} = (0.600)^{1/3}$ $= \$1260$

35. $\dfrac{dx}{dt} = 1 - 0.25x$ $4 \ln(0.25x - 1) = -t + c$ $4 \ln(0.25x - 1) = -t + 4 \ln 2$

      $x = 12 \ ft^3$ for $t = 0$ $\ln \dfrac{0.25x - 1}{2} = -\dfrac{t}{4}$

$\dfrac{dx}{1 - 0.25x} = dt$ $4 \ln(3 - 1) = c, \ c = 4 \ln 2$

      (Note that $0.25x - 1 = 2e^{-0.25t}$

$\dfrac{dx}{0.25x - 1} = -dt$ $0.25x - 1 > 0$) $x = 4(1 + 2e^{-0.25t})$

(To have positive
value for $0.25x - 1$.)

Review Exercises for Chapter 14, p. 498

1. $4xy^3dx + (x^2 + 1)dy = 0$ **3.** $\dfrac{dy}{dx} + 2y = e^{-2x}$ $e^{\int Pdx} = e^{\int 2dx} = e^{2x}$

$\dfrac{4xdx}{x^2 + 1} + \dfrac{dy}{y^3} = 0$ $dy + 2ydx = e^{-2x}dx$ $ye^{2x} = \int e^{-2x}e^{2x}dx = \int dx = x + c$

      $P = 2, \ Q = e^{-2x}$ $y = e^{-2x}(x + c)$

$2 \ln(x^2 + 1) - \dfrac{1}{2y^2} = c$

5. $(x+y)dx + (x+y^3)dy = 0$ **7.** $x\dfrac{dy}{dx} - 3y = x^2$ $yx^{-3} = \int xx^{-3}dx = \int x^{-2}dx$

$xdx + ydx + xdy + y^3dy = 0$ $dy - \dfrac{3}{x}ydx = xdx$ $= -\dfrac{1}{x} + c$

$(ydx + xdy) + xdx + y^3dy = 0$ $P = -\dfrac{3}{x}, \ Q = x$ $y = cx^3 - x^2$

$xy + \dfrac{1}{2}x^2 + \dfrac{1}{4}y^4 = \dfrac{1}{4}c$

$2x^2 + 4xy + y^4 = c$ $e^{\int -3dx/x} = e^{-3 \ln x} = x^{-3}$

9. $dy = 2ydx + y^2dx$ $\dfrac{1}{2} \ln \dfrac{y}{y+2} = x + \dfrac{1}{2} \ln c$ **11.** $y' + 4y = 2$; $dy + 4ydx = 2dx$

$\dfrac{dy}{y^2 + 2y} = dx$; $\dfrac{dy}{y(y+2)} = dx$ $\ln \dfrac{y}{c(y+2)} = 2x$ $P = 4, \ Q = 2, \ e^{\int 4dx} = e^{4x}$

$\dfrac{1}{y(y+2)} = \dfrac{1}{2y} - \dfrac{1}{2(y+2)}$ $y = c(y+2)e^{2x}$ $ye^{4x} = \int 2e^{4x}dx = \dfrac{1}{2}e^{4x} + c$

$\dfrac{dy}{2y} - \dfrac{dy}{2(y+2)} = dx$ $y = \dfrac{1}{2} + ce^{-4x}$

13. $\sin x \dfrac{dy}{dx} + y \cos x + x = 0$

$dy + y \cot x\, dx = -x \csc x\, dx$

$P = \cot x, \quad Q = -x \csc x\, dx$

$e^{\int \cot x\, dx} = e^{\ln \sin x} = \sin x$

$y \sin x = -\int x \csc x \sin x\, dx$

$\qquad = -\int x\, dx = -\dfrac{1}{2} x^2 + \dfrac{c}{2}$

$y = \dfrac{1}{2}(c - x^2)\csc x$

15. $(x^2 + 2x)dy = 2xy\,dx + 5y\,dx$

$(x^2 + 2x)dy = y(2x+5)dx$

$\dfrac{dy}{y} = \dfrac{2x+5}{x^2 + 2x}\,dx \; ; \; \dfrac{2x+5}{x^2 + 2x} = \dfrac{5}{2x} - \dfrac{1}{2(x+2)}$

$\dfrac{dy}{y} = \dfrac{5\,dx}{2x} - \dfrac{dx}{2(x+2)}$

$\ln y = \dfrac{5}{2}\ln x - \dfrac{1}{2}\ln(x+2) + \ln c$

$y = \dfrac{cx^{5/2}}{(x+2)^{1/2}}$

17. $x\,dy - y\,dx = x^3 y\,dx$

$\dfrac{dy}{y} - \dfrac{dx}{x} = x^2\,dx$

$\ln y - \ln x = \dfrac{1}{3}x^3 + \ln c$

$\ln \dfrac{y}{cx} = \dfrac{1}{3}x^3$

$y = cxe^{x^3/3}$

19. $y\,dy = 2(x^2 + y^2)^2\,dx - x\,dx$

$x\,dx + y\,dy = 2(x^2 + y^2)^2\,dx$

$\dfrac{2x\,dx + 2y\,dy}{(x^2 + y^2)^2} = 4\,dx$

$-\dfrac{1}{x^2 + y^2} = 4x + c$

$-1 = (x^2 + y^2)(4x + c)$

$(x^2 + y^2)(4x+c) + 1 = 0$

21. $y' = 2y \cot x$

$\dfrac{dy}{y} = 2 \cot x\, dx$

$\ln y = 2 \ln \sin x + \ln c$

$y = c \sin^2 x$

$x = \dfrac{\pi}{2}$ when $y = 2$

$2 = c \sin^2 \dfrac{\pi}{2}, \quad c = 2$

$y = 2 \sin^2 x$

23. $y' = 4x - 2y$

$dy + 2y\,dx = 4x\,dx$

$P = 2, \quad Q = 4x, \quad e^{\int 2\,dx} = e^{2x}$

$ye^{2x} = \int 4xe^{2x}\,dx = e^{2x}(2x-1) + c$

$\qquad\qquad$ (Formula 44)

$y = 2x - 1 + ce^{-2x}$

$x = 0$ when $y = -2$; $-2 = -1 + c$; $c = -1$

$y = 2x - 1 - e^{-2x}$

25. $\dfrac{dy}{dx} = x - xy = x(1-y)$

$\dfrac{dy}{1-y} = x\,dx \; ; \; -\ln(1-y) = \dfrac{1}{2}x^2 + c$

$x = 2$ when $y = 0$; $-\ln 1 = 2 + c, \quad c = -2$

$-\ln(1-y) = \dfrac{1}{2}x^2 - 2$

$x^2 + 2 \ln(1-y) - 4 = 0$

27. $xy' - 2y = x^3 \cos 4x$

$dy - \dfrac{2}{x}y\,dx = x^2 \cos 4x\, dx$

$P = -\dfrac{2}{x}, \quad Q = x^2 \cos 4x$

$e^{\int -2\,dx/x} = e^{-2 \ln x} = e^{\ln x^{-2}} = x^{-2}$

$\dfrac{y}{x^2} = \int x^{-2}x^2 \cos 4x\, dx = \int \cos 4x\, dx$

$\qquad = \dfrac{1}{4}\sin 4x + c \; ; \; y = x^2 \left(\dfrac{1}{4}\sin 4x + c \right)$

$x = \dfrac{\pi}{4}$ when $y = 0$; $0 = \left(\dfrac{\pi}{4}\right)^2 \left(\dfrac{1}{4}\sin \pi + c\right)$

$c = 0 \; ; \; y = \dfrac{1}{4}x^2 \sin 4x$

29. $\dfrac{dV}{dt} = kA$

$V = \dfrac{4}{3}\pi r^3, \quad A = 4\pi r^2$

$\dfrac{dV}{dt} = 4\pi r^2 \dfrac{dr}{dt}$

$4\pi r^2 \dfrac{dr}{dt} = k(4\pi r^2)$

$dr = k\,dt \; ; \; r = kt + c$

$r = r_0$ when $t = 0, \quad c = r_0$

$r = r_0 + kt$

31. See Example F, Section 14-5

$m = 1\,kg, \quad F = 4\,N, \quad k = 1$

$\dfrac{dv}{dt} = 4 - v \; ; \; \dfrac{dv}{4-v} = dt$

$-\ln(4-v) = t + c$

$v = 0$ when $t = 0, \quad c = -\ln 4$

$-\ln(4-v) = t - \ln 4$

$\ln \dfrac{4}{4-v} = t$

$4 = (4-v)e^t$

$v = 4 - 4e^{-t} = 4(1 - e^{-t})$

When $t = 4\,s$,

$v = 4(1 - e^{-4}) = 3.93\,m/s$

33. See Example C, Section 14-5

$\dfrac{dN}{dt} = kN, \quad N = N_o e^{kt}$

N=100 mg when t=0, $\dot{N}_o = 100$

$N = 100 e^{kt}$

N=95 mg when t=2 years

$95 = 100 e^{2k}, \quad e^k = (0.95)^{1/2}$

$N = 100(0.95)^{t/2} \quad$ OR

$e^k = (0.95)^{1/2} ; \quad k = \dfrac{1}{2}\ln 0.95 = -0.026$

$N = 100 e^{-0.026t}$

35. Let P = population at time t, where t = 0 is 1970

$\dfrac{dP}{dt} = kP \; ; \quad \dfrac{dP}{P} = kdt \; ; \quad \ln P = kt + c$

P=10,000 when t=0, $c = \ln 10,000$

$\ln P = kt + \ln 10,000$

$\ln\dfrac{P}{10,000} = kt \; ; \quad P = 10,000 e^{kt}$

P=12,000 when t=10 years

$12,000 = 10,000 e^{10k}, \quad e^k = 1.2^{1/10}$

$P = 10,000(1.2)^{t/10}$

For t = 20 years : $P = 10,000(1.2)^2 = 14,400$

37. See Example B, Section 14-5

$y = cx^5, \quad c = \dfrac{y}{x^5}$

$y' = 5cx^4 = 5\left(\dfrac{y}{x^5}\right)x^4 = \dfrac{5y}{x}$

$y'\big|_{OT} = -\dfrac{x}{5y} \; ; \quad 5ydy = -xdx$

$\dfrac{5}{2}y^2 = -\dfrac{1}{2}x^2 + \dfrac{1}{2}c$

$5y^2 + x^2 = c$

39. See Example D, Section 14-5

$Ri + \dfrac{1}{C}q = E$

$R\dfrac{dq}{dt} + \dfrac{1}{C}q = E \; ; \quad dq + \dfrac{1}{RC}q\,dt = \dfrac{E}{R}dt$

$P = \dfrac{1}{RC}, \quad Q = \dfrac{E}{R} \; , \quad e^{\int dt/RC} = e^{t/RC}$

$qe^{t/RC} = \int \dfrac{E}{R}e^{t/RC}\,dt = ECe^{t/RC} + c_1$

$q = c_1 e^{-t/RC} + EC$

41. See Fig. 14-7

$r = a\sin\theta, \quad a = \dfrac{r}{\sin\theta}$

$\dfrac{dr}{d\theta} = a\cos\theta = \dfrac{r}{\sin\theta}\cos\theta = r\cot\theta \; ; \quad r\dfrac{d\theta}{dr} = \tan\theta$

$\phi_{OT} = \phi + \dfrac{\pi}{2} \quad$ or $\tan\phi_{OT} = -\dfrac{1}{\tan\phi}$

$\dfrac{dr}{d\theta} = \dfrac{r}{\tan\phi} \; , \quad \tan\phi = r\dfrac{d\theta}{dr}$

$r\dfrac{d\theta}{dr}\Big|_{OT} = -\dfrac{1}{\tan\theta}$

$\tan\theta\,d\theta = -\dfrac{dr}{r}$

$-\ln\cos\theta = -\ln r + \ln c$

$r = c\cos\theta$

Exercises 15-1,15-2 , p. 505

1. $\dfrac{d^2y}{dx^2} - \dfrac{dy}{dx} - 6y = 0$

$D^2y - Dy - 6y = 0$

$m^2 - m - 6 = 0$

$(m-3)(m+2) = 0$

$m = 3, \; -2$

$y = c_1 e^{3x} + c_2 e^{-2x}$

3. $\dfrac{d^2y}{dx^2} + 4\dfrac{dy}{dx} + 3y = 0$

$D^2y + 4Dy + 3y = 0$

$m^2 + 4m + 3 = 0$

$(m+1)(m+3) = 0$

$m = -1, \; -3$

$y = c_1 e^{-x} + c_2 e^{-3x}$

5. $D^2y - Dy = 0$

$m^2 - m = 0$

$m(m-1) = 0$

$m = 0, \; 1$

$y = c_1 e^{0x} + c_2 e^{x}$

$= c_1 + c_2 e^{x}$

7. $3D^2y + 12y = 20Dy$

$3D^2y - 20Dy + 12y = 0$

$3m^2 - 20m + 12 = 0$

$(m-6)(3m-2) = 0$

$m = 6, \; 2/3$

$y = c_1 e^{6x} + c_2 e^{2x/3}$

9. $3y'' + 8y' - 3y = 0$

$3D^2y + 8Dy - 3y = 0$

$3m^2 + 8m - 3 = 0$

$(3m-1)(m+3) = 0$

$m = 1/3, \; -3$

$y = c_1 e^{x/3} + c_2 e^{-3x}$

11. $3y'' + 2y' - y = 0$

$3D^2y + 2Dy - y = 0$

$3m^2 + 2m - 1 = 0$

$(3m-1)(m+1) = 0$

$m = 1/3, \; -1$

$y = c_1 e^{x/3} + c_2 e^{-x}$

13. $2\dfrac{d^2y}{dx^2} - 4\dfrac{dy}{dx} + y = 0$

$2D^2y - 4Dy + y = 0$

$2m^2 - 4m + 1 = 0$

$m = \dfrac{4 \pm \sqrt{16-8}}{4} = 1 \pm \dfrac{1}{2}\sqrt{2}$

$y = c_1 e^{(1+\sqrt{2}/2)x} + c_2 e^{(1-\sqrt{2}/2)x}$

$y = e^{x}(c_1 e^{x\sqrt{2}/2} + c_2 e^{-x\sqrt{2}/2})$

15. $4D^2y - 3Dy - 2y = 0$

$4m^2 - 3m - 2 = 0$

$m = \dfrac{3 \pm \sqrt{9+32}}{8} = \dfrac{3+\sqrt{41}}{8}, \; \dfrac{3-\sqrt{41}}{8}$

$y = c_1 e^{(3+\sqrt{41})x/8} + c_2 e^{(3-\sqrt{41})x/8}$

$= e^{3x/8}(c_1 e^{x\sqrt{41}/8} + c_2 e^{-x\sqrt{41}/8})$

17. $y'' = 3y' + y$

$D^2y - 3Dy - y = 0 \;$; $\; m^2 - 3m - 1 = 0$

$m = \dfrac{3 \pm \sqrt{9+4}}{2} = \dfrac{3+\sqrt{13}}{2}, \; \dfrac{3-\sqrt{13}}{2}$

$y = c_1 e^{(3+\sqrt{13})x/2} + c_2 e^{(3-\sqrt{13})x/2}$

$= e^{3x/2}(c_1 e^{x\sqrt{13}/2} + c_2 e^{-x\sqrt{13}/2})$

19. $y'' + y' = 8y$

$D^2y + Dy - 8y = 0 \;$; $\; m^2 + m - 8 = 0$

$m = \dfrac{-1 \pm \sqrt{1+32}}{2} = \dfrac{-1+\sqrt{33}}{2}, \; \dfrac{-1-\sqrt{33}}{2}$

$y = c_1 e^{(-1+\sqrt{33})x/2} + c_2 e^{(-1-\sqrt{33})x/2}$

$= e^{-x/2}(c_1 e^{x\sqrt{33}/2} + c_2 e^{-x\sqrt{33}/2})$

21. $D^2y - 4Dy - 21y = 0$

$m^2 - 4m - 21 = 0$

$(m-7)(m+3) = 0 \;$; $\; m = 7, \; -3$

$y = c_1 e^{7x} + c_2 e^{-3x}$

$Dy=0$ and $y=2$ when $x=0$

$Dy = 7c_1 e^{7x} \; -3c_2 e^{-3x}$

$0 = 7c_1 - 3c_2 \qquad c_1 = 3/5$

$\left.\begin{array}{l} 2 = c_1 + c_2 \end{array}\right\} \; c_2 = 7/5$

$y = \dfrac{3}{5} e^{7x} + \dfrac{7}{5} e^{-3x}$

$= \dfrac{1}{5}(3e^{7x} + 7e^{-3x})$

23. $D^2y - Dy - 12y = 0$ $y=0$ when $x=0$ and $y=1$ when $x=1$

   $m^2 - m - 12 = 0$ $0 = c_1 + c_2$ $c_1 = \dfrac{e^3}{e^7-1}$, $c_2 = -\dfrac{e^3}{e^7-1}$

   $(m-4)(m+3) = 0$ $1 = c_1e^4 + c_2e^{-3}$

   $m = 4, -3$ ---------------- $y = \dfrac{e^3}{e^7-1}(e^{4x} - e^{-3x})$

   $y = c_1e^{4x} + c_2e^{-3x}$ $e^3 = c_1e^7 + c_2$

   $e^3 = c_1(e^7 - 1)$

25. $y''' - 2y'' - 3y' = 0$ 27. $y''' - 6y'' + 11y' - 6y = 0$

   $D^3y - 2D^2y - 3Dy = 0$ $D^3y - 6D^2y + 11Dy - 6y = 0$

   $m^3 - 2m^2 - 3m = 0$ $m^3 - 6m^2 + 11m - 6 = 0$

   $m(m+1)(m-3) = 0$; $m = 0, -1, 3$ $(m-1)(m-2)(m-3) = 0$; $m = 1, 2, 3$

   $y = c_1 + c_2e^{-x} + c_3e^{3x}$ $y = c_1e^x + c_2e^{2x} + c_3e^{3x}$

Exercises 15-3, p. 510

1. $\dfrac{d^2y}{dx^2} - 2\dfrac{dy}{dx} + y = 0$ 3. $D^2y + 12Dy + 36y = 0$ 5. $\dfrac{d^2y}{dx^2} + 9y = 0$; $D^2y + 9y = 0$

   $D^2y - 2Dy + y = 0$ $m^2 + 12m + 36 = 0$ $m^2 + 9 = 0$; $m = \pm 3j$; $\alpha = 0, \beta = 3$

   $m^2 - 2m + 1 = 0$ $(m+6)^2 = 0$; $m = -6, -6$ $y = e^{0x}(c_1 \sin 3x + c_2 \cos 3x)$

   $(m-1)^2 = 0$; $m = 1, 1$ $y = e^{-6x}(c_1 + c_2x)$ $= c_1 \sin 3x + c_2 \cos 3x$

   $y = e^x(c_1 + c_2x)$

7. $D^2y + Dy + 2y = 0$ 9. $D^2y = 0$ 11. $4D^2y + y = 0$

   $m^2 + m + 2 = 0$ $m^2 = 0$ $4m^2 + 1 = 0$; $m = \pm \frac{1}{2}j$

   $m = \dfrac{-1 \pm \sqrt{1-8}}{2} = \dfrac{-1 \pm j\sqrt{7}}{2}$ $m = 0, 0$ $\alpha = 0, \beta = \frac{1}{2}$

   $\alpha = -\frac{1}{2}, \beta = \frac{1}{2}\sqrt{7}$ $y = e^{0x}(c_1 + c_2x)$ $y = e^{0x}(c_1\sin\frac{1}{2}x + c_2\cos\frac{1}{2}x)$

   $y = e^{-x/2}(c_1\sin\frac{1}{2}\sqrt{7}x + c_2\cos\frac{1}{2}\sqrt{7}x)$ $= c_1 + c_2x$ $= c_1\sin\frac{1}{2}x + c_2\cos\frac{1}{2}x$

13. $16y'' - 24y' + 9y = 0$ 15. $25y'' + 2y = 0$ $y = e^{0x}(c_1 \sin\frac{1}{5}\sqrt{2}x + c_2 \cos\frac{1}{5}\sqrt{2}x)$

   $16D^2y - 24Dy + 9y = 0$ $25D^2y + 2y = 0$ $= c_1 \sin\frac{1}{5}\sqrt{2}x + c_2 \cos\frac{1}{5}\sqrt{2}x$

   $16m^2 - 24m + 9 = 0$ $25m^2 + 2 = 0$

   $(4m-3)^2 = 0$; $m = \frac{3}{4}, \frac{3}{4}$ $m = \pm \frac{1}{5}j\sqrt{2}$

   $y = e^{3x/4}(c_1 + c_2x)$ $\alpha = 0, \beta = \frac{1}{5}\sqrt{2}$

17. $2D^2y + 5y = 4Dy$ 19. $25y'' + 16y = 40y'$

   $2D^2y - 4Dy + 5y = 0$ $25D^2y - 40Dy + 16y = 0$

   $2m^2 - 4m + 5 = 0$ $25m^2 - 40m + 16 = 0$

   $m = \dfrac{4 \pm \sqrt{16-40}}{4} = 1 \pm \frac{1}{2}j\sqrt{6}$ $(5m-4)^2 = 0$; $m = \frac{4}{5}, \frac{4}{5}$

   $\alpha = 1, \beta = \frac{1}{2}\sqrt{6}$ $y = e^{4x/5}(c_1 + c_2x)$

   $y = e^x(c_1 \sin\frac{1}{2}\sqrt{6}x + c_2 \cos\frac{1}{2}\sqrt{6}x)$

21. $2D^2y - 3Dy - y = 0$

$2m^2 - 3m - 1 = 0$

$m = \dfrac{3 \pm \sqrt{9+8}}{4} = \dfrac{3 + \sqrt{17}}{4}, \dfrac{3 - \sqrt{17}}{4}$

(nonrepeated real roots)

$y = c_1 e^{(3+\sqrt{17})x/4} + c_2 e^{(3-\sqrt{17})x/4}$

$\quad = e^{3x/4}(c_1 e^{x\sqrt{17}/4} + c_2 e^{-x\sqrt{17}/4})$

23. $3D^2y + 12Dy = 2y$

$3D^2y + 12Dy - 2y = 0$

$3m^2 + 12m - 2 = 0$

$m = \dfrac{-12 \pm \sqrt{144+24}}{6} = \dfrac{-6 \pm \sqrt{42}}{3}$

(nonrepeated real roots)

$y = c_1 e^{(-6+\sqrt{42})x/3} + c_2 e^{(-6-\sqrt{42})x/3}$

$\quad = e^{-2x}(c_1 e^{x\sqrt{42}/3} \; c_2 e^{-x\sqrt{42}/3})$

25. $y'' + 2y' + 10y = 0$

$D^2y + 2Dy + 10y = 0$

$m^2 + 2m + 10 = 0$

$m = \dfrac{-2 \pm \sqrt{4-40}}{2} = -1 \pm 3j$

$y = e^{-x}(c_1 \sin 3x + c_2 \cos 3x)$

$y=0$ when $x=0$ and

$y=e^{-1}$ when $x=\pi/6$

$0 = 0c_1 + c_2$, $c_2 = 0$

$e^{-1} = e^{-\pi/6}c_1$, $c_1 = e^{\pi/6 - 1}$

$y = e^{\pi/6-1}e^{-x}\sin 3x$

$\quad = e^{(\pi/6-1-x)}\sin 3x$

27. $D^2y - 8Dy + 16y = 0$

$m^2 - 8m + 16 = 0$

$(m-4)^2 = 0$; $m = 4, 4$

$y = e^{4x}(c_1 + c_2 x)$

$Dy=2$ and $y=4$ when $x=0$

$Dy = e^{4x}(c_2 + 4c_1 + 4c_2 x)$

$2 = c_2 + 4c_1$

$4 = c_1$; $c_2 = -14$

$y = e^{4x}(4 - 14x)$

29. $y = c_1 e^{3x} + c_2 e^{-3x}$

$m = 3, -3$; $(m-3)(m+3) = 0$

$m^2 - 9 = 0$; $D^2y - 9y = 0$

$(D^2 - 9)y = 0$

31. $y = c_1 \cos 3x + c_2 \sin 3x$

$\alpha = 0$, $\beta = 3$; $m = \pm 3j$

$m^2 + 9 = 0$; $D^2y + 9y = 0$

$(D^2 + 9)y = 0$

Exercises 15-4, p. 514

1. $D^2y - Dy - 2y = 4$

$m^2 - m - 2 = 0$

$(m-2)(m+1) = 0$, $m=2,-1$

$y_c = c_1 e^{2x} + c_2 e^{-x}$

$y_p = A$, $y_p' = 0$, $y_p'' = 0$

$0 - 0 - 2A = 4$, $A = -2$

$y = c_1 e^{2x} + c_2 e^{-x} - 2$

3. $D^2y + y = x^2$

$m^2 + 1 = 0$, $m = \pm j$

$y_c = c_1 \sin x + c_2 \cos x$

$y_p = A + Bx + Cx^2$

$y_p' = B+2Cx$, $y_p'' = 2C$

$2C + A + Bx + Cx^2 = x^2$

$A+2C=0$, $B=0$, $C=1$, $A=-2$

$y = c_1 \sin x + c_2 \cos x - 2 + x^2$

5. $D^2y + 4Dy + 3y = 2 + e^x$

$m^2 + 4m + 3 = 0$

$(m+1)(m+3)=0$, $m=-1,-3$

$y_c = c_1 e^{-x} + c_2 e^{-3x}$

$y_p = A+Be^x$, $y_p'=Be^x$, $y_p''=Be^x$

$Be^x + 4Be^x + 3(A+Be^x) = 2+e^x$

$3A=2$, $A = \dfrac{2}{3}$; $8B = 1$, $B = \dfrac{1}{8}$

$y = c_1 e^{-x} + c_2 e^{-3x} + \dfrac{2}{3} + \dfrac{1}{8}e^x$

7. $y'' - 3y' = 2e^x + xe^x$

$m^2 - 3m = 0$, $m(m-3) = 0$, $m = 0,3$

$y_c = c_1 + c_2 e^{3x}$, $y_p = Ae^x + Bxe^x$

$y_p' = Ae^x + Be^x + Bxe^x$, $y_p'' = Ae^x+2Be^x+Bxe^x$

$Ae^x+2Be^x+Bxe^x-3(Ae^x+Be^x+Bxe^x) = 2e^x+xe^x$

$(-2A-B)e^x - 2Bxe^x = 2e^x + xe^x$

$-2A-B = 2$, $-2B = 1$, $B = -\dfrac{1}{2}$, $A = -\dfrac{3}{4}$

$y = c_1 + c_2 e^{3x} - \dfrac{3}{4}e^x - \dfrac{1}{2}xe^x$

9. $D^2y - y = \sin x$

$m^2 - 1 = 0$, $m = 1,-1$

$y_c = c_1 e^x + c_2 e^{-x}$, $y_p = A \sin x + B \cos x$

$y_p' = A \cos x - B \sin x$

$y_p'' = -A \sin x - B \cos x$

$-A\sin x - B\cos x - (A\sin x + B\cos x)=\sin x$

$-2A = 1$, $A = -\dfrac{1}{2}$, $-2B = 0$, $B = 0$

$y = c_1 e^x + c_2 e^{-x} - \dfrac{1}{2}\sin x$

11. $\dfrac{d^2y}{dx^2} + 9y = 9x + 5\cos 2x + 10\sin 2x$

$m^2 + 9 = 0, \quad m = \pm 3j$

$y_c = c_1 \sin 3x + c_2 \cos 3x$

$y_p = A + Bx + C\sin 2x + E\cos 2x$

$y_p' = B + 2C\cos 2x - 2E\sin 2x$

$y_p'' = -4C\sin 2x - 4E\cos 2x$

$-4C\sin 2x - 4E\cos 2x$

$\quad + 9(A + Bx + C\sin 2x + E\cos 2x)$

$\quad = 9x + 5\cos 2x + 10\sin 2x$

$9A = 0, A = 0 ; 9B = 9, B = 1 ;$

$5C = 10, C = 2 ; 5E = 5, E = 1$

$y = c_1 \sin 3x + c_2 \cos 3x + 2\sin 2x + \cos 2x + x$

17. $D^2 y - 4y = \sin x + 2\cos x$

$m^2 - 4 = 0, \quad m = 2, -2$

$y_c = c_1 e^{2x} + c_2 e^{-2x}$

$y_p = A\sin x + B\cos x$

$y_p' = A\cos x - B\sin x$

$y_p'' = -A\sin x - B\cos x$

$-A\sin x - B\cos x - 4(A\sin x + B\cos x) =$

$\quad \sin x + 2\cos x$

$-5A = 1, A = -\dfrac{1}{5} ; -5B = 2, B = -\dfrac{2}{5}$

$y = c_1 e^{2x} + c_2 e^{-2x} - \dfrac{1}{5}\sin x - \dfrac{2}{5}\cos x$

21. $D^2 y + 5Dy + 4y = xe^x + 4$

$m^2 + 5m + 4 = 0, \quad (m+1)(m+4) = 0, \quad m = -1, -4$

$y_c = c_1 e^{-x} + c_2 e^{-4x}, \quad y_p = A + Be^x + Cxe^x$

$y_p' = Be^x + Ce^x + Cxe^x, \quad y_p'' = Be^x + 2Ce^x + Cxe^x$

$Be^x + 2Ce^x + Cxe^x + 5(Be^x + Ce^x + Cxe^x)$

$\quad + 4(A + Be^x + Cxe^x) = xe^x + 4$

$4A = 4, A = 1 ; 10B + 7C = 0, 10C = 1,$

$C = \dfrac{1}{10}, \quad 10B + 7\left(\dfrac{1}{10}\right) = 0, \quad B = -\dfrac{7}{100}$

$y = c_1 e^{-x} + c_2 e^{-4x} + 1 - \dfrac{7}{100}e^x + \dfrac{1}{10}xe^x$

13. $\dfrac{d^2y}{dx^2} - \dfrac{dy}{dx} - 30y = 10$

$m^2 - m - 30 = 0 ; (m+5)(m-6) = 0 , m = -5, 6$

$y_c = c_1 e^{-5x} + c_2 e^{6x}$

$y_p = A, \quad y_p' = 0, \quad y_p'' = 0$

$0 - 0 - 30A = 10, \quad A = -\dfrac{1}{3}$

$y = c_1 e^{-5x} + c_2 e^{6x} - \dfrac{1}{3}$

15. $3\dfrac{d^2y}{dx^2} + 13\dfrac{dy}{dx} - 10y = 14e^{3x}$

$3m^2 + 13m - 10 = 0$

$(3m-2)(m+5) = 0, \quad m = \dfrac{2}{3}, -5$

$y_c = c_1 e^{2x/3} + c_2 e^{-5x}$

$y_p = Ae^{3x}, \quad y_p' = 3Ae^{3x}, \quad y_p'' = 9Ae^{3x}$

$3(9Ae^{3x}) + 13(3Ae^{3x}) - 10Ae^{3x} = 14e^{3x}$

$56A = 14, \quad A = \dfrac{1}{4}$

$y = c_1 e^{2x/3} + c_2 e^{-5x} + \dfrac{1}{4}e^{3x}$

19. $D^2 y + y = 4 + \sin 2x$

$m^2 + 1 = 0, \quad m = \pm j$

$y_c = c_1 \sin x + c_2 \cos x$

$y_p = A + B\sin 2x + C\cos 2x$

$y_p' = 2B\cos 2x - 2C\sin 2x$

$y_p'' = -4B\sin 2x - 4C\cos 2x$

$-4B\sin 2x - 4C\cos 2x + A + B\sin 2x$

$\quad + C\cos 2x = 4 + \sin 2x$

$A = 4; -3B = 1, B = -\dfrac{1}{3} ; -3C = 0, C = 0$

$y = c_1 \sin x + c_2 \cos x + 4 - \dfrac{1}{3}\sin 2x$

23. $y'' + 6y' + 9y = e^{2x} - e^{-2x}$

$m^2 + 6m + 9 = 0, \quad (m+3)^2 = 0, \quad m = -3, -3$

$y_c = (c_1 + c_2 x)e^{-3x}, \quad y_p = Ae^{2x} + Be^{-2x}$

$y_p' = 2Ae^{2x} - 2Be^{-2x}, \quad y_p'' = 4Ae^{2x} + 4Be^{-2x}$

$4Ae^{2x} + 4Be^{-2x} + 6(2Ae^{2x} - 2Be^{-2x})$

$\quad + 9(Ae^{2x} + Be^{-2x}) = e^{2x} - e^{-2x}$

$25A = 1, \quad A = \dfrac{1}{25} ; B = -1$

$y = (c_1 + c_2 x)e^{-3x} + \dfrac{1}{25}e^{2x} - e^{-2x}$

25. $D^2y - Dy - 6y = 5 - e^x$

$m^2 - m - 6 = 0$, $(m-3)(m+2) = 0$, $m = 3, -2$

$y_c = c_1 e^{3x} + c_2 e^{-2x}$

$y_p = A + Be^x$, $y_p' = Be^x$, $y_p'' = Be^x$

$Be^x - Be^x - 6(A + Be^x) = 5 - e^x$

$-6A = 5$, $A = -\frac{5}{6}$; $-6B = -1$, $B = \frac{1}{6}$

$y = c_1 e^{3x} + c_2 e^{-2x} - \frac{5}{6} + \frac{1}{6} e^x$

$y = 2$ when $x = 0$

$2 = c_1 + c_2 - \frac{5}{6} + \frac{1}{6}$, $c_1 + c_2 = \frac{8}{3}$

$Dy = 3c_1 e^{3x} - 2c_2 e^{-2x} + \frac{1}{6} e^x$

$Dy = 4$ when $x = 0$

$4 = 3c_1 - 2c_2 + \frac{1}{6}$, $3c_1 - 2c_2 = \frac{23}{6}$

$\left.\begin{array}{l} c_1 + c_2 = \frac{8}{3} \\ 3c_1 - 2c_2 = \frac{23}{6} \end{array}\right\}$ $c_1 = \frac{11}{6}$, $c_2 = \frac{5}{6}$

$y = \frac{1}{6}(11e^{3x} + 5e^{-2x} + e^x - 5)$

27. $y'' + y = x + \sin 2x$

$m^2 + 1 = 0$, $m = \pm j$, $y_c = c_1 \sin x + c_2 \cos x$

$y_p = A + Bx + C \sin 2x + E \cos 2x$

$y_p' = B + 2C \cos 2x - 2E \sin 2x$

$y_p'' = -4C \sin 2x - 4E \cos 2x$

$-4C \sin 2x - 4E \cos 2x + A + Bx + C \sin 2x$
    $+ E \cos 2x = x + \sin 2x$

$A = 0$; $B = 1$; $-3C = 1$, $C = -\frac{1}{3}$; $-3E = 0$, $E = 0$

$y = c_1 \sin x + c_2 \cos x + x - \frac{1}{3} \sin 2x$

$y = 0$ when $x = \pi$

$0 = c_1 \sin \pi + c_2 \cos \pi + \pi - \frac{1}{3} \sin 2\pi$, $c_2 = \pi$

$Dy = c_1 \cos x - \pi \sin x + 1 - \frac{2}{3} \cos 2x$; $Dy = 1$ when $x = \pi$

$1 = c_1 \cos \pi - \pi \sin \pi + 1 - \frac{2}{3} \cos 2\pi$

$1 = -c_1 + 1 - \frac{2}{3}$, $c_1 = -\frac{2}{3}$

$y = -\frac{2}{3} \sin x + \pi \cos x + x - \frac{1}{3} \sin 2x$

Exercises 15-5, p. 521

1. $D^2x + 100x = 0$

$m^2 + 100 = 0$, $m = \pm 10j$

$x = c_1 \sin 10t + c_2 \cos 10t$

$x = 4$ when $t = 0$

$4 = c_1(0) + c_2$, $c_2 = 4$

$x = c_1 \sin 10t + 4 \cos 10t$

$Dx = 10c_1 \cos 10t - 40 \sin 10t$

$Dx = 0$ when $t = 0$; $0 = 10c_1 - 40(0)$, $c_1 = 0$

$x = 4 \cos 10t$

3. $D^2x + bDx + 100x = 0$

$m^2 + bm + 100 = 0$

$m = \dfrac{-b \pm \sqrt{b^2 - 400}}{2}$

For critical damping,
$b^2 - 400 = 0$, $b = 20$.

(b=-20 not physically
significant)

5. $F = 4$ lb, $x = \frac{1}{8}$ ft

$4 = k(\frac{1}{8})$, $k = 32$ lb/ft

$m = \frac{4}{32} = \frac{1}{8}$ slug

$\frac{1}{8} D^2x + 32x = 0$

$m^2 + 256 = 0$, $m = \pm 16j$

$x = c_1 \sin 16t + c_2 \cos 16t$

$x = \frac{1}{4}$ ft for $t = 0$

$\frac{1}{4} = c_1(0) + c_2$, $c_2 = \frac{1}{4}$

$x = c_1 \sin 16t + \frac{1}{4} \cos 16t$

$Dx = 16c_1 \cos 16t - 4 \sin 16t$

$Dx = 0$ for $t = 0$

$0 = 16c_1 - 4(0)$, $c_1 = 0$

$x = \frac{1}{4} \cos 16t$

7. See Exercises 5. $k = 32$ lb/ft, $m = \frac{1}{8}$ slug

$\frac{1}{8} D^2 x + 32x = 4 \sin 2t$

$D2x + 256x = 32 \sin 2t$

$x_c = c_1 \sin 16t + c_2 \cos 16t$

$x_p = A \sin 2t + B \cos 2t$

$Dx_p = 2A \cos 2t - 2B \sin 2t$

$D^2 x_p = -4A \sin 2t - 4B \cos 2t$

$-4A \sin 2t - 4B \cos 2t + 256(A \sin 2t + B \cos 2t)$
$\quad = 32 \sin 2t$

$252A = 32$, $A = \frac{8}{63}$; $252B = 0$, $B = 0$

$x = c_1 \sin 16t + c_2 \cos 16t + \frac{8}{63} \sin 2t$

$x = \frac{1}{4}$ ft for $t = 0$, $\frac{1}{4} = c_1(0) + c_2 + \frac{8}{63}(0)$, $c_2 = \frac{1}{4}$

$x = c_1 \sin 16t + \frac{1}{4} \cos 16t + \frac{8}{63} \sin 2t$

$Dx = 16c_1 \cos 16t - 4 \sin 16t + \frac{16}{63} \cos 2t$

$Dx = 0$ for $t = 0$; $0 = 16c_1 - 4(0) + \frac{16}{63}$, $c_1 = -\frac{1}{63}$

$x = -\frac{1}{63} \sin 16t + \frac{1}{4} \cos 16t + \frac{8}{63} \sin 2t$

9. $m = 0.820$ kg, $x = 0.250$ m

$F = 0.820(9.80) = 8.036$ N

$8.036 = k(0.250)$, $k = 32.144$ N/m

$0.820 D^2 x + 32.144x = 0$

$D^2 x + 39.2x = 0$

$m^2 + 39.2 = 0$, $m = \pm 6.26j$

$x = c_1 \sin 6.26t + c_2 \cos 6.26t$

$x = 0.150$ m for $t = 0$

$0.150 = c_1(0) + c_2$, $c_2 = 0.150$

$x = c \sin 6.26t + 0.150 \cos 6.26t$

$Dx = 6.26 c_1 \cos 6.26t - 0.939 \sin 6.26t$

$Dx = 0$ for $t = 0$

$0 = 6.26 c_1 - 0.939(0)$, $c_1 = 0$

$x = 0.150 \cos 6.26t$

11. $LD^2 q + RDq + \frac{q}{C} = 0$ $\quad (D = \frac{d}{dt})$

$L = 0.2$ H, $R = 8 \,\Omega$, $C = 1 \,\mu F$

$0.2 D^2 q + 8Dq + 10^6 q = 0$

$D^2 q + 40Dq + 5 \times 10^6 q = 0$

$m^2 + 40m + 5 \times 10^6 = 0$

$m = \dfrac{-40 \pm \sqrt{1600 - 20 \times 10^6}}{2}$

$\alpha = -20$, $\beta = 2240$

$q = e^{-20t}(c_1 \sin 2240t + c_2 \cos 2240t)$

$q = 0$ when $t = 0$; $0 = c_1(0) + c_2$, $c_2 = 0$

$q = c_1 e^{-20t} \sin 2240$

$Dq = c_1 e^{-20t}(-20 \sin 2240t + 2240 \cos 2240t)$

$i = Dq = 0.5$ A when $t = 0$

$0.5 = c_1[-20(0) + 2240]$, $c = 2.23 \times 10^{-4}$

$q = 2.23 \times 10^{-4} e^{-20t} \sin 2240t$

13. $LD^2 q + RDq + \frac{q}{C} = E$ $\quad (D = \frac{d}{dt})$

$L = 0.1$ H, $R = 0$, $C = 100 \,\mu F$, $E = 100$ V

$0.1 D^2 q + 10^4 q = 100$, $D^2 q + 10^5 q = 1000$

$m^2 + 10^5 = 0$, $m = \pm 316j$

$q_c = c_1 \sin 316t + c_2 \cos 316t$

$q_p = A$, $Dq_p = 0$, $D^2 q_p = 0$

$0 + 10^5 A = 10^3$, $A = 0.01$

$q = c_1 \sin 316t + c_2 \cos 316t + 0.01$

$q = 0$ for $t = 0$

$0 = c_1(0) + c_2 + 0.01$, $c_2 = -0.01$

$q = c_1 \sin 316t - 0.01 \cos 316t + 0.01$

$Dq = 316 c_1 \cos 316t + 3.16 \sin 316t$

$Dq = 0$ for $t = 0$; $0 = 316 c_1$, $c_1 = 0$

$q = -0.01 \cos 316t + 0.01$

$\quad = \frac{1}{100}(1 - \cos 316t)$

15. $LD^2q + RDq + \dfrac{q}{C} = E$

$L = 0.5$ H, $R = 10\ \Omega$, $C = 200\ \mu$F, $E = 100 \sin 200t$

$0.5D^2q + 10\ Dq + 0.5 \times 10^4 q = 100 \sin 200t$

$D^2q + 20Dq + 10^4 q = 200 \sin 200t$

$m^2 + 20m + 10^4 = 0$

$m = \dfrac{-20 \pm \sqrt{400 - 4 \times 10^4}}{2} = -10 \pm 99.5j$

$q = e^{-10t}(c_1 \sin 99.5t + c_2 \cos 99.5t)$

$q_p = A \sin 200t + B \cos 200t$

$Dq_p = 200A \cos 200t - 200B \sin 200t$

$D^2 q_p = -4 \times 10^4 A \sin 200t - 4 \times 10^4 B \cos 200t$

$-4 \times 10^4 A \sin 200t - 4 \times 10^4 B \cos 200t$

$+ 20(200A \cos 200t - 200B \sin 200t)$

$+ 10^4(A \sin 200t + B \cos 200t) = 200 \sin 200t$

$-3 \times 10^4 A - 4 \times 10^3 B = 200,\quad -3 \times 10^4 B + 4 \times 10^3 A = 0$

$\left.\begin{matrix} -300A - 40B = 2 \\ 4A - 30B = 0 \end{matrix}\right\}$ $A = -\dfrac{15}{2290},\quad B = -\dfrac{2}{2290}$

$q = e^{-10t}(c_1 \sin 99.5t + c_2 \cos 99.5t) - \dfrac{1}{2290}(15 \sin 200t + 2 \cos 200t)$

17. See Exercise 15.

Steady state solution for the charge is

$q = -\dfrac{1}{2290}(15 \sin 200t + 2 \cos 200t)$

Since $i = dq/dt$

$i = -\dfrac{1}{2290}(3000 \cos 200t - 400 \sin 200t)$

$= \dfrac{20}{229}(2 \sin 200t - 15 \cos 200t)$

19. $LD^2q + RDq + \dfrac{q}{C} = E$; $L = 1$ H, $R = 5\ \Omega$, $C = 150\ \mu$F, $E = 100 \sin 400t$

$D^2q + 5Dq + 6667q = 100 \sin 400t$; For steady state solution, find q_p.

$q_p = A \sin 400t + B \cos 400t$, $Dq_p = 400A \cos 400t - 400B \sin 400t$

$D^2 q_p = -1.6 \times 10^5 A \sin 400t - 1.6 \times 10^5 B \cos 400t$

$-1.6 \times 10^5 A \sin 400t - 1.6 \times 10^5 B \cos 400t + 5(400A \cos 400t - 400B \sin 400t)$

$+ 6667(A \sin 400t + B \cos 400t) = 100 \sin 400t$

$\left.\begin{matrix} -153333A - 2000B = 100 \\ 2000A - 153333B = 0 \end{matrix}\right\}$ $A = -6.5 \times 10^{-4},\ B = -8.5 \times 10^{-6}$

$q_p = -6.5 \times 10^{-4} \sin 400t - 8.5 \times 10^{-6} \cos 400t$

Review Exercises for Chapter 15, p. 522

1. $2D^2y + Dy = 0$

$2m^2 + m = 0$

$m(2m + 1) = 0$, $m = 0, -\dfrac{1}{2}$

$y = c_1 + c_2 e^{-x/2}$

3. $y'' + 2y' + y = 0$

$D^2y + 2Dy + y = 0$

$m^2 + 2m + 1 = 0$

$(m+1)^2 = 0$, $m = -1, -1$

$y = (c_1 + c_2 x)e^{-x}$

5. $D^2y - 2Dy + 5y = 0$

$m^2 - 2m + 5 = 0$

$m = \dfrac{2 \pm \sqrt{4 - 20}}{2} = 1 \pm 2j$

$y = e^x(c_1 \sin 2x + c_2 \cos 2x)$

7. $\dfrac{d^2y}{dx^2} = 0.08y - 0.2\dfrac{dy}{dx}$

$D^2y + 0.2Dy - 0.08y = 0$

$m^2 + 0.2m - 0.08 = 0$

$(m-0.2)(m+0.4) = 0$, $m = 0.2, -0.4$

$y = c_1 e^{0.2x} + c_2 e^{-0.4x}$

9. $y'' - y' - 56y = 28$

$m^2 - m - 56 = 0$

$(m-8)(m+7) = 0$, $m = 8, -7$

$y_c = c_1 e^{8x} + c_2 e^{-7x}$

$y_p = A$, $y_p' = 0$, $y_p'' = 0$; $0 - 0 - 56A = 28$, $A = -\dfrac{1}{2}$

$y = c_1 e^{8x} + c_2 e^{-7x} - \dfrac{1}{2}$

11. $\dfrac{d^2y}{dx^2} + 6\dfrac{dy}{dx} + 9y = 3x$

$m^2 + 6m + 9 = 0$

$(m+3)^2 = 0, \ m = -3, -3$

$y_c = (c_1 + c_2 x)e^{-3x}$

$y_p = A + Bx, \ y_p' = B, \ y_p'' = 0$

$0 + 6B + 9(A + Bx) = 3x$

$6B + 9A = 0, \ 9B = 3, \ B = \dfrac{1}{3}, \ A = -\dfrac{2}{9}$

$y = (c_1 + c_2 x)e^{-3x} - \dfrac{2}{9} + \dfrac{1}{3}x$

15. $9D^2y - 18Dy + 8y = 16 + 4x$

$9m^2 - 18m + 8 = 0$

$(3m-2)(3m-4) = 0, \ m = \dfrac{2}{3}, \ \dfrac{4}{3}$

$y_c = c_1 e^{2x/3} + c_2 e^{4x/3}$

$y_p = A + Bx, \ y_p' = B, \ y_p'' = 0$

$9(0) - 18B + 8(A + Bx) = 16 + 4x$

$-18B + 8A = 16, \ 8B = 4, \ B = \dfrac{1}{2}, \ A = \dfrac{25}{8}$

$y = c_1 e^{2x/3} + c_2 e^{4x/3} + \dfrac{25}{8} + \dfrac{1}{2}x$

21. $\dfrac{d^2y}{dx^2} + \dfrac{dy}{dx} + 4y = 0$

$m^2 + m + 4 = 0$

$m = \dfrac{-1 \pm \sqrt{1-16}}{2} = -\dfrac{1}{2} \pm \dfrac{1}{2}\sqrt{15}j$

$y = e^{-x/2}(c_1 \sin \tfrac{1}{2}\sqrt{15}x + c_2 \cos \tfrac{1}{2}\sqrt{15}x)$

$y = 0$ when $x = 0$; $\ 0 = c_1(0) + c_2, \ c_2 = 0$

$Dy = e^{-x/2}(\tfrac{1}{2}\sqrt{15}c_1 \cos \tfrac{1}{2}\sqrt{15}x - \tfrac{1}{2}c_1 \sin \tfrac{1}{2}\sqrt{15}x)$; $Dy = \sqrt{15}$ when $x = 0$

$\sqrt{15} = \dfrac{1}{2}\sqrt{15}\,c_1, \ c_1 = 2$; $y = 2e^{-x/2}\sin \tfrac{1}{2}\sqrt{15}x$

13. $4y'' + 3y = 3x^2$

$4m^2 + 3 = 0, \ m = \pm \dfrac{1}{2}\sqrt{3}j$

$y_c = c_1 \sin \tfrac{1}{2}\sqrt{3}x + c_2 \cos \tfrac{1}{2}\sqrt{3}x$

$y_p = A + Bx + Cx^2, \ y_p' = B + 2Cx, \ y_p'' = 2C$

$4(2C) + 3(A + Bx + Cx^2) = 3x^2$

$8C + 3A = 0, \ 3B = 0, \ B = 0, \ 3C = 3, \ C = 1, \ A = -\dfrac{8}{3}$

$y = c_1 \sin \tfrac{1}{2}\sqrt{3}x + c_2 \cos \tfrac{1}{2}\sqrt{3}x - \dfrac{8}{3} + x^2$

17. $y'' + y' - y = 2e^x$

$m^2 + m - 1 = 0, \ m = \dfrac{-1 \pm \sqrt{1+4}}{2} = -\dfrac{1}{2} \pm \dfrac{1}{2}\sqrt{5}$

$y = e^{-x/2}(c_1 e^{x\sqrt{5}/2} + c_2 e^{-x\sqrt{5}/2})$

$y_p = Ae^x, \ y_p' = Ae^x, \ y_p'' = Ae^x$

$Ae^x + Ae^x - Ae^x = 2e^x, \ A = 2$

$y = e^{-x/2}(c_1 e^{x\sqrt{5}/2} + c_2 e^{-x\sqrt{5}/2}) + 2e^x$

19. $D^2y + 9y = \sin x$

$m^2 + 9 = 0, \ m = \pm 3j, \ y_c = c_1 \sin 3x + c_2 \cos 3x$

$y_p = A \sin x + B \cos x, \ y_p' = A \cos x - B \sin x$

$y_p'' = -A \sin x - B \cos x$

$-A \sin x - B \cos x + 9(A \sin x + B \cos x) = \sin x$

$8A = 1, \ A = \dfrac{1}{8}$; $8B = 0, \ B = 0$

$y = c_1 \sin 3x + c_2 \cos 3x + \dfrac{1}{8}\sin x$

23. $(D^2 + 4D + 4)y = 4 \cos x$

$m^2 + 4m + 4 = 0$

$(m+2)^2 = 0, \ m = -2, -2$

$y_c = (c_1 + c_2 x)e^{-2x}$

$y_p = A \sin x + B \cos x$

$y_p' = A \cos x - B \sin x$

$y_p'' = -A \sin x - B \cos x$

$-A \sin x - B \cos x + 4(A \cos x - B \sin x) + 4(A \sin x + B \cos x)$

$= 4 \cos x$; $\ 3A - 4B = 0$ $\quad A = \dfrac{16}{25}, \ B = \dfrac{12}{25}$
$\qquad\qquad\qquad\quad 4A + 3B = 4$

$y = (c_1 + c_2 x)e^{-2x} + \dfrac{16}{25}\sin x + \dfrac{12}{25}\cos x$

$y = 0$ when $x = 0$; $\ 0 = c_1 + \dfrac{16}{25}(0) + \dfrac{12}{25}, \ c_1 = -\dfrac{12}{25}$

$y = (-\dfrac{12}{25} + c_2 x)e^{-2x} + \dfrac{16}{25}\sin x + \dfrac{12}{25}\cos x$

$Dy = e^{-2x}(\dfrac{24}{25} - 2c_2 x + c_2) + \dfrac{16}{25}\cos x - \dfrac{12}{25}\sin x$

$Dy = 1$ when $x = 0$; $\ 1 = \dfrac{24}{25} + c_2 + \dfrac{16}{25}, \ c_2 = -\dfrac{15}{25}$

$y = \dfrac{1}{25}[(-12 - 15x)e^{-2x} + 16 \sin x + 12 \cos x] = \dfrac{1}{25}[16 \sin x + 12 \cos x - 3(4+5x)e^{-2x}]$

25. $F = 64$ lb, $x = 2$ ft

$64 = k(2)$, $k = 32$ lb/ft

$m = \dfrac{64 \text{ lb}}{32 \text{ ft/s}^2} = 2$ slug

$2D^2x + 16Dx + 32x = 0$

$D^2x + 8Dx + 16x = 0$

$m^2 + 8m + 16 = 0$

$(m+4)^2 = 0$, $m = -4, -4$

$x = (c_1 + c_2 t)e^{-4t}$

$x = 2$ ft when $t = 0$

$2 = (c_1 + 0)(1)$, $c_1 = 2$

$x = (2 + c_2 t)e^{-4t}$

$Dx = e^{-4t}(-8 - 4c_2 t + c_2)$

$Dx = 0$ when $t = 0$

$0 = (-8 + c_2)$, $c_2 = 8$

$x = (2 + 8t)e^{-4t}$

critically damped

27. $m = 1.25$ kg, $x = 0.200$ m

$F = 1.25(9.80) = 12.25$ N

$12.25 = k(0.200)$, $k = 61.25$ N/m

$1.25D^2x + 4Dx + 61.25x = 0$

$D^2x + 3.2Dx + 49x = 0$

$m^2 + 3.20m + 49 = 0$

$m = \dfrac{-3.20 \pm \sqrt{3.20^2 - 4(49)}}{2} = -1.60 \pm 6.81j$

$x = e^{-1.60t}(c_1 \sin 6.81t + c_2 \cos 6.81t)$

$x = 0$ when $t = 0$, $c_2 = 0$

$x = c_1 e^{-1.60t} \sin 6.81t$

$Dx = c_1 e^{-1.60t}(-1.60 \sin t\, 6.81t + 6.81 \cos 6.81t)$

$Dx = 4.50$ m/s when $t = 0$

$4.50 = c_1(6.81)$, $c_1 = 0.661$

$x = 0.661 e^{-1.60t} \sin 6.81t$

29. $LD^2q + RDq + \dfrac{q}{C} = E \qquad (D = \dfrac{d}{dt})$

$L = 0.5$ H, $R = 6\ \Omega$, $C = 20$ mF, $E = 24 \sin 10t$

$0.5D^2q + 6Dq + 50q = 24 \sin 10t$, $D^2q + 12Dq + 100q = 48 \sin 10t$

$m^2 + 12m + 100 = 0$, $m = \dfrac{-12 \pm \sqrt{144 - 400}}{2} = -6 \pm 8j$

$q_c = e^{-6t}(c_1 \sin 8t + c_2 \cos 8t)$, $q_p = A \sin 10t + B \cos 10t$

$Dq_p = 10A \cos 10t - 10B \sin 10t$, $D^2q_p = -100A \sin 10t - 100B \cos 10t$

$-100A \sin 10t - 100B \cos 10t + 12(10A \cos 10t - 10B \sin 10t)$

$\quad +100(A \sin 10t + B \cos 10t) = 48 \sin 10t$

$-120B = 48$, $B = -0.4$; $120A = 0$, $A = 0$

$q = e^{-6t}(c_1 \sin 8t + c_2 \cos 8t) - 0.4 \cos 10t$

$q = 0$ when $t = 0$, $0 = c_2 - 0.4$, $c_2 = 0.4$

$q = e^{-6t}(c_1 \sin 8t + 0.4 \cos 8t) - 0.4 \cos 10t$

$Dq = e^{-6t}(-6c_1 \sin 8t - 2.4 \cos 8t + 8c_1 \cos 8t - 3.2 \sin 8t) + 4 \sin 10t$

$Dq = 0$ when $t = 0$; $0 = -2.4 + 8c_1$, $c_1 = 0.3$

$q = e^{-6t}(0.3 \sin 8t + 0.4 \cos 8t) - 0.4 \cos 10t$

31. $LD^2q + RDq + \dfrac{q}{c} = E$

$L = 4\ H,\ R = 20\ \Omega,\ C = 100\ \mu F,\ E = 100\ V$

$4D^2q + 20Dq + 10^4q = 100$

$D^2q + 5Dq + 2500q = 25$

$m^2 + 5m + 2500 = 0$

$m = \dfrac{-5 \pm \sqrt{25 - 10000}}{2} = -\dfrac{5}{2} \pm 49.9j$

$q_c = e^{-5t/2}(c_1 \sin 49.9t + c_2 \cos 49.9t)$

$q_p = A,\ q_p' = 0,\ q_p'' = 0$

$0 + 5(0) + 2500A = 25,\ A = 0.01$

$q = e^{-5t/2}(c_1 \sin 49.9t + c_2 \cos 49.9t)$
$\quad + 0.01$

$q = 10\ mC = 0.01\ C$ when $t = 0$

$0.01 = c_2 + 0.01,\ c_2 = 0$

$q = c_1 e^{-5t/2} \sin 49.9t + 0.01$

$i = c_1 e^{-5t/2}(-2.5 \sin 49.9t + 49.9 \cos 49.9t)$

$i = 0$ when $t = 0$, $0 = 49.9c_1,\ c_1 = 0$

$i = 0$

33. $EI\dfrac{d^2y}{dx^2} = M,\ M = 2000x - 40x^2$

$EID^2y = 2000x - 40x^2$

$D^2y = \dfrac{1}{EI}(2000x - 40x^2)$

$Dy = \dfrac{1}{EI}(1000x^2 - \dfrac{40}{3}x^3) + c_1$

$y = \dfrac{1}{EI}(\dfrac{1000}{3}x^3 - \dfrac{10}{3}x^4) + c_1 x + c_2$

$y = 0$ for $x = 0$ and $x = L$

$0 = \dfrac{1}{EI}(0 - 0) + 0 + c_2,\ c_2 = 0$

$0 = \dfrac{1}{EI}(\dfrac{1000}{3}L^3 - \dfrac{10}{3}L^4) + c_1 L$

$c_1 = \dfrac{1}{EI}(\dfrac{10}{3}L^3 - \dfrac{1000}{3}L^2)$

$y = \dfrac{1}{EI}(\dfrac{1000}{3}x^3 - \dfrac{10}{3}x^4) + \dfrac{1}{EI}(\dfrac{10}{3}L^3 - \dfrac{1000}{3}L^2)x$

$\quad = \dfrac{10}{3EI}(100x^3 - x^4 + L^3x - 100L^2x)$

Exercises 16-1, p. 528

1. $\dfrac{dy}{dx} = x + 1$

x	y	x+1	dy	y(correct)
0.0	1.00	1.0	0.20	1.00
0.2	1.20	1.2	0.24	1.22
0.4	1.44	1.4	0.28	1.48
0.6	1.72	1.6	0.32	1.78
0.8	2.04	1.8	0.36	2.12
1.0	2.40	2.0	0.40	2.50

$y = \dfrac{1}{2}x^2 + x + c$

$y = 1$ when $x = 0$

$c = 1$

$y = \dfrac{1}{2}x^2 + x + 1$

3. $\dfrac{dy}{dx} = y(0.4x + 1)$

x	y	y(0.4x+1)	dy	y(correct)
-0.2	2.0000	1.8400	0.1840	2.0000
-0.1	2.1840	2.0966	0.2097	2.1971
0.0	2.3937	2.3937	0.2394	2.4233
0.1	2.6330	2.7384	0.2738	2.6836
0.2	2.9069	3.1394	0.3139	2.9836
0.3	3.2208	3.6073	0.3607	3.3306
0.4	3.5815	4.1546	0.4155	3.7338
0.5	3.9970	4.7964	0.4796	4.2003

$\dfrac{dy}{y} = (0.4x + 1)dx$, $\ln y = 0.2x^2 + x + \ln c$

$y = 2$ when $x = -0.2$

$\ln 2 = 0.008 - 0.2 + \ln c$, $c = 2.4233$

$y = 2.4233e^{0.2x^2+x}$

5.
x	y	x+1	dy	y(correct)
0.0	1.00	1.0	0.10	1.000
0.1	1.10	1.1	0.11	1.105
0.2	1.21	1.2	0.12	1.220
0.3	1.33	1.3	0.13	1.345
0.4	1.46	1.4	0.14	1.480
0.5	1.60	1.5	0.15	1.625
0.6	1.75	1.6	0.16	1.780
0.7	1.91	1.7	0.17	1.945
0.8	2.08	1.8	0.18	2.120
0.9	2.26	1.9	0.19	2.305
1.0	2.45	2.0	0.20	2.500

$\dfrac{dy}{dx} = x + 1$

From Exercise 1

$y = \dfrac{1}{2}x^2 + x + 1$

7. $\dfrac{dy}{dx} = y(0.4x + 1)$

x	y	y(0.4x+1)	dy	y(correct)
-0.20	2.0000	1.8400	0.0920	2.0000
-0.15	2.0920	1.9665	0.0983	2.0952
-0.10	2.1903	2.1027	0.1051	2.1971
-0.05	2.2955	2.2496	0.1125	2.3063
0.00	2.4079	2.4079	0.1204	2.4233
0.05	2.5283	2.5789	0.1289	2.5489
0.10	2.6573	2.7636	0.1382	2.6836
0.15	2.7955	2.9632	0.1482	2.8282
0.20	2.9436	3.1791	0.1590	2.9836
0.25	3.1026	3.4128	0.1706	3.1508

x	y	y(0.4x+1)	dy	y(corr.)
0.30	3.2732	3.6660	0.1833	3.3306
0.35	3.4565	3.9404	0.1970	3.5242
0.40	3.6535	4.2381	0.2119	3.7328
0.45	3.8654	4.5612	0.2281	3.9576
0.50	4.0935	4.9122	0.2456	4.2003

From Exercise 3

$y = 2.4233e^{0.2x^2+x}$

197

9. $\dfrac{dy}{dx} = xy + 1$

x	y	xy+1	dy
0.0	0.0000	1.0000	0.1000
0.1	0.1000	1.0100	0.1010
0.2	0.2010	1.0402	0.1040
0.3	0.3050	1.0915	0.1092
0.4	0.4142	1.1657	0.1166
0.5	0.5307	1.2654	0.1265
0.6	0.6573	1.3944	0.1394
0.7	0.7967	1.5577	0.1558
0.8	0.9525	1.7620	0.1762
0.9	1.1287	2.0158	0.2016
1.0	1.3303	2.3303	0.2330

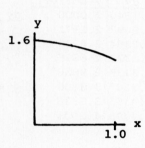

11. $\dfrac{dy}{dx} = e^{xy}$

x	y	e^{xy}	dy
0.0	0.0000	1.0000	0.2000
0.2	0.2000	1.0408	0.2082
0.4	0.4082	1.1773	0.2355
0.6	0.6436	1.4713	0.2943
0.8	0.9379	2.1177	0.4235
1.0	1.3614	3.9018	0.7804
1.2	2.1418	13.068	2.6136

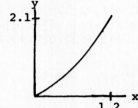

13. $\dfrac{dy}{dx} = \cos(x + y)$

x	y	cos(x+y)	dy
0.0	1.5708	0.0000	0.0000
0.1	1.5708	-0.0998	-0.0100
0.2	1.5608	-0.1889	-0.0189
0.3	1.5419	-0.2678	-0.0268
0.4	1.5151	-0.3376	-0.0338
0.5	1.4814	-0.3991	-0.0399
0.6	1.4415	-0.4535	-0.0453
0.7	1.3961	-0.5015	-0.0501
0.8	1.3460	-0.5440	-0.0544
0.9	1.2916	-0.5817	-0.0582
1.0	1.2334	-0.6152	-0.0615

15. $\dfrac{dy}{dx} = x + y$

x	y	x+y	dy	x	y	x+y	dy
0.00	1.0000	1.0000	0.0200	0.40	1.5719	1.9719	0.0394
0.02	1.0200	1.0400	0.0208	0.42	1.6113	2.0313	0.0406
0.04	1.0408	1.0808	0.0216	0.44	1.6520	2.0920	0.0418
0.06	1.0624	1.1224	0.0224	0.46	1.6938	2.1538	0.0431
0.08	1.0849	1.1649	0.0233	0.48	1.7369	2.2169	0.0443
0.10	1.1082	1.2082	0.0242	0.50	1.7812	2.2812	0.0456
0.12	1.1323	1.2523	0.0250	0.52	1.8268	2.3468	0.0469
0.14	1.1574	1.2974	0.0259	0.54	1.8738	2.4138	0.0483
0.16	1.1833	1.3433	0.0269	0.56	1.9220	2.4820	0.0496
0.18	1.2102	1.3902	0.0278	0.58	1.9717	2.5517	0.0510
0.20	1.2380	1.4380	0.0288	0.60	2.0227	2.6227	0.0525
0.22	1.2667	1.4867	0.0297				
0.24	1.2965	1.5365	0.0307				
0.26	1.3272	1.5872	0.0317				
0.28	1.3590	1.6390	0.0328				
0.30	1.3917	1.6917	0.0338				
0.32	1.4256	1.7456	0.0349				
0.34	1.4605	1.8005	0.0360				
0.36	1.4965	1.8565	0.0371				
0.38	1.5336	1.9136	0.0383				

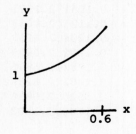

17. $y' = \ln(xy)$

x	y	$\ln(xy)$	dy
1.0	1.0000	0.0000	0.0000
1.1	1.0000	0.0953	0.0095
1.2	1.0095	0.1918	0.0192
1.3	1.0287	0.2907	0.0291
1.4	1.0578	0.3926	0.0393

x	y	$\ln(xy)$	dy
1.5	1.0970	0.4981	0.0498
1.6	1.1469	0.6070	0.0607
1.7	1.2076	0.7192	0.0719

$y = 1.2076$ for $x = 1.7$

19. $\dfrac{di}{dt} + 2i = \sin t$

t	i	$\sin t - 2i$	di
0.0	0.0000	0.0000	0.0000
0.1	0.0000	0.0998	0.0100
0.2	0.0100	0.1787	0.0179
0.3	0.0279	0.2398	0.0240
0.4	0.0518	0.2857	0.0286
0.5	0.0804	0.3186	0.0319

$i = 0.0804$ A for $t = 0.5$ s

$di + 2i\,dt = \sin t\,dt$; $e^{\int 2dt} = e^{2t}$

$ie^{2t} = \int e^{2t}\sin t\,dt = \dfrac{e^{2t}(2\sin t - \cos t)}{4+1} + c$

$i = \dfrac{1}{5}(2\sin t - \cos t) + ce^{-2t}$ (Formula 49)

$i = 0$ for $t = 0$, $0 = \dfrac{1}{5}(0-1) + c$, $c = \dfrac{1}{5}$

$i = \dfrac{1}{5}(2\sin t - \cos t + e^{-2t})$

$i = 0.0898$ A for $t = 0.5$ s

Exercises 16-2, p. 533

1. $y' = y^2$, $(0,1)$

$dy = (1)dx$

$y_1 = x + c_1$, $1 = 0 + c_1$, $c_1 = 1$, $y_1 = x + 1$

$dy = (x+1)^2 dx$

$y_2 = \dfrac{1}{3}(x+1)^3 + c_2$, $1 = \dfrac{1}{3} + c_2$, $c_2 = \dfrac{2}{3}$

$y_2 = \dfrac{1}{3}(x+1)^3 + \dfrac{2}{3} = \dfrac{1}{3}x^3 + x^2 + x + 1$

$y_2(0.1) = \dfrac{1}{3}(0.1)^3 + (0.1)^2 + 0.1 + 1 = 1.1103$

$\dfrac{dy}{y^2} = dx$

$-\dfrac{1}{y} = 0 + c$, $c = -1$

$-\dfrac{1}{y} = x - 1$

$y = \dfrac{1}{1-x}$

$y(0.1) = 1.1111$

3. $y' = 2x(1+y)$, $(0,0)$

$dy = 2x(1+0)dx = 2x\,dx$

$y_1 = x^2 + c_1$, $0 = 0 + c_1$, $c_1 = 0$, $y_1 = x^2$

$dy = 2x(1+x^2)dx = (2x + 2x^3)dx$

$y_2 = x^2 + \dfrac{1}{2}x^4 + c_2$, $0 = 0 + 0 + c_2$, $c_2 = 0$

$y_2 = x^2 + \dfrac{1}{2}x^4$

$dy = 2x(1 + x^2 + \dfrac{1}{2}x^4)dx = (2x + 2x^3 + x^5)dx$

$y_3 = x^2 + \dfrac{1}{2}x^4 + \dfrac{1}{6}x^6 + c_3$, $c_3 = 0$

$y_3 = x^2 + \dfrac{1}{2}x^4 + \dfrac{1}{6}x^6$; $y_3(1) = 1 + \dfrac{1}{2} + \dfrac{1}{6} = 1.67$

$\dfrac{dy}{1+y} = 2x\,dx$

$\ln(1+y) = x^2 + c$

$\ln(1+0) = 0 + c$, $c = 0$

$\ln(1+y) = x^2$, $1 + y = e^{x^2}$

$y = e^{x^2} - 1$

$y(1) = 1.72$

5. $y' = x + y^2$, (0,0)

   $dy = x\,dx$

   $y_1 = \frac{1}{2}x^2 + c_1$, $c_1 = 0$, $y = \frac{1}{2}x^2$

   $dy = [x + (\frac{1}{2}x^2)^2]dx = (x + \frac{1}{4}x^4)dx$

   $y_2 = \frac{1}{2}x^2 + \frac{1}{20}x^5 + c_2$, $c_2 = 0$

   $dy = [x + (\frac{1}{2}x^2 + \frac{1}{20}x^5)^2]dx$

   $\quad = (x + \frac{1}{4}x^4 + \frac{1}{20}x^7 + \frac{1}{400}x^{10})dx$

   $y_3 = \frac{1}{2}x^2 + \frac{1}{20}x^5 + \frac{1}{160}x^8 + \frac{1}{4400}x^{11} + c_3, c_3 = 0$

   $y_3 = \frac{1}{2}x^2 + \frac{1}{20}x^5 + \frac{1}{160}x^8 + \frac{1}{4400}x^{11}$

7. $y' = y - \cos 2x$, (0,1)

   $dy = (1 - \cos 2x)dx$

   $y_1 = x - \frac{1}{2}\sin 2x + c_1$, $c_1 = 1$

   $y_1 = x - \frac{1}{2}\sin 2x + 1$

   $dy = (x - \frac{1}{2}\sin 2x + 1 - \cos 2x)dx$

   $y_2 = \frac{1}{2}x^2 + \frac{1}{4}\cos 2x + x - \frac{1}{2}\sin 2x + c_2$

   $1 = 0 + \frac{1}{4} + 0 - 0 + c_2$, $c_2 = \frac{3}{4}$

   $y_2 = \frac{3}{4} + x + \frac{1}{2}x^2 + \frac{1}{4}\cos 2x - \frac{1}{2}\sin 2x$

9. $e^x = 1 + x + \frac{1}{2}x^2 + \frac{1}{6}x^3 + \frac{1}{24}x^4 + \ldots$

   $2e^x - x - 1 = 2(1 + x + \frac{1}{2}x^2 + \frac{1}{6}x^3 + \frac{1}{24}x^4 + \ldots) - x - 1$

   $\qquad\qquad\quad = 1 + x + x^2 + \frac{1}{3}x^3 + \frac{1}{12}x^4 + \ldots$

   From Example A: $y = 1 + x + x^2 + \frac{1}{3}x^3 + \frac{1}{24}x^4$

11. $\frac{dv}{dt} = 2 - v^3$, $v = 0$ when $t = 0$ (from rest)

   $dv = 2\,dt$, $v_1 = 2t + c_1$, $c_1 = 0$, $v_1 = 2t$, $dv = [2 - (2t)^3]dt$

   $v_2 = 2t - 2t^4 + c_2$, $c_2 = 0$, $v_2 = 2t - 2t^4$; $v_2(0.3) = 2(0.3) - 2(0.3)^4 = 0.58$ ft/s

Exercises 16-3, p. 539

1. $f(t) = 1$

   $L(f) = f(t) = \int_0^\infty e^{-st}(1)dt = \lim_{h\to\infty}\int_0^h e^{-st}dt$

   $\quad = \lim_{h\to\infty} -\frac{1}{s}e^{-st}\Big|_0^h$

   $\quad = \lim_{h\to\infty} -\frac{1}{s}e^{-hs} + \frac{1}{s} = \frac{1}{s}$

5. $f(t) = e^{3t}$

   Transform 3: $a = -3$

   $L(f) = \frac{1}{s - 3}$

7. $f(t) = t^3 e^{-2t}$

   Transform 12: $n-1 = 3$, $n = 4$, $a = 2$

   $L(f) = \frac{3!}{(s + 2)^4} = \frac{6}{(s + 2)^4}$

3. $f(t) = \sin at$

   $L(f) = f(t) = \int_0^\infty e^{-st}\sin at\,dt$

   $\quad = \lim_{h\to\infty}\int_0^h e^{-st}\sin at\,dt$

   $\quad = \lim_{h\to\infty}\frac{e^{-st}(-s\sin at - a\cos at)}{s^2 + a^2}\Big|_0^h$

   $\quad = \lim_{h\to\infty}\frac{e^{-sh}(-s\sin ah - a\cos ah)}{s^2 + a^2}$

   $\quad - \frac{(-s\sin 0 - a\cos 0)}{s^2 + a^2} = \frac{a}{s^2 + a^2}$

9. $f(t) = \cos 2t - \sin 2t$

   Transforms 5 and 6: $a = 2$

   $L(f) = \frac{s}{s^2 + 4} - \frac{2}{s^2 + 4} = \frac{s - 2}{s^2 + 4}$

11. $f(t) = 3 + 2t \cos 3t$

Transforms 1 and 18: $a = 3$

$$L(f) = 3(\frac{1}{s}) + 2[\frac{s^2 - 9}{(s^2 + 9)^2}]$$

$$= \frac{3}{s} + \frac{2(s^2 - 9)}{(s^2 + 9)^2}$$

13. $y'' + y'$, $f(0) = 0$, $f'(0) = 0$

$$L(y'') + L(y') = s^2 L(f) - sf(0) - f'(0) + sL(f) - f(0)$$

$$= s^2 L(f) - s(0) - 0 + sL(f) - 0$$

$$= s^2 L(f) + sL(f)$$

15. $2y'' - y' + y$, $f(0) = 1$, $f'(0) = 0$

$$2L(y'') - L(y') + L(y) = 2[s^2 L(f) - sf(0) - f'(0)] - [sL(f) - f(0)] + L(f)$$

$$= 2s^2 L(f) - 2s - 2(0) - sL(f) + 1 + L(f) = (2s^2 - s + 1)L(f) - 2s + 1$$

17. $F(s) = \frac{2}{s^3}$

Transform 2: $n = 3$

$f(t) = 2(\frac{t^2}{2!}) = t^2$

19. $F(s) = \frac{1}{s + 5}$

Transform 3: $a = 5$

$f(t) = e^{-5t}$

21. $F(s) = \frac{1}{s^3 + 3s^2 + 3s + 1} = \frac{1}{(s+1)^3}$

Transform 12: $n = 3$, $a = 1$

$$F(s) = \frac{1}{2}[\frac{2}{(s+1)^3}]$$

$$f(t) = \frac{1}{2} t^2 e^{-t}$$

23. $F(s) = \frac{s + 2}{(s^2 + 9)^2}$; Transforms 16 and 15: a=3

$$F(s) = \frac{s}{(s^2+9)^2} + \frac{2}{(s^2+9)^2} = \frac{1}{6}[\frac{6s}{(s^2+9)^2}] + \frac{1}{27}[\frac{2(27)}{(s^2+9)^2}]$$

$$f(t) = \frac{1}{6} t \sin 3t + \frac{1}{27}(\sin 3t - 3t \cos 3t) = \frac{1}{54}(9t \sin 3t + 2 \sin 3t - 6t \cos 3t)$$

25. $F(s) = \frac{2s - 1}{s^3 - s} = \frac{2s - 1}{s(s^2 - 1)} = \frac{A}{s} + \frac{B}{s + 1} + \frac{C}{s - 1}$

$2s - 1 = A(s+1)(s-1) + Bs(s-1) + Cs(s+1)$

$s=0$: $-1 = -A$, $A = 1$
$s=-1$: $-3 = 2B$, $B = -3/2$
$s=1$: $1 = 2C$, $C = 1/2$

$$F(s) = \frac{1}{s} - \frac{3}{2(s + 1)} + \frac{1}{2(s - 1)}$$

Transforms 1 and 3:

$$f(t) = 1 - \frac{3}{2} e^{-t} + \frac{1}{2} e^t$$

27. $F(s) = \frac{s^2 - 2s + 3}{(s-1)^2 (s+1)}$

$$= \frac{A}{s-1} + \frac{B}{(s-1)^2} + \frac{C}{s+1}$$

$s^2 - 2s + 3 = A(s-1)(s+1)$
$\qquad\qquad + B(s+1) + C(s-1)^2$

$s=-1$: $6 = 4C$, $C = 3/2$
$s=1$: $2 = 2B$, $B = 1$
s^2: $1 = A + C$, $A = -1/2$

$$F(s) = -\frac{1}{2(s-1)} + \frac{1}{(s-1)^2} + \frac{3}{2(s+1)}$$

Transforms 3 and 11:

$$f(t) = -\frac{1}{2} e^t + t e^t + \frac{3}{2} e^{-t}$$

$$= \frac{1}{2}(2t - 1)e^t + \frac{3}{2} e^{-t}$$

29. $F(s) = \frac{3s^2 + 2s + 18}{(s+3)(s^2+4)}$

$$= \frac{A}{s+3} + \frac{Bs + C}{s^2 + 4}$$

$3s^2 + 2s + 18 = A(s^2 + 4) + Bs(s + 3) + C(s + 3)$

$s=-3$: $39 = A(13)$, $A = 3$
s^2: $3 = A + B$, $B = 0$ $\qquad$ $F(s) = \frac{3}{s + 3} + \frac{2}{s^2 + 4}$
s : $2 = 3B + C$, $C = 2$

Transforms 3 and 6: $f(t) = 3e^{-3t} + \sin 2t$

Exercises 16-4, p. 542

1. $y' + y = 0$, $y(0) = 1$
   $L(y') + L(y) = L(0)$
   $sL(y) - 1 + L(y) = 0$
   $L(y) = \dfrac{1}{s+1}$
   $f(t) = e^{-t}$

3. $2y' - 3y = 0$, $y(0) = -1$
   $2L(y') - 3L(y) = L(0)$
   $2[sL(y) + 1] - 3L(y) = 0$
   $L(y) = \dfrac{-2}{2s-3} = \dfrac{-1}{s - 3/2}$
   $f(t) = -e^{3t/2}$

5. $y' + 3y = e^{-3t}$, $y(0) = 1$
   $L(y') + 3L(y) = L(e^{-3t})$
   $sL(y) - 1 + 3L(y) = \dfrac{1}{s+3}$
   $(s+3)L(y) = 1 + \dfrac{1}{s+3}$
   $L(y) = \dfrac{1}{s+3} + \dfrac{1}{(s+3)^2}$
   $f(t) = e^{-3t} + te^{-3t}$
   $\qquad = (1 + t)e^{-3t}$

7. $y'' + 4y = 0$, $y(0) = 0$, $y'(0) = 1$
   $L(y'') + 4L(y) = L(0)$; $s^2L(y) - s(0) - 1 + 4L(y) = 0$
   $L(y) = \dfrac{1}{s^2 + 4}$; $f(t) = \dfrac{1}{2}\sin 2t$

9. $y'' + 2y' = 0$, $y(0) = 0$, $y'(0) = 2$
   $L(y'') + 2L(y') = L(0)$
   $s^2L(y) - s(0) - 2 + 2[sL(y) - 0] = 0$
   $L(y) = \dfrac{2}{s^2 + 2s} = \dfrac{2}{s(s+2)}$
   $f(t) = 1 - e^{-2t}$

11. $y'' - 4y' + 5y = 0$, $y(0) = 1$, $y'(0) = 2$
    $L(y'') - 4L(y') + 5L(y) = L(0)$
    $s^2L(y) - s(1) - 2 - 4[sL(y) - 1] + 5L(y) = 0$
    $L(y) = \dfrac{s-2}{s^2 - 4s + 5} = \dfrac{s-2}{(s-2)^2 + 1}$
    $f(t) = e^{2t}\cos t$

13. $y'' + y = 1$, $y(0) = 1$, $y'(0) = 1$
    $L(y'') + L(y) = L(1)$
    $s^2L(y) - s(1) - 1 + L(y) = \dfrac{1}{s}$
    $(s^2 + 1)L(y) = 1 + s + \dfrac{1}{s}$
    $L(y) = \dfrac{1}{s^2+1} + \dfrac{s}{s^2+1} + \dfrac{1}{s(s^2+1)}$
    $f(t) = \sin t + \cos t + (1 - \cos t)$
    $\qquad = 1 + \sin t$

15. $y'' + 2y' + y = e^{-t}$, $y(0) = 1$, $y'(0) = 2$
    $L(y'') + 2L(y') + L(y) = L(e^{-t})$
    $s^2L(y) - s(1) - 2 + 2[sL(y) - 1] + L(y) = \dfrac{1}{s+1}$
    $(s^2 + 2s + 1)L(y) = s + 4 + \dfrac{1}{s+1}$
    $L(y) = \dfrac{s}{(s+1)^2} + \dfrac{4}{(s+1)^2} + \dfrac{1}{(s+1)^3}$
    $f(t) = e^{-t}(1 - t) + 4te^{-t} + \dfrac{1}{2}t^2 e^{-t}$
    $\qquad = e^{-t}(\dfrac{1}{2}t^2 + 3t + 1)$

17. $2\dfrac{dv}{dt} = 6 - v$, $v(0) = 0$
    $2v' + v = 6$; $2L(v') + L(v) = L(6)$
    $2[sL(v) - 0] + L(v) = \dfrac{6}{s}$
    $(2s + 1)L(v) = \dfrac{6}{s}$
    $L(v) = \dfrac{6}{s(2s+1)} = \dfrac{3}{s(s + 1/2)}$
    $v = 6(1 - e^{-t/2})$

19. $R\dfrac{dq}{dt} + \dfrac{q}{C} = E$, $q(0) = 0$
    $R = 50\ \Omega$, $C = 4\ \mu F$, $E = 40\ V$
    $50q' + 2.5{\times}10^5 q = 40$
    $5q' + 2.5{\times}10^4 q = 4$
    $5L(q') + 2.5{\times}10^4 L(q) = L(4)$
    $5[sL(q) - 0] + 2.5{\times}10^4 L(q) = \dfrac{4}{s}$
    $L(q) = \dfrac{4}{s(5s + 2.5{\times}10^4)} = \dfrac{0.8}{s(s + 5{\times}10^3)}$
    $\qquad = \dfrac{0.8}{5{\times}10^3}[\dfrac{5{\times}10^3}{s(s + 5{\times}10^3)}]$
    $q = \dfrac{0.8}{5{\times}10^3}(1 - e^{-5{\times}10^3 t}) = 1.6{\times}10^{-4}(1 - e^{-5{\times}10^3 t})$

21. $Lq'' + \dfrac{q}{C} = E$ $(q' = \dfrac{dq}{dt})$, $q(0) = 0$, $q'(0) = 0$

$L = 10$ H, $C = 40$ μF, $E = 100 \sin 50t$

$10q'' + 2.5 \times 10^4 q = 100 \sin 50\,t$

$q'' + 2500q = 10 \sin 50t$

$L(q'') + 2500L(q) = L(10 \sin 50t)$

$s^2 L(q) - s(0) - 0 + 2500L(q) = \dfrac{10(50)}{s^2 + 50^2}$

$(s^2 + 2500)L(q) = \dfrac{500}{s^2 + 50^2}$

$L(q) = \dfrac{500}{(s^2 + 50^2)^2} = \dfrac{5}{50^2}[\dfrac{2(50)(50)^2}{(s^2 + 50^2)^2}]$; $q = 0.002(\sin 50t - 50t \cos 50t)$

$i = \dfrac{dq}{dt} = 0.002(50 \cos 50t + 2500t \sin 50t - 50 \cos 50t) = 5t \sin 50t$

23. $D^2 y + 9y = 18 \sin 3t$, $y(0) = 0$, $y'(0)=0$

$L(y'') + 9L(y) = L(18 \sin 3t)$

$s^2 L(y) - s(0) - 0 + 9L(y) = \dfrac{18(3)}{s^2 + 9}$

$L(y) = \dfrac{54}{(s^2 + 9)^2} = \dfrac{2(3^3)}{(s^2 + 9)^2}$

$y = \sin 3t - 3t \cos 3t$

Review Exercises for Chapter 16, p. 543

1. $y' = x^2 - y^2$

x	y	x^2-y^2	dy
0.0	1.0000	-1.0000	-0.1000
0.1	0.9000	-0.8000	-0.0800
0.2	0.8200	-0.6324	-0.0632
0.3	0.7568	-0.4827	-0.0483
0.4	0.7085	-0.3420	-0.0342
0.5	0.6743	-0.2047	-0.0205

3. $y' = \sqrt{x + y}$

x	y	$\sqrt{x+y}$	dy
0.0	1.0000	1.0000	0.2000
0.2	1.2000	1.1832	0.2366
0.4	1.4366	1.3552	0.2710
0.6	1.7077	1.5191	0.3038
0.8	2.0115	1.6768	0.3354
1.0	2.3469	1.8294	0.3659
1.2	2.7128	1.9781	0.3956
1.4	3.1084	2.1233	0.4247
1.6	3.5330	2.2656	0.4531
1.8	3.9861	1.4054	0.4811
2.0	4.4672	2.5431	0.5086

5. $y' = \sin x + \tan y$

x	y	sinx+tany	dy
0.0	0.5000	0.5463	0.0546
0.1	0.5546	0.7193	0.0719
0.2	0.6266	0.9225	0.0923
0.3	0.7188	1.1705	0.1170
0.4	0.8359	1.4958	0.1496
0.5	0.9854	1.9881	0.1988
0.6	1.1843	3.0215	0.3022
0.7	1.4864	12.4663	1.2466

7. $y = x^2 - y^2$

x	y	x^2-y^2	dy	x	y	x^2-y^2	dy
0.00	1.0000	-1.0000	-0.0200	0.26	0.7954	-0.5651	-0.0113
0.02	0.9800	-0.9600	-0.0192	0.28	0.7841	-0.5365	-0.0107
0.04	0.9608	-0.9215	-0.0184	0.30	0.7734	-0.5082	-0.0102
0.06	0.9424	-0.8845	-0.0177	0.32	0.7633	-0.4802	-0.0096
0.08	0.9247	-0.8486	-0.0170	0.34	0.7536	-0.4524	-0.0090
0.10	0.9077	-0.8139	-0.0163	0.36	0.7446	-0.4248	-0.0085
0.12	0.8914	-0.7802	-0.0156	0.38	0.7361	-0.3975	-0.0079
0.14	0.8758	-0.7475	-0.0149	0.40	0.7282	-0.3702	-0.0074
0.16	0.8609	-0.7155	-0.0143	0.42	0.7208	-0.3431	-0.0069
0.18	0.8466	-0.6843	-0.0137	0.44	0.7139	-0.3160	-0.0063
0.20	0.8329	-0.6537	-0.0131	0.46	0.7076	-0.2891	-0.0058
0.22	0.8198	-0.6237	-0.0125	0.48	0.7018	-0.2621	-0.0052
0.24	0.8073	-0.5942	-0.0119	0.50	0.6965	-0.2352	-0.0047

9. $y' = y + x^2$, $(0,1)$

$dy = (1 + x^2)dx$

$y_1 = x + \frac{1}{3}x^3 + c_1$, $c_1 = 1$

$y_1 = 1 + x + \frac{1}{3}x^3$

$dy = (1 + x + \frac{1}{3}x^3 + x^2)dx$

$y_2 = x + \frac{1}{2}x^2 + \frac{1}{12}x^4 + \frac{1}{3}x^3 + c_2$, $c_2 = 1$

$y_2 = 1 + x + \frac{1}{2}x^2 + \frac{1}{3}x^3 + \frac{1}{12}x^4$

$dy = (1 + x + \frac{1}{2}x^2 + \frac{1}{3}x^3 + \frac{1}{12}x^4 + x^2)dx$

$y_3 = x + \frac{1}{2}x^2 + \frac{1}{2}x^3 + \frac{1}{12}x^4 + \frac{1}{60}x^5 + c_3$

$c_3 = 1$

$y_3 = 1 + x + \frac{1}{2}x^2 + \frac{1}{2}x^3 + \frac{1}{12}x^4 + \frac{1}{60}x^5$

11. $y' = 1 + x^2y$, $(1,1)$

$dy = (1 + x^2)dx$

$y_1 = x + \frac{1}{3}x^3 + c_1$, $1 = 1 + \frac{1}{3} + c_1$, $c_1 = -\frac{1}{3}$

$y_1 = x + \frac{1}{3}x^3 - \frac{1}{3}$

$dy = [1 + x^2(x + \frac{1}{3}x^3 - \frac{1}{3})]dx$

$= (1 - \frac{1}{3}x^2 + x^3 + \frac{1}{3}x^5)dx$

$y_2 = x - \frac{1}{9}x^3 + \frac{1}{4}x^4 + \frac{1}{18}x^6 + c_2$

$1 = 1 - \frac{1}{9} + \frac{1}{4} + \frac{1}{18} + c_2$, $c_2 = -\frac{7}{36}$

$y_2 = -\frac{7}{36} + x - \frac{1}{9}x^3 + \frac{1}{4}x^4 + \frac{1}{18}x^6$

13. $4y' - y = 0$, $y(0) = 1$

$4L(y') - L(y) = L(0)$

$4[sL(y) - 1] - L(y) = 0$

$L(y) = \frac{4}{4s - 1} = \frac{1}{s - 1/4}$

$y = e^{t/4}$

15. $y' - 3y = e^t$, $y(0) = 0$

$L(y') - 3L(y) = L(e^t)$

$sL(y) - 0 - 3L(y) = \frac{1}{s - 1}$

$(s - 3)L(y) = \frac{1}{s - 1}$

$L(y) = \frac{1}{(s - 3)(s - 1)} = \frac{1}{2(s - 3)} - \frac{1}{2(s - 1)}$

$y = \frac{1}{2}(e^{3t} - e^t)$

17. $y'' + y = 0$, $y(0) = 0$, $y'(0) = -4$

$L(y'') + L(y) = L(0)$

$s^2L(y) - s(0) + 4 + L(y) = 0$

$L(y) = \frac{-4}{s^2 + 1}$, $y = -4\sin t$

19. $y'' + 9y = 3t$, $y(0) = 0$, $y'(0) = -1$

$L(y'') + 9L(y) = 3L(t)$

$s^2L(y) - s(0) + 1 + 9L(y) = \frac{3}{s^2}$

$L(y) = \frac{3}{s^2(s^2 + 9)} - \frac{1}{s^2 + 9}$

$y = \frac{1}{9}(3t - \sin 3t) - \frac{1}{3}\sin 3t = \frac{1}{3}t - \frac{4}{9}\sin 3t$

21. $y' = 1 + ye^{-x}$, $(0,1)$

$dy = (1 + e^{-x})dx$

$y_1 = x - e^{-x} + c_1$

$1 = 0 - 1 + c_1$, $c_1 = 2$

$y_1 = 2 + x - e^{-x}$

$dy = [1 + (2 + x - e^{-x})e^{-x}]dx$

$= (1 + 2e^{-x} + xe^{-x} - e^{-2x})dx$

$y_2 = x - 2e^{-x} + e^{-x}(-x - 1) + \frac{1}{2}e^{-2x} + c_2$

$1 = 0 - 2 + (-1) + \frac{1}{2} + c_2$, $c_2 = \frac{7}{2}$

$y = \frac{7}{2} + x - (3 + x)e^{-x} + \frac{1}{2}e^{-2x}$

$y(0.5) = 3.5 + 0.5 - 3.5e^{-0.5} + 0.5e^{-1}$

$= 2.0611$

23. $y' = 1 + ye^{-x}$

x	y	1-ye^{-x}	dy
0.0	1.0000	2.0000	0.2000
0.1	1.2000	2.0858	0.2086
0.2	1.4086	2.1532	0.2153
0.3	1.6239	2.2030	0.2203
0.4	1.8442	2.2362	0.2236
0.5	2.0678	2.2542	0.2254
0.6	2.2932	2.2586	0.2259
0.7	2.5191	2.1510	0.2251
0.8	2.7442	2.2330	0.2233
0.9	2.9675	2.2065	0.2206
1.0	3.1882	2.1729	0.2173

25. $2\dfrac{di}{dt} + i = 12$, $i(0) = 0$

$2L(i') + L(i) = L(12)$

$2[sL(i) - 0] + L(i) = \dfrac{12}{s}$

$L(i) = \dfrac{12}{s(2s+1)} = \dfrac{12(1/2)}{s(s+1/2)}$

$i = 12(1 - e^{-t/2})$

$1(0.3) = 12(1 - e^{-0.3/2})$

$\qquad = 1.67\,A$

29. $2\dfrac{di}{dt} + i = 12$

$\dfrac{di}{dt} = 6 - 0.5i$

t	i	6-0.5i	di
0.00	0.0000	6.0000	0.3000
0.05	0.3000	5.8500	0.2925
0.10	0.5925	5.7038	0.2852
0.15	0.8777	5.5612	0.2781
0.20	1.1557	5.4221	0.2711
0.25	1.4269	5.2866	0.2643
0.30	1.6912	5.1544	0.2577

$i(0.5) = 1.69\,A$

33. $m = 5\,kg$, $F = 50\,N$, $x = 1\,m$

$50 = k(1)$, $k = 50\ N/m$

$5D^2y + 50y = 0$, $y(0) = 1$, $y'(0) = 0$

$D^2y + 10y = 0$

$L(D^2y) + 10L(y) = L(0)$

$s^2L(y) - s(1) - 0 + 10L(y) = 0$

$L(y) = \dfrac{s}{s^2 + 10}$; $y = \cos 3.16t$

27. $LD^2q + RDq + \dfrac{q}{C} = 0$ $(D = \dfrac{d}{dt})$, $q(0) = 400\ \mu C$, $q'(0) = 0$

$L = 0.25\ H$, $R = 4\ \Omega$, $C = 100\ \mu F$

$0.25D^2q + 4Dq + 10^4q = 0$

$D^2q + 16Dq + 4\times10^4q = 0$

$L(D^2q) + 16L(Dq) + 4\times10^4L(q) = 0$

$[s^2L(q) - s(4\times10^{-4}) - 0] + 16[sL(q) - 4\times10^{-4}]$

$\quad + 4\times10^4L(q) = 0$

$(s^2 + 16s + 4\times10^4)L(q) = 4\times10^{-4}(s + 16)$

$L(q) = 4\times10^{-4}[\dfrac{s + 16}{(s^2 + 16s + 64) + 39936}]$

$\quad = 4\times10^{-4}[\dfrac{s+8}{(s+8)^2+200^2} + \dfrac{8}{(s+8)^2+200^2}]$

$q = 4\times10^{-4}(e^{-8t}\cos200t + \dfrac{8}{200}e^{-8t}\sin 200t)$

$\quad = 10^{-4}e^{-8t}(4\cos 200\,t + 0.16\sin 200t)$

31. $m = 0.25\ slug$, $F = 8\ lb$, $x = 6\ in. = 0.5\ ft$

$8 = 0.5k$, $k = 16\ lb/ft$

$0.25D^2y + 16y = \cos 8t$ $(D = d/dt)$

$D^2y + 64y = 4\cos 8t$, $y(0) = 0$, $Dy(0) = 0$

$L(D^2y) + 64L(y) = 4L(\cos 8t)$

$s^2L(y) - s(0) - 0 + 64L(y) = 4(\dfrac{s}{s^2 + 64})$

$L(y) = \dfrac{4s}{(s^2 + 64)^2} = \dfrac{1}{4}[\dfrac{2(8s)}{(s^2 + 64)^2}]$

$y = \dfrac{1}{4}t\sin 8t$

35. $F(s) = \dfrac{b - a}{(s + a)(s + b)} = \dfrac{A}{s + a} + \dfrac{B}{s + b}$

$b - a = A(s + b) + B(s + a)$

$s=-a$: $b - a = A(-a + b)$, $A = 1$

$s=-b$: $b - a = B(-b + a)$, $B = -1$

$F(s) = \dfrac{1}{s + a} - \dfrac{1}{s + b}$

$L^{-1}(F) = e^{-at} - e^{-bt}$